Digitalization in Construction

This book highlights the latest trends and advances in applications of digital technologies in construction engineering and management. A collection of chapters is presented, explicating how advanced technological solutions can innovatively address challenges and improve outcomes in the construction industry. Promising technologies that are highlighted include digital twins, virtual reality, augmented reality, artificial intelligence, robotics, blockchain, and distributed ledger technologies. The first section presents recent applications of extended reality technologies for construction education and advanced project control. The subsequent chapters explore Artificial Intelligence (AI), blockchain, and BIM-enabled digitalization in construction through a series of case studies, reviews, and technical studies. Innovative technologies and digitalized solutions are proposed for improved design, planning, training, monitoring, inspection, and operations management in Architectural, Engineering and Construction (AEC) contexts. In addition to the technological perspectives and insights presented, pressing issues such as decarbonization, safety, and sustainability in the built environment are also discussed.

This book provides foundational knowledge and in-depth technical studies on emerging technologies for students, academics, and industry practitioners. The research demonstrates how the effective use of new technologies can enhance work methods, transform organizational structures, and bring profound advantages to construction project participants.

Spon Research

Publishes a stream of advanced books for built environment researchers and professionals from one of the world's leading publishers. The ISSN for the Spon Research programme is ISSN 1940-7653 and the ISSN for the Spon Research E-book programme is ISSN 1940-8005

The Connectivity of Innovation in the Construction Industry
Edited by Malena Ingemansson Havenvid, Åse Linné, Lena E. Bygballe and Chris Harty

Contract Law in the Construction Industry Context
Carl J. Circo

Corruption in Infrastructure Procurement
Emmanuel Kingsford Owusu and Albert P. C. Chan

Improving the Performance of Construction Industries for Developing Countries
Programmes, Initiatives, Achievements and Challenges
Edited by Pantaleo D Rwelamila and Rashid Abdul Aziz

Work Stress Induced Chronic Diseases in Construction
Discoveries Using Data Analytics
Imriyas Kamardeen

Life-Cycle Greenhouse Gas Emissions of Commercial Buildings
An Analysis for Green-Building Implementation Using A Green Star Rating System
Cuong N. N. Tran, Vivian W. Y. Tam and Khoa N. Le

Data-driven BIM for Energy Efficient Building Design
Saeed Banihashemi, Hamed Golizadeh and Farzad Pour Rahimian

Successful Development of Green Building Projects
Tayyab Ahmad

BIM and Construction Health and Safety
Uncovering, Adoption and Implementation
Hamed Golizadeh, Saeed Banihashemi, Carol Hon and Robin Drogemuller

Digitalisation in Construction
Recent Trends and Advances
Edited by Chansik Park, Farzad Pour Rahimian, Nashwan Dawood, Akeem Pedro, Dongmin Lee, Rahat Hussain, Mehrtash Soltani

Digitalization in Construction

Recent Trends and Advances

Edited by Chansik Park,
Farzad Pour Rahimian, Nashwan Dawood,
Akeem Pedro, Dongmin Lee, Rahat Hussain,
and Mehrtash Soltani

Routledge
Taylor & Francis Group

LONDON AND NEW YORK

First published 2024
by Routledge
4 Park Square, Milton Park, Abingdon, Oxon OX14 4RN

and by Routledge
605 Third Avenue, New York, NY 10158

Routledge is an imprint of the Taylor & Francis Group, an informa business

British Library Cataloguing-in-Publication Data
A catalogue record for this book is available from the British Library

Library of Congress Cataloguing-in-Publication Data
Names: Park, Chansik, editor. | Rahimian, Farzad Pour, editor. | Dawood, Nashwan, editor. | Pedro, Akeem, editor. | Lee, Dongmin (Professor of architecture and building science) editor. | Hussain, Rahat, editor. | Soltani, Mehrtash, editor.
Title: Digitalization in construction : recent trends and advances / edited by Chansik Park, Farzad Pour Rahimian, Nashwan Dawood, Akeem Pedro, Dongmin Lee, Rahat Hussain, Mehrtash Soltani.
Description: Abingdon, Oxon ; New York, NY : Routledge, 2024. |
Series: Spon research | Includes bibliographical references and index.
Identifiers: LCCN 2023026759 (print) | LCCN 2023026760 (ebook) | ISBN 9781032517896 (hardback) | ISBN 9781032528892 (paperback) | ISBN 9781003408949 (ebook)
Subjects: LCSH: Building information modeling. | Construction industry--Management--Data processing.
Classification: LCC TH438.13 .D54 2024 (print) | LCC TH438.13 (ebook) | DDC 624.0285--dc23/eng/20231019
LC record available at https://lccn.loc.gov/2023026759
LC ebook record available at https://lccn.loc.gov/2023026760

ISBN: 978-1-032-51789-6 (hbk)
ISBN: 978-1-032-52889-2 (pbk)
ISBN: 978-1-003-40894-9 (ebk)

DOI: 10.1201/9781003408949

Typeset in Times New Roman
by MPS Limited, Dehradun

Contents

Preface

The rapid proliferation of digital technologies is heralding unprecedented transformation across diverse industries and sectors. This book highlights the latest trends and advances in applications of digital technologies in construction engineering and management. A collection of articles is presented, explicating how advanced technological solutions can innovatively address challenges and improve outcomes in the construction industry. Promising technologies that are highlighted include digital twins, virtual reality, augmented reality, artificial intelligence, computer vision, robotics, blockchain, and distributed ledger technologies.

The drive towards digital transformation is essential for the construction industry, which tends to lag behind other industries in terms of labor productivity, safety, and cost efficiency. Unforeseen disruptions such as the COVID-19 pandemic have shown the need for construction collaboration, training, monitoring, and contracting innovation. The immense potential of paradigm-shifting technologies such as artificial intelligence, computer vision, digital twins, and robotics has been recognized. Their integration into construction workflows is expected to bring tremendous value to the building industry. However, many challenges and questions regarding the digitalization of construction remain unaddressed. Many emerging technological tools and solutions in construction are still at a low level of maturity, and there is a lack of documented work on how they can be implemented in practice. Issues such as tool interoperability and concerns regarding the high cost of entry, worker upskilling, and safety remain underexamined. This book fills this gap by providing a body of research outputs which showcase state-of-the-art advances and applications of innovative technologies in construction.

The book comprises 18 chapters. The first chapter provides an overview of the state of the construction industry through a collection of review papers and qualitative studies. The subsequent chapters explore applications of innovative technologies through case studies, review articles, conceptual articles, and technical papers. Innovative technologies and digitalized solutions are proposed for improved design, planning, training, monitoring, inspection, and operations management in Architectural, Engineering, and Construction (AEC) contexts. In addition to the technological perspectives and insights presented, the last few chapters explore pressing issues such as health and safety in the built environment.

Chansik Park, Farzad Pour Rahimian, Nashwan Dawood,
Akeem Pedro, Dongmin Lee, Rahat Hussain, Mehrtash Soltani

Editors' Biographies

Chansik Park is a Professor at the School of Architecture and Building Science and a former Dean of the Graduate School of Construction Engineering of Chung-Ang University in South Korea. Professor Park has published over 100 papers in peer-reviewed journals and conferences internationally and has served editorial board member and reviewer of many international journals including *Automation in Construction* and *International Journal of Project Management*. He has been the recipient of numerous academic and professional awards, including the prestigious Elsevier Atlas Award in recognition of 'outstanding achievement and significant positive contribution to society'.

Farzad Rahimian is a Professor of Digital Engineering and Manufacturing at Teesside University, UK. He is the Editor-in-Chief of the *Journal of Smart and Sustainable Built Environment* and Associate Editor of Automation in Construction. He is the author and editor of a number of books including *Industry 4.0 Solutions for Building Design and Construction* (Routledge, 2019) and *Industry 4.0 Solutions for Building Design and Construction* (Routledge, 2021)

Nashwan Dawood is a Professor of Digital Construction in the Centre for Sustainable Engineering at Teesside University. His expertise is in sustainable infrastructure and in modelling ways to ensure carbon reduction in construction, housing and engineering projects. He has long-standing experience in undertaking research projects with major industrial partners in the areas of 5D modelling, serious game engine technology, Building Information Modelling, and the application of digital technologies for energy-efficient buildings.

Akeem Pedro is a Research Associate at Chung-Ang University in Seoul, South Korea. His research focuses on Technology Enhanced Learning, Interactive Learning Environments, Knowledge Engineering and Informatics for construction safety and health. He has published and co-authored over 20 papers in international journals and conferences, garnering over 900 citations.

Dongmin Lee is an Assistant Professor in the Department of Architecture and Building Science at Chung-Ang University. His research interests lie in the integration of construction equipment, method, planning, scheduling, and control to support a better human-robot collaborative working environment. Dr Lee has published around 40 papers in various journals and conferences internationally. He has also served as a reviewer of many international as well as national journals.

Rahat Hussain is currently working towards PhD degree at Chung-Ang University in Seoul, South Korea. His current research works focus on construction informatics, technology-enhanced education, metaverse, visualization technologies for construction safety, education, and management.

Mehrtash Soltani earned his PhD from the University of Malaya in Malaysia. With over a decade of research expertise in building materials, waste management, and construction safety management, he currently serves as a research associate at ConTi Lab, located at Chung-Ang University in Seoul, South Korea.

Contributors

Abbas Muhammad Sibtain, Chung-Ang University, South Korea

Akanmu Abiola Abosede, Myers Lawson School of Construction, Virginia Tech, Blacksburg, VA, USA

Aminudin Eeydzah, School of Civil Engineering, Faculty of Engineering, Universiti Teknologi Malaysia, Skudai Johor, Malaysia

Arashpour Mehrdad, Department of Civil Engineering, Monash University, Australia

Bairaktarova Diana, Department of Engineering Education, Virginia Tech, Blacksburg, VA, USA

Chauhan Jatin Kumar, School of Natural and Built Environment, Queen's University Belfast, United Kingdom

Daniel Emmanuel, University of Wolverhampton, Wolverhampton, United Kingdom

Dawood Nashwan, School of Computing, Engineering and Digital Technologies, Teesside University, Middlesbrough, UK

Dudhee Vishak, School of Computing, Engineering and Digital Technologies, Teesside University, Middlesbrough, UK

Dzuwa Christopher, Faculty of Engineering and the Built Environment, University of Johannesburg, South Africa

Elghaish Faris, School of Natural and Built Environment, Queen's University Belfast, United Kingdom

Gonsalves Nihar James, Myers-Lawson School of Construction, Virginia Tech, Blacksburg, VA, USA

Haahr Meinhardt Thorlund, University College of Northern Denmark (UCN), Aalborg, Denmark

Hosseini M Reza, School of Architecture and Building, Deakin University, Australia

Hussain Rahat, Chung-Ang University, South Korea

Jiancheng Shan, Faculty of Engineering, Computing and Science, Swinburne University of Technology Sarawak Campus, Kuching, Sarawak, Malaysia

Kocaturk Tuba, School of Architecture and Built Environment, Deakin University, Australia

Laishram Boeing, Department of Civil Engineering, IIT Guwahati, Assam, India

Le Thai-Hoa, Department of Civil Engineering, National Taiwan University, Taiwan

Ledari Masoomeh Bararzadeh, Energy Department, Sharif University of Technology, Tehran, Iran

Lee Doyeop, Chung-Ang University, South Korea

Lin Jacob J., Department of Civil Engineering, National Taiwan University, Taiwan

Matarneh Sandra, Faculty of Engineering, Al-Ahliyya Amman University, Amman, Jordan

Miranda Arturo De Jesús, University of Wolverhampton, Wolverhampton, United Kingdom

Molusiwa Ramabodu, Faculty of Engineering and the Built Environment, University of Johannesburg, South Africa

Musonda Innocent, Faculty of Engineering and the Built Environment, University of Johannesburg, South Africa

Nguyen Linh, Myongji University, South Korea

Ogunseiju Omobolanle, School of Building of Construction, College of Design, Georgia Tech, Atlanta, Georgia, USA

Onososen Adetayo, Faculty of Engineering and the Built Environment, University of Johannesburg, South Africa

Park Chan-Sik, Chung-Ang University, South Korea

Park Man-Woo, Myongji University, South Korea

Pedro Akeem, Chung-Ang University, South Korea

Pham Hung, Eindhoven University of Technology, Netherlands

Pidgeon Andrew Gerard, School of Computing, Engineering and Digital Technologies, Teesside University, Middlesbrough, UK

Rahimian Farzad Pour, School of Computing, Engineering and Digital Technologies, Teesside University, Middlesbrough, UK

Saar Chai Chang, Faculty of Engineering, Computing and Science, Swinburne University of Technology Sarawak Campus, Kuching, Sarawak, Malaysia

Sharma Nandini, Department of Civil Engineering, IIT Guwahati, Assam, India

Shih Yun Ting, Department of Civil Engineering, National Taiwan University, Taiwan

Soltani Mehrtash, Chung-Ang University, South Korea

Sood Rhijul, Department of Civil Engineering, IIT Guwahati, Assam, India

Vanderhorst Hamlet Reynoso, University of Wolverhampton, Wolverhampton, United Kingdom

Vukovic Vladimir, School of Computing, Engineering and Digital Technologies, Teesside University, Middlesbrough, UK

Yang Jaehun, Chung-Ang University, South Korea

Zaidi Syed Farhan Alam, Chung-Ang University, South Korea

1   Multi-user Virtual Reality-Based Design Review of Students' Construction Designs

Meinhardt Thorlund Haahr

University College of Northern Denmark (UCN), Aalborg

1.1   Introduction

A construction process is considered successful if the construction is completed on time, at the agreed price and quality and with a high degree of customer satisfaction. Unfortunately, there are many examples of this not being achieved (Shirkavand et al., 2016). McKinsey reported that 98% of megaprojects are associated with overruns or delays, with an average cost increase being 80% of original budget and least but not last with an average delay of 20 months (McKinsey Global Institute, 2015). Among others, they report the cause being, lack of interdisciplinary collaboration and accumulation of unresolved issues. According to the report conducted by The Danish Construction Federation, in collaboration with Chalmers University in Gothenburg and the company BIMobject, the Danish private construction sector can save up to 10.5 billion DKK (approximately 1.41 billion Euro) by proper use of digital tools and improving communication in the design phase of a construction project (Malmgreen, 2020). Among other things, they conclude that economic profit can be gained in the transition between design and construction phases. Here, several ambiguities often arise between consultants and contractors, leading to misunderstandings and disagreements between the partners.

1.1.1   Communication

Communication, especially information delivery, depends on the context in which the information is presented, the background of the sender and receiver as well as their natural environment (Berlo, 1960). For information to be delivered, it must be conceptualized and coded in the form of language, text and/or a medium. The prerequisite for encoding and decoding to be successful is familiarity with the codes that apply in the culture, in which the communication takes place (Hall, 1980). Design and construction of buildings and especially its building services are interdisciplinary and complex processes, as each building is unique, the partners are many, with different educational backgrounds and change from project to project. Finally, several of the stakeholders, such as users and the client, are not necessarily building professionals.

1.1.2   Design Review

During the design process, design reviews are conducted in order to detect omissions and defects, also referred to as issues. Design and execution of building is a complex process, as each building is unique, and the partners are many and different from project to project. The current design and review processes in the design phase are predominantly

DOI: 10.1201/9781003408949-1

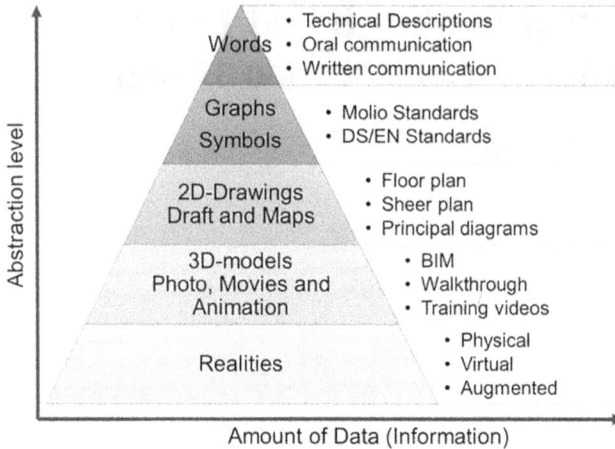

Figure 1.1 Abstraction pyramid – inspired by Kjems (2000).

communicated via media such as Building Information Model (BIM), design review in the construction phase is predominantly communicated via media such as 2D drawings and descriptions and to a lesser extent 3D models. (Haahr et al., 2019; Johansson and Roupé, 2019). According to space cognition theory (Golledge, 1991), which is grounded in visuospatial skills, processing and working memory, information delivery in current media is abstract (Figure 1.1). They place high demands on the individual actor's ability to encode and decode information, and as consequence a higher cognitive workload for the actor (Johansson and Roupé, 2019; Kwiatek et al., 2019).

1.1.3   *Virtual Reality*

Virtual reality (VR) offers a solution to this problem. By using VR, it is possible to achieve a more natural and intuitive design review in a size ratio of 1:1, similar to that used in erected buildings. The goal is that the technology will result in actors arriving at the same conclusions as they would in a real-world investigation. According to Windham and Liu, VR significantly improves the (1) understanding of space, (2) detecting and understanding of potential issues and (3) understanding of functionality (Windham and Liu, 2018). Furthermore, VR gives a common base of reference in a 1:1 scale, which gives actors from the Architectural Engineering and Construction (AEC) Industry improved ability to (1) understand the building as whole, (2) decode the intended design compared to 2D drawings and 3D models in BIM and lastly (3) detect clashes and design errors before commenced work (Johansson and Roupé, 2019). A literature study conducted by the author concludes, that existing studies argue that VR can improve design review process in the designing phase of a building (Haahr, 2021). Furthermore, VR is mature for use in the industry and that existing research lacks investigation with many actors in a multidisciplinary environment, which can give more transferable results.

One disadvantage with existing VR is that most commercialized applications only involve a single user experience. As opposed to multi-user, the common base of reference is gained individually. From the author's own observation, collaboration with single-user experience (1) leads to actors spending shorter time in an immersive VR experience or

(2) rapidly taking off VR equipment, when reviewing a design. Furthermore, VR is associated with expensive and troublesome setups, which causes resistance to the adoption of the technology. Fortunately, Prospect offers an easy-to-use and multi-user solution intended for interdisciplinary immersive collaboration, and is supported by low-priced VR equipment, Oculus Quest 2 (PROSPECT BY IRISVR, 2022).

The aim of this study is to investigate the value of multi-user VR-supported design review of buildings in a multidisciplinary collaboration.

1.2   Method

The physical frame around this study was at an annual event at UCN, called The Digital Days (DDD) (de Digitale Dage, 2022). The purpose of DDD is to bring together students from different education within the AEC industry to design a building, where the main goal is to learn digital tools centred around BIM. For the event, students from different disciplines within AEC industry, work in groups, aiming to use digital applications to design building projects.

1.2.1   Case

The case was a multi-story building with a real client. Each group received the official material of the Schematic Design, including a Revit model, see Figure 1.2. Within three days each group move from Schematic Design to Detailed Design, including considerations and a rough design of Architecture, MEP and Structural bearing system. Each group had a supervisor from the local AEC industry, which supported the group with his own experience and knowledge from the AEC industry. The groups compete in making the best project, by exploring methods, digital applications and collaboration. Before and during the event, various presentations are held by companies, which are to inspire the students in their work.

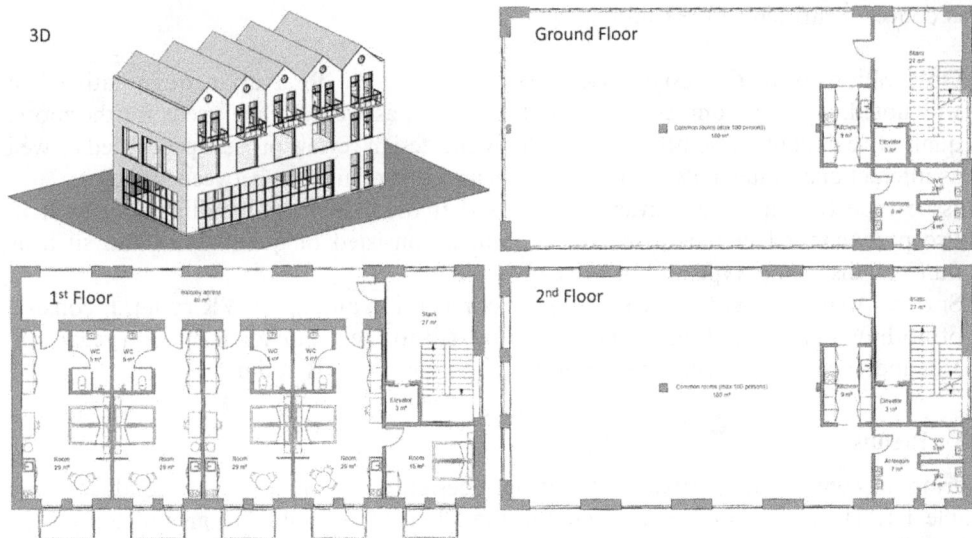

Figure 1.2 An extract of the Revit model and shows a 3D view, ground floor, first floor and second floor.

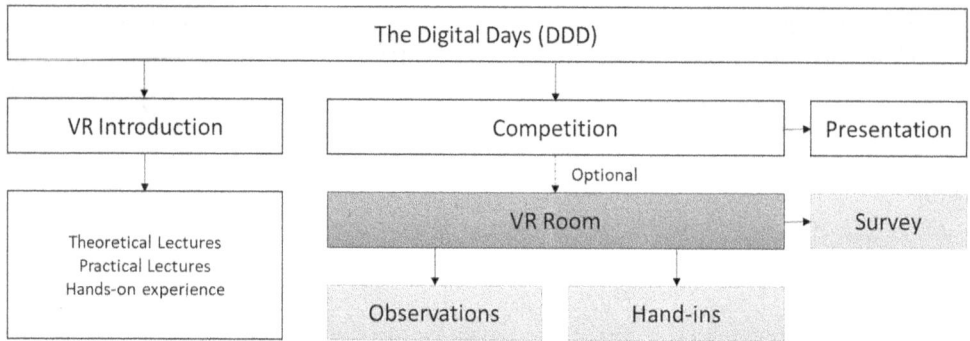

Figure 1.3 An overview of data collected during this study.

1.2.2   *Data Collection*

This year had particular focus on VR, which included two intensive days of VR prior to the DDD competition. This included different lectures in the use of VR, both theoretical and practical from two Danish companies. The theoretical lectures focused on the technology and existing studies on VR in the AEC industry, whereas the practical lectures focused on the practical use-case and value of VR based on their experience within the AEC industry. To get some hands-on experience, the students were introduced to different applications for issue detecting and handling, among here Navisworks for clash detection and Prospect for immersive and collaborative design review.

   For this study, data were collected during and after the three days of competition, as shown in Figure 1.3. During the DDD competition, a VR room was set up for voluntary design review. The VR room consisted of four sets of Oculus Quest 2 (OQ) for use in Prospect, furthermore, students could also participate with they're desktop (DT). Data was collected through three sources.

1 Observation during the design review sessions: During the observation, quantitative data was noted, such as number of OQ and DT users as well as time used. Furthermore, qualitative data was also noted, such as how the design review process proceeded as well as interactions, statements, announcements and body language between VR users.
2 Issue report from design reviews: Groups that used the issue-handling tool built in Prospect, handed in their issue report, which consisted of qualitative data, such as number, discipline, type and priority of issues.
3 Survey: A survey was conducted after the competition and use of VR room. It consists of both qualitative and quantitative data related to their individual opinion regarding VR and experience during the VR introduction and/or VR room.

1.3   Results

During the competition period, a total of eight design review sessions were conducted, see Table 1.1. The sessions were conducted in the VR room by different groups during the DDD. In total 25 users participated in a total time of 201 minutes, 18 OQ and 7 DT users. On average, rounded to whole persons, in total three users, two OQ and one DT, in a 25-minute session.

Table 1.1 Overview of multi-user VR-supported design review session

Session	Oculus Quest	Desktop	Total	Time	Process			Issues	
No.		No. of Users		Minutes	Documentation		Approach	No.	Per Minute
1	4	2	6	39	BCF	Digital	Structured	19	0.49
2	2	0	2	15	No	Analog	Semi		
3	2	1	3	20	Screenshot	Digital	Semi		
4	1	1	2	25	BCF	Digital	Structured	6	0.24
5	2	0	2	15	No	Analog	Non		
6	1	1	2	40	BCF	Digital	Structured	15	0.38
7	3	1	4	22	BCF	Digital	Structured	8	0.36
8	3	1	4	25	Hand notes	Analog	Non		
Total	18	7	25	201				48	
Average	2	1	3	25				12	0.37

Session 1 was the greatest and one of the longest sessions, with a total of six users, four OQ users and two DTs, and 39 min long. The longest session was 40 minutes, which was conducted in session 6. On the other hand, in four sessions only two users were participating, in which two sessions had two OQ users and the other two sessions had one OQ and one DC user. In the case of two OQ users, the sessions lasted 15 minutes, which was the two shortest sessions conducted.

1.3.1 Design Review Process

As mentioned in Section 1.2, the VR-supported design reviews were optional for the students participating in DDD, where the main objective of DDD is to learn digital solutions based on hands-on experience and own reflections. Therefore, no instruction was given on how to perform design reviews in VR. As shown in Table 1.1, in six out of eight sessions, at least one DT user was participating. Interestingly enough, in all six sessions, the role of the DT user was to document the issues detected by the OQ user(s). In all eight sessions, the role of the OQ user was to review the models and detect issues.

The documentation for registered issues was handled in four different ways, Building Collaboration Format (BCF), screenshots, hand notes or no documentation. In four out of eight sessions, the detection of issues was documented in the issue-handling tool in Prospect, in which supports BCF. This allows the users to create a so-called issue, which among other things saves the location of the issue in the 3D model, screenshot including any animation (left and right, Figure 1.4) and/or measurements (left, Figure 1.5) created in VR, description, status and to assign responsibility to a project participant. In session 3 the documentation consisted of DT users taking screenshots in Prospect and reporting and describing the issues in a Word document. In session 8, the documentation consisted of DT user taking analogy hand notes on a piece of paper. Common for both sessions 2 and 5 was that no documentation was conducted. Interestingly, these were also the only two sessions where no DT user participated. Even though, this does not mean that the design review session was useless. In both sessions, the users gain an improved understanding of the project and issues were also detected. For instance, in session 5, when looking at a bearing beam, one OQ user states "This one is too high up". Furthermore, during another conversation, one OQ user states "The suspended ceiling is floating and

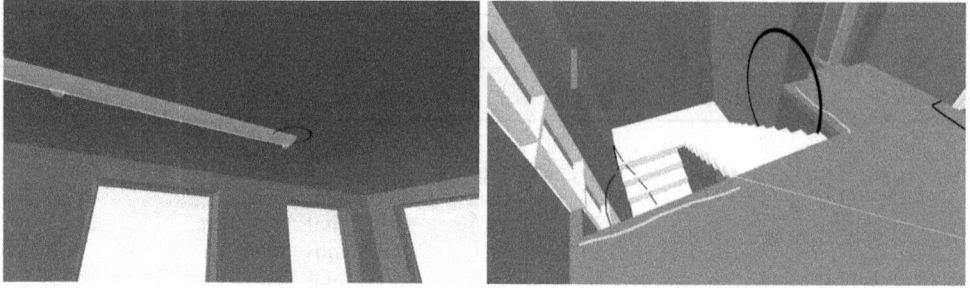

Figure 1.4 Examples of issue registration in Prospect. (Left) Hard class between ventilation duct and ceiling in session 5. (Right) Omission of stair rail from session 6. The animation shows where stair railing is to be placed.

Figure 1.5 Examples of issue registration from session 1in Prospect. (Left) An issue from poor accessibility in the kitchen, including a measurement supporting the claim. (Right) An issue of wrong choice of material in the bathroom floor.

also it must be lowered", and the other one answers "The models off-set is completely wrong". Lastly, when finishing the design review one OQ user states "We made a mess", where he implies that there are a lot of problems they have not noticed when designing and reviewing the project in Revit.

Three different approaches were observed, structured, semi-structured and non-structured. In four out of eight sessions, a structured approach was conducted. In the structured approaches, the participants had made a detailed plan of the design review session in advance, which included they had prioritized and scheduled walking path as well as a prepared list of point of interest to be reviewed in a joint force. For instance, in session 1 they start with the first room on the ground floor and begin with the first thing on the list, which in this particular case was collisions. In two out of eight sessions, the approaches were non-structured. In the non-structured approaches, there was no plan for the design review session. The participants started walking in random directions and walked around looking announcing what caught the eye. In the semi-structured approaches, there was not necessarily a plan in advance, but during the design review session, the participants, especially the DT, felt a need to start structuring the process. This was mainly because it was difficult for the DT to keep track of the issues being reported.

1.3.2   Issue Report

As shown in Table 1.1, in four out of eight sessions, the issue report was documented using issue-handling tool in Prospect. In Figures 1.4–1.6, six examples of issues from the issue report are shown. In this section, these issues are analysed and classified. In total 48 issues were registered in the four sessions. In session 1 a total amount of 19 issues was registered and in relation to effectiveness 0.49 issues were registered per minute, which was the highest score in relation to total amount and effectiveness. In session 4 total amount of six issues were registered and in relation to effectiveness, 0.24 issues were registered per minute, which was the lowest score in relation to total amount of effectiveness. On average a total of 12 issues were registered whereas 0.37 issues were registered per minute.

Going through all the issue reports, the issues were categorized in relation to type, discipline and priority. For each category issues were classified, as shown in Table 1.2.

1.3.2.1   Type of Issue

Four different types of issues were identified and classified as follows.

Omission was used when something is missing from the model. A total of 12 issues were classified as omissions. Among these issues the description was "Missing suspended ceiling" or "Missing stair rail", an example is shown in Figure 1.4 (right).

Accessibility was used for one single issue, which was described as "Kitchen Accessibility", this example is shown in Figure 1.5 (left). This issue was also noted in the observation in session 1, as one user moved into the kitchen and stated that there was not

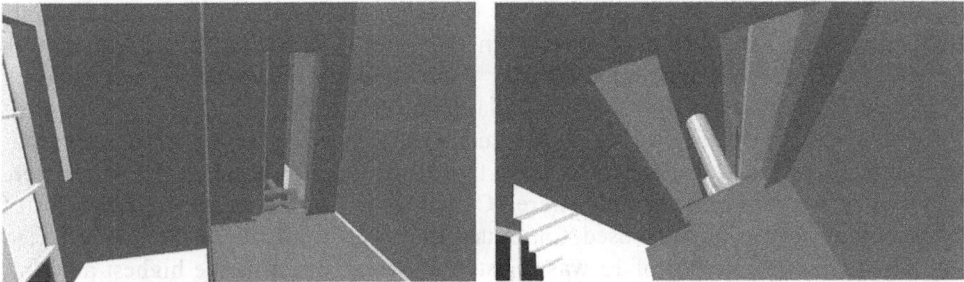

Figure 1.6 Examples of issue registration in Prospect. (Left) Collisions with elevator shaft and ventilation in session 1. (Right) Collision with elevator shafts and ventilation in section session 4.

Table 1.2 Shows issue report from Prospect, categorized by type, discipline and priority

Type	No.	Per cent	Discipline	No.	Per cent	Priority	No.	Per cent
Omission	12	25%	Architecture	35	73%	Trivial	12	25%
Accessibility	1	2%	MEP	11	23%	Low	4	8%
Hard Clash	5	10%	Structural	2	4%	Medium	12	25%
Defect	30	63%				High	13	27%
						Critical	7	15%
Total	48	100%		48	100%		48	100%

enough space for proper passage in the kitchen. This was followed up by a measurement in VR, that confirms his claim. This issue was then registered in the issue report as "Kitchen Accessibility".

Hard clash was used when two or more objects collide in the model, an example is shown in the left picture of Figure 1.4. A total of five issues were classified as hard clash. In all five cases, the issue consisted of MEP colliding with other BIM objects, such as "Collision between ventilation and bearing beam" and "Plumbing colliding with above floorplan".

Defect was used when an object has been modelled poorly or misplaced and not necessarily is clashing. A total of 40 issues were classified as defect, which consisted of 63%. Examples were "Floating Wall", "Movement of elevator shaft" and "wrong material", an example is shown in Figure 1.5 (right) and Figure 1.6 (right and left).

1.3.2.2 Discipline

The issues were classified into three different disciplines, Architectural, Mechanical Electrical and Plumbing (MEP) and Structural. With a total of 35 issues, Architectural issues were by far the highest amount registered. Followed by MEP and Structural, which consisted of respectively 11 and 2 issues. In some cases, it was not clear who was responsible for correcting the issue. In this case, this was discussed during the design review sessions.

1.3.2.3 Priority of Issue

The issues were also classified by their priority. (1) Trivial was used for issues that were easy to fix and needed no interdisciplinary coordination, such as "Ventilation supply fitting placement" and "missing glazing on the window". A total of 12 issues were registered as trivial. (2) Low was used for issues that were easy to fix, but couldn't be decided by a single person such as "Placement of suspended ceiling". A total of four issues were registered as low. (3) Medium was used for issues that were somewhat more difficult to fix and affected at least two disciplines. A total of 12 issues were registered as medium and examples were "Ceiling level height" and "Misplacement of stair landing". (4) High was used for issues that were difficult to fix and affected multiple disciplines. A total of 13 was registered as high and was the highest amount. Examples were "Ventilation duct rout collides with elevator shaft" and "Floating Bearing column". (5) Critical was used for issues that affected multiple disciplines and in solving the issue, may have caused new issues. Examples were "Movement of elevator shafts", see Figure 1.6 (left and right), and "collision between ventilation and bearing beam".

1.3.2.4 Model-Confirmation and Solution-Discussing

In the design review sessions, the participants typically reviewed the whole model. In all eight sessions, there were incidents where participants started addressing solutions of the registered issues during the design review session, in order to get an agreement on how to move on. Furthermore, there were many examples of participants confirming their model, such as the following statements from an OQ user "The bearing beam is good enough" or the conversation between an OC and DT user; OQ said "Is the suspended ceiling not too faar down?" and the DT user replied, "Let me check that ... no, it is good". It is noteworthy to mention, that confirming the model was very valuable, however, these confirmations are not shown in the issue report. Lastly, it was also observed that accessibility was a subject that was often discussed in many sessions, however, as shown in Table 1.2, only one issue was registered in issue-handling tool.

1.3.3 *Communication and Interaction*

During all eight sessions, observations were made of the participants' interactions, statements, announcements and body language. It is important to mention, that when a user in Prospect moved around, the other users could see his orientation and movement. Furthermore, for OQ users, the movement of hands was also visible for other users, making it possible to communicate through body language to a degree. In terms of oral communication, the OQ and DT have built-in speakers and microphone, making it possible to communicate orally, however, this was not necessary in this study, since all users in the design review sessions were in the same room, making it possible to communicate normally.

During the design review sessions, it was noted that in general the communication and interaction between the participants, especially communication between OQ users, was reduced to simple and concrete expressions, such as the oral conversation between three OQ users in session 1:

OQ1 speech:	*"Try entering here"*
OQ1–3 actions:	Move with their controllers
OQ2 action:	Turns his head facing upwards
OQ2 speech:	*"Hold on, that is completely wrong"*
OQ2 action:	rising his hands upwards
OQ1 and 3 action:	turn their head facing upward.
OQ3 speech:	*"There is also something missing"*
OQ1 speech:	*"OK!, this is really bad"*

For an outsider, who was not participating in the VR design review session 1, it is almost impossible to understand what they were communicating. In the first sentence, OQ1 requested the other users to enter a place "here". For the OQ users, here was not difficult to understand, since it was a simple action of following another user using the controllers. After entering the place referred to, OQ2 detected something when looking upwards. He then pointed by using his controller and expressed that "something" was completely wrong. Both OQ1 and OQ3 reacted, by looking in the direction OQ2 pointed. Both users seemed to easily understand what OQ1 was refereeing to. This led to OQ3 also expressing that something was also missing, which OQ1 and OQ2 seemed to "silently" agree with. Lastly, OQ1 stated that this was bad. The above-mentioned example took place in session 1. The place referred to by OQ1 was the Common Room on Ground Floor, see Figure 1.2. The "that" referred to by OQ2 was a ventilation colliding with a bearing beam, see Figure 1.7 (left). The "something" referred to by OQ3 was a missing suspended ceiling. The "This" referred to a bigger problem including multiple issues that needed attending to by multiple and interdisciplinary actions.

Another example of simple and concrete expressions is during session 7, where three OQ and one DT users were participating. In advance of this particular conversation, the

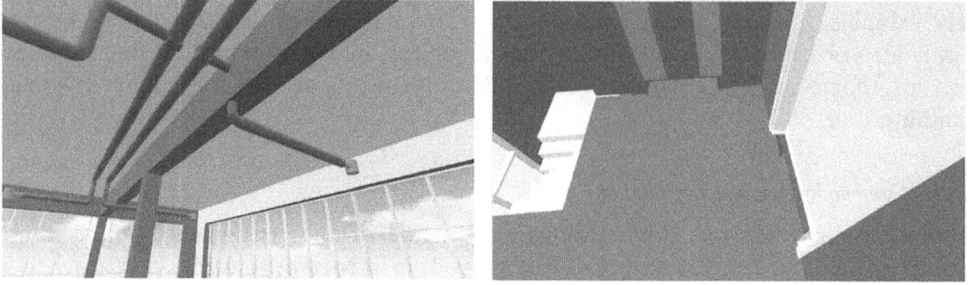

Figure 1.7 Examples of issue registration in Prospect. (Left) An issue during session 1 of a hard clash between a bearing beam and ventilation. (Right) An issue during session 7 of the floor deck is missing underneath the door.

OQ user in question wandered alone around to a distant place of the model, while the rest of the users were busy registering other issues. During this time the OQ user in question discovered the issue shown in Figure 1.7 (right), in which the floor deck was missing at a specific location. However, at the moment of time, DT user was busy. After a while, the conversation started.

DT speech:	*"Were was it the floor deck was missing?"*
OQ1 speech:	*"How should I explain ..."*
OQ1 action:	stops and thinks a few seconds
OQ1 speech:	*"I'll move to the place in question"*
OQ1 action:	*starts move towards the place in question*
DT, OQ2 and OQ3 action:	*starts to follow OQ1.*

For an outsider, who was not participating in the VR design review session 7, it is almost impossible to know where the place in question is located. All the concrete information we get is that the floor deck is missing somewhere in the model, which is a fairly broad demarcation. Even though OQ1 knows of the place in question himself, it is too difficult for him to explain orally. According to Figure 1.1, an explanation through oral communication requires the coding and decoding between the sender (OQ1) and receivers (DT, OQ2 and OQ3) on a more abstract level. However, after some consideration, the OQ1 user decides to move to the place in question. With no opposition, the rest of the group move to the location. Even though, the oral communication happens in the VR environment, the act of explaining it orally still requires high abstraction. On the other hand, by moving to the place in question in VR, the act of delivering information from sender (OQ1) to receivers (DT, OQ2 and OQ3) is reduced to a simple movement, by pressing the joystick, hence the communication is on the lowest level of abstraction.

The two above-mentioned examples were but one of many. Other statement examples are as such; "I'm standing right behind you", "There's nothing here", "That one is too high up", "We made a mess" and "It doesn't connect here". Furthermore, in general lots of body language was observed, such as pointing or waving around to address a single issue or an area with issues and giving direction when guiding another user around in the model.

1.3.4 Survey

After the two-day's introduction to VR and optional VR-supported design review sessions during the DDD, a survey was conducted. The purpose was to get the students' individual viewpoint about the multi-user VR-supported design review. A total of 37 respondents completed the questionnaire, whereas 25 of them participated in the design review sessions. Even more so two respondents were present at the VR sessions, but did not participate in the sessions as an OC or DT user. Lastly, ten of the respondents only participated in the two days of introduction to VR. Twenty-two of the respondents have experience in the AEC industry were among them 17 have primary experience within the construction phase. The other 15 of them have no experience prior to their current study, as shown in Figure 1.8. In relation to respondents that participated in the VR sessions, the demography of work experience is somewhat similar to the total. Therefore, even though the respondent in this study are students, many of them still have a significant amount of practical experience to support their viewpoint.

1.3.4.1 Strengths and Drawbacks

The result in this section is based on all 37 respondents. In the survey, the respondents were asked to describe strengths and weaknesses in relation to VR in the AEC industry. The results are coded and summarized in Table 1.3. In general, the strengths are that VR improves the understanding of the project, especially compared to 2D and desktop version of 3D models. For instance, respondent number (RN) 34;

RN 34 states: "*It is possible to discover possible errors in buildings, as well as get an understanding of how the building will appear in the real world. There is a big difference from a screen in 3D and then standing in the building in a virtual world*".

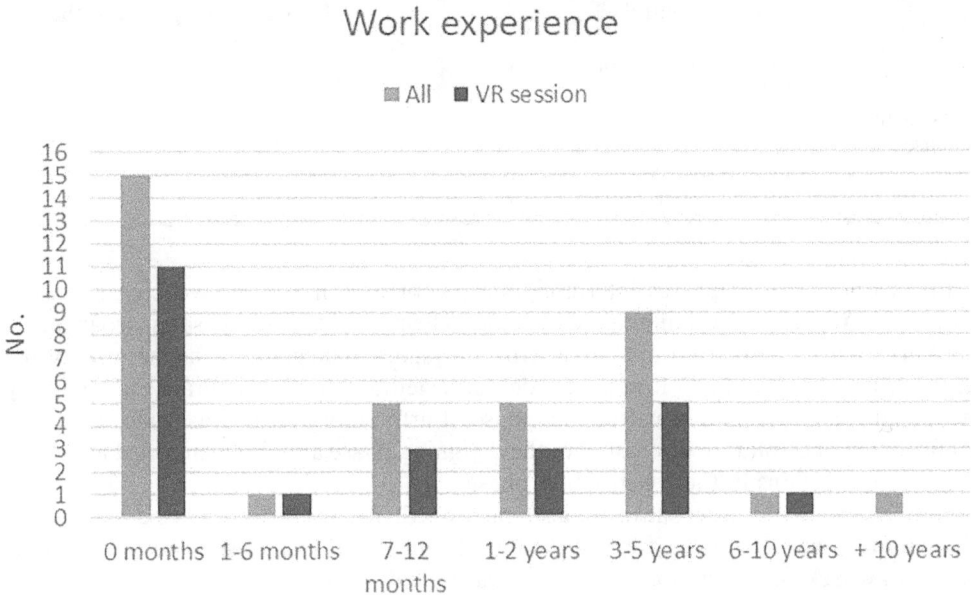

Figure 1.8 Amount of work experience by the respondents.

Table 1.3 Shows the strengths and drawbacks of VR appointed by the respondents

	Strengths	*No.*		*Drawbacks*	*No.*
General	Improves understanding	12	General	Difficult with new technology	12
	Improved visualisation	8		Value vs cost and use of time	9
	Improved working process	5		Time-consuming and difficult to set up	8
	Superior to 3D	4		Clumsy to use	2
	Improved interdisciplinary cooperation	3		Motion sickness	1
	Superior to 2D	3			
Presentation	Client	12	AEC	Expensive to implement	14
	Building contractor	4		Lack of skills in the industry	12
	Investors	3		Resistance of new technology adoption	7
	Sale	3		Conservatism in AEC-industry	3
				Requires broad agreement among the AEC-industry	2
				Difficult to implement for small and medium-sized enterprises	2
Design phase	Quality assurance	24	Technology	Inferior Clash detection compared to clash detection algorithms	4
	Design Review	22		Requires a model with high level of LOD	4
	Collision control	4		Difficult to keep track with many users	1
	Model-confirmation	4			
	Solution-discussing	3			
Construction phase	Quality assurance	22			
	Design Review	16			
	Building instructions	4			

Furthermore, VR improves interdisciplinary cooperation and the working processes in general. More specific building contractors, clients, investors and sales departments would particularly benefit from VR. In terms of quality management, the majority of the respondents point out that VR improves design review and quality assurance in especially the designing phase and construction phase. Furthermore, strengths such as model confirmation and solution discussion in the designing phase are also mentioned as well as building instructions in the construction phase.

On the other hand, drawbacks are also mentioned. The most drawbacks described are difficulty with new technology, expensive to implement and lack of skills in the industry. These drawbacks are of a more general in nature and are supported by drawbacks such as time consuming and difficult to setup, resistance of new technology adoption, conservativism in the AEC industry. For instance;

RN 20 "*… in order to achieve a proper implementation of VR, it requires a large knowledge base to, which we currently don't have*".

Following on from that, nine respondents also point out that it is not known whether or not the value of the use of VR is good enough when compared to costs and time consumption.

RN 29 states: "*I think it is difficult for companies to know whether it is a good investment, and I also think that the investment and retraining of employees there is the biggest challange*".

In relation to the technology, four respondents point out that a model needs a certain Level of Detail (LOD), for the AEC industry to gain from using VR. Furthermore, four respondents mentioned that VR was inferior to clash detection compared with clash detection algorithms such as in Navisworks.

RS 28 states. "*It can be difficult to persuade people to use it when, for example, when you can do clash detection in Navisworks, without the use of VR glasses*".

1.3.4.2   VR versus Traditional Design Review

The result in this section is based on all 37 respondents. When comparing design review carried out in 2D drawings and 3D view with VR, the bar chart in Figure 1.9. speaks quite unambiguously. To a high degree, the respondents either strongly agree or agree, that compared with both traditional 2D drawings and 3D views, VR gave them (1) a better understanding of the project as a whole, (2) more spatial understanding of the project, (3) a better understanding of the interdisciplinary work and (4) a better opportunity to communicate issues.

Although it is worth mentioning that there were significantly more respondents that are neutral or disagreed when it comes to a better understanding of the interdisciplinary work. For instance, compared with 3D view, a total of four respondents (14.8%) disagreed and five respondents (18.5%) are neutral. It is also worth mentioning that in relation to communicating in 3D view versus communicating in VR, for each answer one respondent answered "I don't know", "Disagree" and "Neutral".

1.3.4.3   Value of VR in Design Review Sessions

The result in this section is based on the 27 respondents, who used were presented during the design review sessions. This section investigates the respondents' own experience on the design review sessions described in Sections 3.1 to 3.3, see also Table 1.1.

The results shown in Figure 1.10 speak quite unambiguously, respondents that participated in design review sessions aider strongly agree or agree, that (1) reviewing the project in general gave them value, (2) a lot of issues were detected when reviewing the model in VR, (3) without VR, they would not detect some issues and (4) reviewing the project with multiple VR users gave them value. In the survey, there were many examples supporting these results such as. In relation to value of VR

RN 12: "*We found errors that we couldn't find in 3D Revit, and as an architect it gave me a better overview*".

Compared to 2D drawings, reviewing in VR gives me:

■ Strongly   ■ Agree ■ Neutral   ■ Disagree   ■ Strongly   ■ I don't know
Agree   Disagree

a better understanding of the project as a
whole

a more spatial understanding of the project

a better understanding of the
interdisciplinary

a better opportunity to communicate issues
(when using many VR users)

100%   0%   100%

Compared to 3D-view, reviewing in VR gives me:

■ Strongly   ■ Agree ■ Neutral   ■ Disagree   ■ Strongly   ■ I don't know
Agree   Disagree

a better understanding of the project as a
whole

a more spatial understanding of the project

a better understanding of the
interdisciplinary

a better opportunity to communicate issues
(when using many VR users)

100%   0%   100%

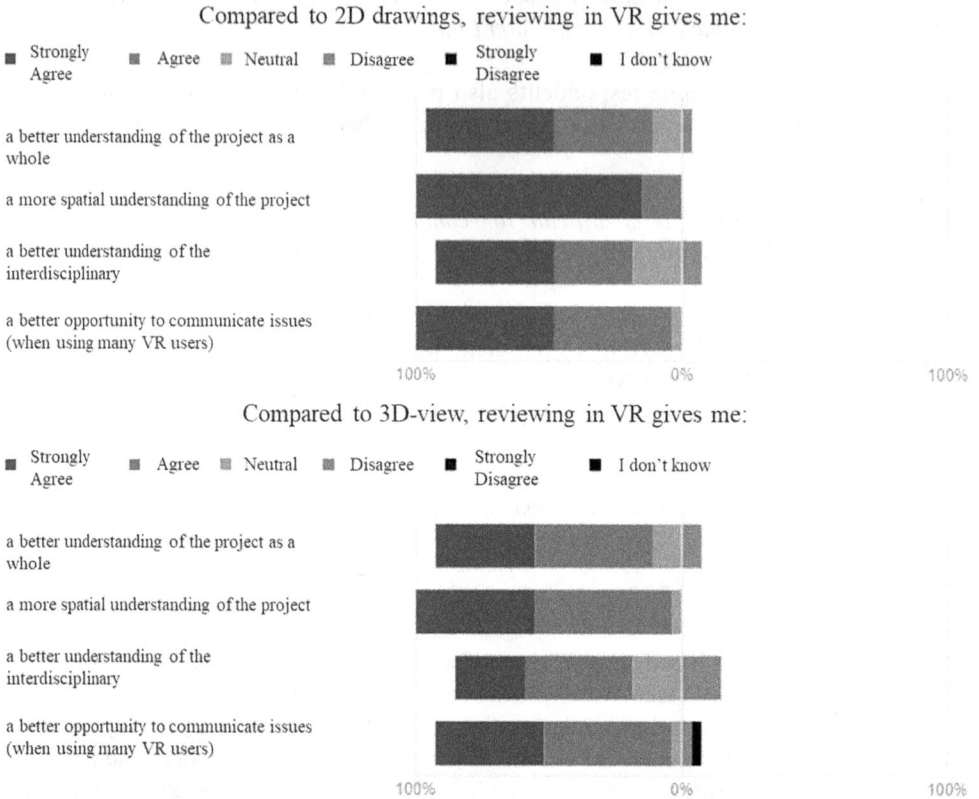

Figure 1.9 The degree to which the respondents agree or disagree to the statement listed above the charts.

RN 15: *"Being able to stand inside the building yourself gives a better understanding of the entire project, and it is much easier to see and to perform a quality assurance on the building".*

RN 23: *"It was great to be able to walk around inside our building. It gave a better understanding of room size, corridor widths, stairs, ceiling height, etc. in 1:1".*

RN 25: *"It added value to the project because we could see and discuss things that were not visible in our 2D/3D drawings".*

Respondents also mentioned many examples of issues they would not discover without a VR design review session such as

RN 2: *"There were many technical installations that collided. There were building parts that collided or were not connected".*

RN 12: *"we had some ventilation pipes that we had made holes in the walls to implement. - We then moved these ventilation pipes, and it turned out when we performed quality assurance in VR that there was still a hole in the wall".*

RN 14: *" ... the elevator shaft collided with a ventilation pipe. It would have been difficult to see in 3D".*

Value of Virtual Reality in Design Review Sessions

■ Strongly   ■ Agree   ■ Neutral   ■ Disagree   ■ Strongly   ■ I don't know
Agree   Disagree

Reviewing the project in VR gave us value

A lot of issues was discovered when
reviewing the model in VR

Without VR, we would not discover some
issues

Reviewing the project with multiple VR users
gave us value

100%   0%   100%

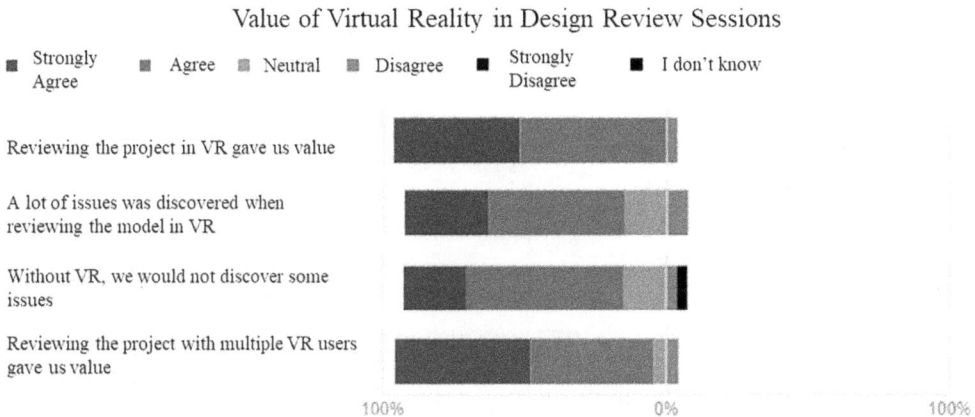

Figure 1.10 The degree to which the respondents agree to the statement listed above the charts.

RN 18: *"We could detect some other issues in VR than we could with Navisworks"*.

In relation to multiple users in VR, there were also many examples supporting the results. Among other things respondents described

RN 3 *"You can coordinate the troubleshooting and discuss the errors on the spot"*.

RN 11; *"Because everyone got a common understanding of how the building looked at the same time as being able to see the things that were wrong"*

RN 17: *"It gave understanding to several students at once, and therefore we could pass on the answers better to the rest of the group"*.

RN 19 *"Quality assurance became very fluid as we could ask questions immediately and also point to the area in the meantime"*.

As shown in Figure 1.11 in general respondents agree to strongly agree to the questions regarding multiple users in VR. In relation to communication and interaction with multiple users being easy, a total of one Strongly disagrees, four disagree and two were neutral. In addition to that, the answers were somewhat similar in relation to interaction and usage of body language to communicate. Lastly, 91.3% of the respondents agreed or strongly agreed that they used the virtual model to support their explanation.

1.4   Discussion

Due to the nature of the setup in the DDD, where the primary aim was for students to learn, the process in the design review sessions varied a lot. Seemingly it appears, that a well-organized and structured session supported by BCF supported issue-handling tool was the most efficient design review session. However, technically this study only has documentation for the effect of design review for sessions, in which BCF supported issue-handling tool was used. This claim is also supported by Gade et al., who conclude that for VR sessions, even ordinary agenda, with topics for discussions is not enough (Gade et al., 2014). They point out that routes, specific junctions or tasks had to be included. On the other hand, one could argue, that in order to do a strictly planned design review session,

Multiple VR user Design Review

■ Strongly   ■ Agree   ■ Neutral   ■ Disagree   ■ Strongly   ■ I don't know
Agree   Disagree

It was easy to communicate to another VR user

It was easy to interact

I used my body language to communicate
(pointing etc.)

I used the virtual model in order to support my
explanation

100%   0%   100%

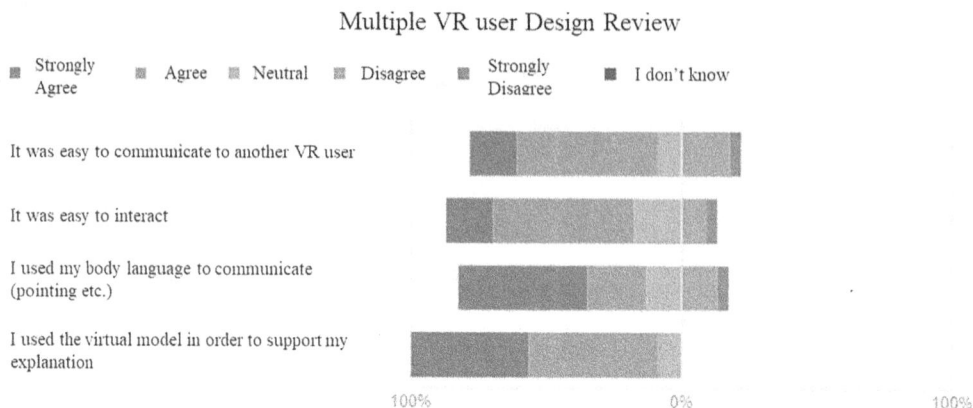

Figure 1.11 The degree to which the respondents agree to the statement listed above the charts.

the points of interest have to be known beforehand, which is a somewhat paradoxical problem. In that sense, a structured design review session should be followed up by an unstructured exploration.

In order to evaluate the effect of multi-user VR-supported design review, the ratio of issues per minute is calculated in Table 1.1. This calculation presumes that the design review is only effective if an issue is registered. However, as stated in Section 3.2.4, in the design review session a significant amount of time was also spent on model confirmation and discussing a solution of the issues registered. This is also supported in the survey, where respondents were asked to describe strengths and drawbacks of VR. Four respondents describe model confirmation and three respondents that solution-discussing as a strength of VR. For instance RN 5 states "... *as well as discuss solutions in the model together with different actors*". Obviously, these activities are of value to the building project, therefore the ratio of issue per minute is not an accurate expression for measuring the effect of the design review.

In this chapter a total of eight multi-user VR-supported design review sessions were studied, whereas in four sessions only two users participated. Furthermore, in one of these sessions, there was only one OQ was participating. On the other hand, the other four sessions 3–6 users were participating. In that sense, this study includes variety of data, that potentially tell different truths. According to the survey in Section 3.4.3, see also Figure 1.10. question 4, a total of 21 (91.3%) of respondents strongly agree or agree that reviewing the design with multi-VR users gave value for their building project. Among other things, they mentioned that multi-user in VR gives a common understanding of the project and makes it possible to coordinate and troubleshoot on the spot. By observation of the sessions and analysis of the data, there are also indications that the number of users has a negative effect on the quality of the sessions. For instance, in sessions with many users, one or two participants lose focus and tend to go goofing around. This is also supported by Section 3.4.1, where one respondent, RN 6, associates multi-user in VR with a drawback. RN 6 states: " *It is difficult to find out where people are and are you look at the same thing? ... several actors in the same model can seem messy and difficult*". All in all, this study indicates that multi-users in VR-design review are of value and at the same time that too many users can be troublesome. However, this study does not have enough data to conclude the optimal number of users.

In the survey in Section 3.4.1, see also Table 1.3, some respondents stated that VR-supported design review was not as efficient as Navisworks, as it automatically detects clashes in relatively short time. At the same time, in the survey, Section 3.4.3 contradicts this, as some respondents state that they detected many issues that they could not in Navisworks. Furthermore, this statement is supported statistically by question three in Figure 1.10, as the majority of users agree that without VR they would not have detected some issues. For all sessions, a clash detection was planned to be run in Navisworks, which could have made the participants less focused on hard clashes. The consequence could be the reason why only 10% of the issues detected were hard clashes. However, stating that clash detection in ex. Navisworks is sufficient in itself, but is a bit overvalued. Because the other 90% of the issues could not be detected by the automatic clash detected algorithm. Korman et al. state that there are certain types of clashes, which are difficult for algorithms to detect (Korman et al., 2003). Furthermore, despite significant endeavours at improving clash detection algorithms, the tools throw up a large number of irrelevant clashes, which require a lot of time and resources to sort (Akponeware and Adamu, 2017). For example, Wang et al. find that 78% of automated clashes detected in the project they studied were insufficient (Wang et al., 2016).

Many of the drawbacks identified in the survey in Section 3.4.1 are general, see also Table 1.3. These are drawbacks such as difficulty with technology, difficult to setup, lack of skills in the industry, resistance to new technology adoption and conservatism in the industry. One could argue that all the above-mentioned drawbacks could be improved by more knowledge and skills within VR. As a note on the final open comment, RN 25 states "There just needs to be a lot more of it, if that's the direction the construction industry is going, I think students should be ready for that kind of digitization". By "it", the respondent refers to education in VR. This statement is supported by several trade associations within the Danish AEC industry. In a report made by Quartz, they investigate the most trending technologies within the MEP industry, among which VR is one of these technologies. They conclude that in order to obtain the expected improvement with the new technology, it's necessary for the people involved to acquire the necessary skills (Qvartz, 2017).

Observation of the participants during the multi-user design review sessions shows that communication and interaction between the VR participants was reduced to simple and concrete expressions, that relate to the reality the participants are situated in. This is a linguistic phenomenon, in which it denotes the relationship between the communicated content and the communication situation. In their conclusion Kragh and Strudsholm state that, Deixis is a central element in less complex contexts, and is often replaced by unambiguous and non-situational references in more complex contexts (Kragh and Strudsholm, 2015). The results in Section 3.3 Communication and Interaction indicate that VR allows for a setting that is less complex, enabling the participants to communicate and interact in the same frame of reference. This is also supported by the survey in Section 3.4.3, see also Figure 1.11. Most respondents agree to strongly agree that it is easy to communicate and interact in a VR session with multiple VR users. Furthermore, that they use body language and the virtual model to support their communication.

In the survey the respondents are asked about their work-related experience prior to their study, see also Figure 1.8. As shown in the figures in Sections 3.4.2 to 3.4.3, most respondents strongly agree or agree on the questions, hence this study in general indicates that work experience does not affect the individual opinion of the respondents. Furthermore, no significant difference caught the eye, when the respondents were asked to elaborate on their answers, no significant difference was found based on work experience.

1.5   Conclusion

The aim of this study was to investigate the value of multi-user VR-supported design review of buildings in a multidisciplinary collaboration. The study was based on both qualitative and quantitative studies, where students in AEC education were observed, while performing VR-supported design review. Furthermore, issue reports were analysed and lastly their own opinion and experience were surveyed in a questionnaire. The main findings are:

- VR-based Design Review Process: The study indicates that a well-organized and structured session, where they point out that route, specific junctions or tasks had to be included, supported by BCF-supported issue-handling tool was the most efficient design review session. However, it is also important to follow up by an unstructured exploration, as all points of interest are not necessarily known.
- Issue Report: Based on the issue reports conducted in Prospect, issues were categorized into type, profession and prioritize. Sixty-three per cent of the issues were identified as defects and 73% were related to architecture. In relation to prioritized, the issues were more distributed. With the highest ratio, 0.49 issues were detected per minute, whereas the lowest was 0.24. However, it is also worth mentioning that during the design review session participants used time on model-checking and solution-discussing, which also gave value to the design review.
- Communication and interaction: Observation revealed that communication between participants in VR was reduced to simple and concrete expressions. The survey unambiguously tells that the participants agree that VR makes it easier to communicate and interact with other participants. This indicates that VR enables a setting for a common frame of reference, improving the interdisciplinary collaboration.
- Value of multi-user VR: In general the three above points argue positively for the value of multi-user VR. This is significantly further supported by the participants' own opinions and experiences.

This study was based on students, in which most of them have somewhere in-between seven months and five years of experience in the AEC industry. In that sense, the data can be seen as somewhat indicative for practice. However, the nature of the setup is limited compared to actual operation, therefore, a similar study in actual construction project will give more transferrable results and an insight on the value design review processes supported by multi-user VR real settings. Furthermore, when measuring the effect of multi-user VR-supported design review the ratio of issues per minute is not sufficient. A future study should also include a way of measuring the effect of model confirmation and solution-discussing during the design review session.

References

Akponeware, A.O., Adamu, Z.A., 2017. Clash detection or clash avoidance? An investigation into coordination problems in 3D BIM. Buildings 7. 10.3390/buildings7030075

Berlo, D.K., 1960. The Process of Communication: An Introduction to Theory and Practice. New York.

de Digitale Dage, 2022. De Digitale Dage 2022 afholdes d. 6–8 April 2022. [WWW Document]. De Digitale Dage. https://dedigitaledage.dk/ (accessed September 15, 2022).

Gade, P., Jensen, C.G., Nissen, S., Christensen, D.H., Knak, H., Gregersen, M.L., Bertel, F.G., Holm, J., Wulf, T., Andersen, M., 2014. Virtual reality – De Digitale Dage. Aalborg.

Golledge, R.G., 1991. Cognition of physical and built environments. in: Garling, T., Evans, G.W. (Eds.), Environment, Cognition, and Action: An Integrated Approach. Oxford University Press, Inc, pp. 35–62.

Haahr, M.T., 2021. XR-GRANSKNING AF INSTALLATIONER: Extended reality-teknologier til granskning, produktion samt kvalitetskontrol af tekniske installationer i byggeriet – et indledende scoping review. UCN PERSPEKTIV 9, 45–57.

Haahr, M.T., Svidt, K., Jensen, R.L., 2019. How can virtual reality and augmented reality support the design review of building services, in: Proceedings of the 19th International Conference on Construction Applications of Virtual Reality (CONVR2019). Bangkok, Thailand, pp. 84–93.

Hall, S., 1980. Encoding/decoding, in: Hall, S., Hobson, D., Lowe, A., Willis, P. (Eds.), *Culture, Media, Language*. London, pp. 128–138.

Johansson, M., Roupé, M., 2019. BIM and virtual reality (VR) at the construction site. in: Proceedings of the 19th International Conference on Construction Applications of Virtual Reality, Bangkok, Thailand, pp. 1–10.

Kjems, E., 2000. Er tiden inde til at tage det næste skridt ind i VR verdenen? *Dansk Vejtidsskrift* 5, 3.

Korman, T.M., Fischer, M.A., Tatum, C.B., 2003. Knowledge and reasoning for MEP coordination. *J Constr Eng Manag* 129, 627–634. 10.1061/(ASCE)0733-9364(2003)129:6(627)

Kragh, K.J., Strudsholm, E., 2015. Deiksis i sprog og kontekst. *Scandinavian Studies in Language* 6.

Kwiatek, C., Sharif, M., Li, S., Haas, C., Walbridge, S. 2019. Impact of augmented reality and spatial cognition on assembly in construction. *Autom Constr* 108. 10.1016/j.autcon.2019.102935.

Malmgreen, H., 2020. Ny analyse: Alt for mange fejl og mangler koster dyrt i den danske byggebranche [WWW Document]. Finans. https://finans.dk/erhverv/ECE11920119/ny-analyse-alt-for-mange-fejl-og-mangler-koster-dyrt-i-den-danske-byggebranche/?ctxref=ext (accessed September 25, 2022).

McKinsey Global Institute, 2015. The construction productivity imperative.

PROSPECT BY IRISVR, 2022. Prospect for Meta Quest 2 [WWW Document]. Prospect. https://irisvr.com/oculus-quest (accessed September 15, 2022).

Qvartz, 2017. INSTALLATION 4.0 - Teknologiske Tendenser Med Betydning For Installationsbranchen Frem Mod 2025.

Shirkavand, I., Lohne, J., Lædre, O., 2016. Defects at Handover in Norwegian Construction Projects. *Procedia Soc Behav Sci* 226, 3–11. 10.1016/j.sbspro.2016.06.155

Wang, J., Wang, X., Shou, W., Chong, H.-Y., Guo, J., 2016. Building information modeling-based integration of MEP layout designs and constructability. *Autom Constr* 61, 134–146. 10.1016/j.autcon.2015.10.003

Windham, A., Liu, F., 2018. Comparing Perceptions of Occupant low and space functionality in virtual reality and an actual space, in: 18th International Conference on Construction Applications of Virtual Reality. Auckland, New Zealand, pp. 266–274.

2 Towards Personalized Mixed Reality-Based Learning Experience in Construction Education

Omobolanle Ogunseiju[1], Nihar James Gonsalves[2], Abiola Abosede Akanmu[3], and Diana Bairaktarova[4]

[1]*School of Building of Construction, College of Design, Georgia Tech, Atlanta, GA, USA;* [2]*Myers-Lawson School of Construction, Virginia Tech, Blacksburg, VA, USA;* [3]*Department of Smart Design and Construction, Myers Lawson School of Construction, Virginia Tech, Blacksburg, VA, USA;* [4]*Department of Engineering Education, Virginia Tech, Blacksburg, VA, USA*

2.1 Introduction

As the construction industry continues to technically advance, there is a high adoption rate of sensing technologies such as laser scanners, cameras, and radiofrequency identification devices (Ogunseiju et al., 2021a). Compared to other sensing technologies, laser scanners are one of the most adopted, but institutions are lagging in equipping students with the technical capabilities of laser scanning (Ogunseiju et al., 2021a). This could be due to limited access to the construction site for hands-on learning, the high up-front cost of these technologies, and unformalized competencies. One way to address this is by adopting alternative and captivating learning environments such as mixed reality (MR), where students can perform hands-on implementation of laser scanners for authentic and constructive learning experiences (Yusoff et al., 2010) without exposure to jobsite hazards.

In a previous chapter, the authors (Ogunseiju et al., 2021b) presented an MR environment for learning sensing technologies, with an example of laser scanning. However, to implement such learning environments as a pedagogical tool in construction education, a usability study was conducted. The usability study employed eye-tracking and think-aloud protocol as measures of evaluating the quality of the learning experience in the environment. This is because learning is usually informed by the perception of stimuli and information in the environment (Andrzejewska and Stolińska, 2016). While the think-aloud protocol provides immediate feedback on the environment, learning, and cognitive activities can often be inferred by exploring learners' eye-tracking data (Andrzejewska and Stolińska, 2016). Since learning can vary based on individual needs and abilities (Toker et al., 2014), detecting cognitive stages, and human-computer interactions can provide opportunities for designing dynamic and intelligent virtual learning environments in construction. However, eye-tracking data is highly multi-dimensional, and manual comparison of different eye metrics can be computationally expensive and time-consuming. For instance, Lee et al. (2019) explained that while it is visually possible to assess scan paths for two images, it becomes more challenging to compare multiple scan paths. Zemblys et al. (2018) explained event detection from eye-tracking data is challenging when conducted manually. Hence, advanced techniques such as machine learning are required to further provide deep insights into eye gaze patterns.

DOI: 10.1201/9781003408949-2

In this chapter, we present machine learning models that employ eye-tracking data for detecting learning stages and users' interaction difficulty during laser scanning activities in an MR environment. The findings revealed that eye movement data possess significant information about learning stages and interaction difficulties. The classification models will help detect users who require additional support to acquire the necessary technical skills for deploying laser scanners in the construction industry and inform the specific training needs of users to enhance seamless interaction with the learning environment. The findings of this study provide evidence of the potential of a smart MR environment for improved learning experiences in construction education.

2.2   Background

This section presents a review of MR environments in education and eye-tracking as a usability measure of virtual environments. This section also provides a review of machine learning for providing insights into eye-tracking data.

2.2.1   *MR/AR Environments in Construction Education*

MR and augmented reality (AR) environments have been widely adopted in the construction industry for enhancing safety (Moore and Gheisari, 2019), facility management (Ensafi et al., 2021), and design communications (Chalhoub and Ayer, 2018). Likewise in construction education, MR has been employed as a learning environment for simulating construction activities in the classroom. For example, Shanbari et al. (2016) employed MR to project environmental contents and constraints of construction processes. The study revealed the efficacy of MR in improving the memorability of roofing and masonry work. Bosché et al. (2016) presented the potential of MR for simulating difficult construction conditions such as work-at-height. With an example of a wood framing lab, Wu et al. (2018) presented the potential of MR for acquiring tacit knowledge and revealed its efficacy for design comprehension and for technical skills acquisition for building wood-frame structures. Despite the opportunities of MR environments in construction education, reports for acquiring hands-on experiences of currently adopted technologies are scarce. As such, the authors in a previous chapter (Ogunseiju et al., 2021b) presented the potential of an MR environment for learning different sensing technologies currently adopted in the industry. It is also imperative to conduct a usability test for evaluating its capability to enable users to achieve specified goals with effectiveness, productivity, and satisfaction.

2.2.2   *Eye-Tracking for Usability Studies*

Several measures such as usability questionnaires, think-aloud protocol, and eye-tracking are often used for procuring data during usability studies. While usability questionnaires and think-aloud protocols often provide subjective data, eye-tracking is often employed for assessing usability evaluations of virtual environments and user interfaces involving cognitive activities and human-computer interactions. For example, visual attention from eye-tracking was used to analyze end users' satisfaction with construction design (Mohammadpour et al., 2015). Similarly, eye-tracking data revealed users' preferences and performance during engineering design reviews (Satter and Butler, 2015). Eye movement data such as fixations have also been explored for inferring cognitive processes during steel installation in an AR learning environment

(Wang et al., 2018). In construction safety, Ye and König (2019) employed eye-tracking data such as scan paths and heatmaps for hazard identification in a virtual environment. However, inferring high-level cognitive activities and interaction difficulties is imperative for reducing cognitive load and designing intelligent and dynamic virtual learning environments in construction education, and this has not been explored in these studies.

2.2.3   *Machine Learning for Understanding Eye-Tracking Data*

Machine learning has been effective in detecting users' cognitive states, performance levels, and learning stages from eye-tracking data. For example, Eivazi and Bednarik (2011) employed machine learning techniques on eye-tracking features and labeled high-level behaviors to predict cognitive states for intelligent user interfaces. By assessing think-aloud protocol data, five cognitive states: cognition and evaluation, intention, planning, and concurrent move were classified (Eivazi and Bednarik, 2011). The study revealed that while cognition was difficult to predict, the performance level of participants during the cognitive activity was predicted more accurately.

By assessing eye-tracking data, Toker et al. (2014) employed machine learning techniques for predicting users' learning phase during a visualization task. Users repeated experimental tasks twice, and the model was trained to predict the 'before skill acquisition' phase (first trial) and 'after skill acquisition' phase (second trial) owning to a statistical difference in the task completion time during both trials. Likewise, Conati et al. (2020) predicted users' cognitive abilities using eye-tracking and users' interaction data. Notably, fixation was the common eye-tracking metrics adopted in the reviewed studies, although the adopted machine learning classifiers varied. For example, Eivazi and Bednarik (2011) adopted Support Vector Machine (SVM), Toker et al. (2014) adopted Logistic regression (LR), and Conati et al. (2020) adopted Boosted LR and Random Forest (RF) for developing their classification models. In the same vein, Lagun et al. (2011) employed both LR and SVM for detecting cognitive impairments during behavioral tests, and Salminen et al. (2019) adopted Neural Network (NN) and RF for inferring users' confusion levels from eye-tracking data.

Similarly, Schechter et al. (2021) employed Feed-Forward Neural Network (FFNN) and Light Gradient Boosting Machine (LGBM) on eye-tracking data to predict confusion during silent reading. However, the developed algorithm revealed low accuracy, which the authors related to an insufficient training dataset and a poor eye-tracker. However, the main obstacle to exploring machine learning for eye-tracking data is getting the appropriate data inputs and machine learning methods (Bednarik et al., 2012, Eivazi and Bednarik, 2011). To this end, literature was reviewed to investigate eye-tracking features and user-level data commonly adopted for machine learning systems.

Dalrymple et al. (2019) employed machine learning to understand how eye movement patterns vary according to age. Conati et al. (2020) explored only fixations for training their classification model, which generated a high accuracy. However, the more the eye-tracking data inputs, the better the classification's performance. For example, Lagun et al. (2011) reported an increased classification performance when all eye-tracking data was employed for training the classifiers than when the baseline feature (Novelty preference) was used. However, different eye-tracking data inputs such as fixation duration, and fixation start time often possess different underlying information and should be selected based on the embedded cognitive information and impacts on improving classification performance during machine learning. For example, fixation duration has been revealed to carry significant information and insights to usability studies. Goldberg and Helfman (2010) explained that a longer fixation duration may infer confusion of the perceived information. Similarly, fixation duration increases as the mental effort and information processing for a task increase or becomes more demanding (Shojaeizadeh et al., 2019, He and Mccarley, 2010). Also, fixation positions can reveal cognitive processes in the mind and indicate users' tasks. Since learning was performed in two different scenes, fixation positions can provide more insights into the classification model. The fixation distance provides further validation of the fixation positions as this is computed as the Euclidian distance of the fixation positions. In addition, fixation start time creates a continuous time sequence that could better improve the classifier's performance. Salminen et al. (2019) revealed that data inputs creating sequences and connections with preceding data can improve the performance of classification models. Hence, to achieve a robust training dataset, this study employed fixation duration, fixation start time and fixation positions, and fixation distance (Table 2.1).

Table 2.1 Machine learning techniques for eye-tracking data

Sources	Eye-Tracking Features and Data Labels	Objectives	Machine Learning Classifier and Sample Size	Evaluation Measures
(Shojaeizadeh et al., 2019)	Fixation, saccade, blink, pupil dilation, eye movement ratio Data label: Task load	Task load detection system	RF (highest accuracy) SVM (low accuracy) 48 participants	Accuracy
(Salminen et al., 2019)	Fixation duration, fixation positions (X and Y coordinates), fixation distance, dispersion, and fixation start, demographics (age and gender), and Areas of Interest (AOI) Data label: confusion levels	Predicting user's confusion	RF (highest accuracy) NN (low accuracy) 29 participants	Accuracy, precision, recall, and F1 scores
(Eivazi and Bednarik, 2011)	Mean fixation duration, total fixation duration, average and total path distance, fixation counts, fixation rate, and visited rates, task completion time behavioral codes: Data label: cognition states	visual attention patterns and problem-solving behavior predictions	SVM 14 participants	Accuracy

(*Continued*)

Table 2.1 (Continued)

Sources	Eye-Tracking Features and Data Labels	Objectives	Machine Learning Classifier and Sample Size	Evaluation Measures
(Toker et al., 2014)	Fixation count, fixation duration, saccade length, relative and absolute saccade angles, Data label: learning acquisition phases	Predict users' skill acquisition phase	LR 62 participants	Accuracy
(Lagun et al., 2011)	Novelty preference, fixation duration, Re-fixations, saccade orientation, pupil diameter, Data label: performance measures	Detecting mild cognitive impairment	LR (highest accuracy), SVM, Naive Bayes (NB), 50 participants	Accuracy, sensitivity, specificity, Area under (ROC) curve
(Bednarik et al., 2012)	Fixations: fixation duration, saccades, and pupil diameter. Data label: Intentions	Intention prediction	SVM 13 participants	Classifier accuracy, Area Under Curve (AUC)
(Schechter et al., 2021)	Fixations, and pupil diameter, Data label: self-reported confusion ratings	Confusion prediction	FFNN and LGBM 10 participants	Accuracy
(Conati et al., 2020)	Fixation rate and duration, relative and absolute saccade angles, pupil width and head distance, mouse click rates, interface action features. Data label: cognitive abilities	Predicting users' cognitive abilities	Boosted Logistic Regression, Random Forest 166 participants	Accuracy
(Kandemir and Kaski, 2012)	Fixation duration, saccade length, and pupil area Data label: Intentions	Predicting relevance on user interface	Gaussian Process 12 participants	Area under ROC, precision, recall curve, accuracy
(Dalrymple et al., 2019)	Fixations, image features. Data label: toddler's ages	Predicting toddler's ages	Deep learning and SVM 73 participants	Accuracy, sensitivity, specificity, Area under ROC

2.2.4 Theoretical Framework

The development of the holographic scenes was guided by the cognitive apprenticeship theory. The theory posits that cognitive development happens when learners' previous knowledge is enhanced through the reification of cognitive activities (Hennessy, 1993). Collins et al. (2018) explain that learning can be enhanced through modeling, scaffolding, coaching, fading, articulation, and reflection on their problem-solving skills. Hennessy (1993) further explained that in a classroom setting where cognitive apprenticeship is adopted, students seek help from adults or other more knowledgeable peers, but with the advent of digital learning environment, scaffolding can be achieved via guided participation in social activity. In particular, '*Scaffolding refers to the help which thereby enables learners to engage more successfully in activity at the expanding limits of their competence, and which they would not have been quite able to manage alone, i.e. within the 'zone of proximal development*' (Vygotsky, 1980). The zone of proximal development often referred to as ZPD is the distance between the 'potential level of development' (determined by the tasks that can be solved with 'help') and the 'actual level of development' (which can be determined by the tasks solved independently) (Veresov, 2004). Guided by this theory, and to ensure knowledge is scaffolded, the learning environment was developed as three learning scenes: (1) explore jobsite scene, (2) sensor tutorial scene, and (3) sensor implementation scene. The explore jobsite scene provides a learning platform for situating students in their domain. Students can explore different construction activities and identify the risks and construction resources for these activities, and selectively explore tasks, operations, resources involved, and workspaces. After interaction with the explore jobsite scene, students can proceed to the sensor tutorial scene. It is assumed that at this stage (sensor tutorial scene), students are at the level of potential development of skills for implementing sensing technologies on construction projects. According to Lunsford (2017), Vygotsky posits that students often possess the inherent ability to learn which can be fostered through assistance and learning strategies. Hence, 'help' was provided via learning in a less-congested environment (that is exclusion of other construction activities irrelevant to the learning-in-progress), and a guided interface that outlines the required steps for accomplishing the cognitive tasks (Figure 2.1a). After interacting with the sensing technologies in the sensor tutorial scene, students can proceed to the sensor implementation scene where

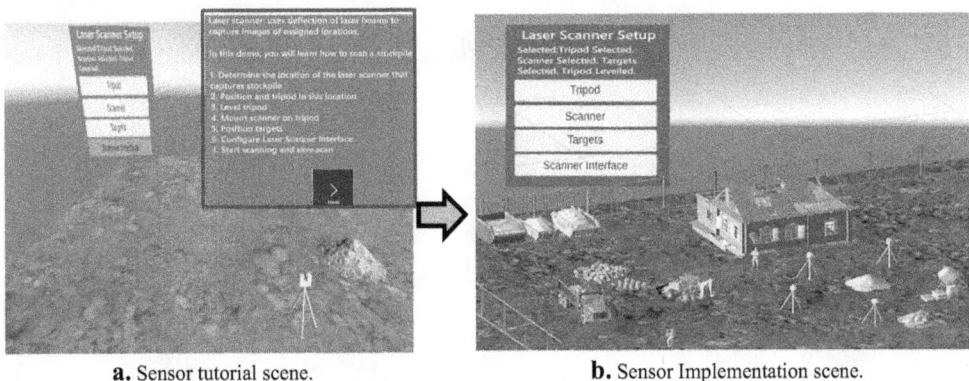

a. Sensor tutorial scene.   **b.** Sensor Implementation scene.

Figure 2.1 Sensor tutorial and sensor implementation learning scenes. a. Sensor tutorial scene. b. Sensor implementation scene.

similar cognitive tasks for each sensing technology can be performed without the 'help' in the Sensor tutorial scene (Figure 2.1b). The authors posited that at this stage (sensor implementation scene), students are at the stage of actual development of technical skills for implementing laser scanning on construction projects.

2.2.5   Research Gaps

Despite the opportunities offered by machine learning, eye-tracking data from usability studies in construction education are still often analyzed manually which might become time-consuming especially when multiple eye-tracking data must be compared. Machine learning helps computers learn and understand patterns, and make predictions from multi-dimensional data without explicit programming (Simon, 2013, Lee et al., 2019). Also, machine learning can extract relevant relationships between eye movements and human cognition for inferring deep insights such as cognitive stages and interaction difficulties with a virtual learning environment. Based on the significance of the sensor tutorial and sensor implementation scenes, and the distinct design features in each scene, this study presents the utilization of machine learning to automatically learn features from eye-tracking data for predicting when learners are in potential development and actual development stages and predicting interaction difficulties in the learning environment.

2.3   Materials and Methods

This section details the adopted methodology in this study. As outlined in Figure 2.2, data collection was performed during the usability evaluation of an MR environment for learning laser scanning. During the experimental tasks, eye-tracking data were collected through the embedded eye-tracker in the AR head-mounted device (HMD), and a think-aloud protocol was employed to procure immediate feedback on the interaction difficulties in the learning environment. The eye-tracking data and think-aloud protocol were then analyzed and utilized for developing the classification models.

2.3.1   Experiment Procedure

2.3.1.1   Participants

The usability study was conducted during the Fall 2021 semester. Based on similar studies (Ramsier, 2019, Wang and Dunston, 2008, Hercegfi et al., 2019), a total of 18 participants

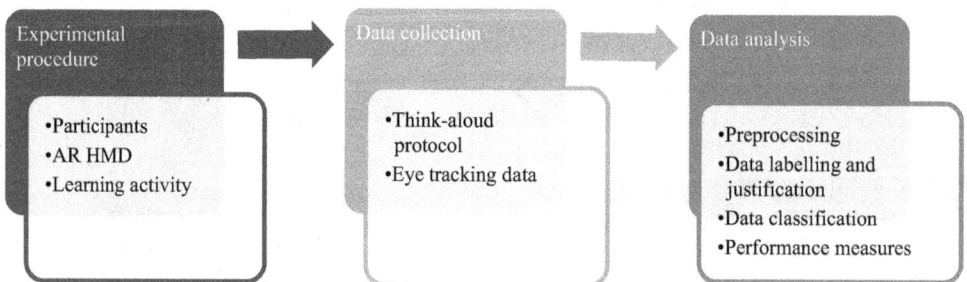

Figure 2.2 Methodology overview.

were involved in the usability evaluation. This sample size is sufficient as more usability issues are often detected with the first few participants and observing more participants reveals only fewer new usability problems (Hwang and Salvendy, 2010). The study also adopted a mixed research method where quantitative (eye-tracking) and qualitative (think-aloud protocol) data were procured.

Participants of the usability study were 18 students (15 males and 3 females) of XXXX University with an average age and standard deviation of 28 years (±7). Based on the objective of the learning environment, which is to equip construction engineering and management students with the competencies for deploying sensing technologies in the industry, only participants with construction backgrounds were recruited using convenient sampling method. The participants were required to sign two consent forms before proceeding with the experimental tasks: one pertaining to in-person research during COVID-19, and the other entails consent to participate in the study. Participants were then introduced to the AR HMD and briefed on the experimental tasks.

2.3.1.2 AR HMD

The AR HMD adopted for this study was the HoloLens 2, which is a see-through device that allows natural communication between users, by overlaying the virtual environment on the real world. The HoloLens 2 displays on a field of view of 52 degrees, a resolution of 47 pixels per degree at the rate of 60 frames per second. The device also has an embedded eye-tracker, which was leveraged for the study.

2.3.1.3 Learning Environment and Activity

After collecting the consent forms, participants were immersed in the learning environment through the AR HMD. All participants were required to calibrate their eye-gaze for procuring accurate eye movement data. The experimental activity entailed the performance of similar laser scanning tasks in the 'sensor tutorial' and 'sensor implementation' scenes. The laser scanner in the learning environment was modeled after Faro Focus M70 and Trimble X8. Guided by the menu interface, the laser scanning activity entails basic steps of the laser scanning activity for taking scans of a stockpile in the simulated construction site. The 'laser scanner setup' menu interface (Figure 2.3a) provides a sequential guide for setting up the laser scanner components, while the 'laser scanner interface' provides a sequential guide for interacting with the settings of the laser scanner.

Hence, guided by the 'laser scanner setup', participants were required to interact with the laser scanner components (tripod, scanner, and targets 1, 2, and 3) such as positioning and leveling the tripod (Figure 2.3a), mounting the scanner on the tripod, and positioning targets in the desired location (Figure 2.3b). The targets are an important aspect of laser scanning as they allow for stitching of two or more scans. After interacting with the laser scanner components, participants were required to set up the scan settings (such as scan coverage, resolution, quality, color, and profile) by interacting with the 'scanner interface' menu (Figure 2.3c). For example, on selecting the coverage button, the participants could set up the required scan coverage (Figure 2.3d) by interacting with the horizontal and vertical sliders. To provide an engaging learning experience, the selected coverage was visualized as a 'red light beam' (see Figure 2.3d). This guided their knowledge of the extent of the selected scan coverage.

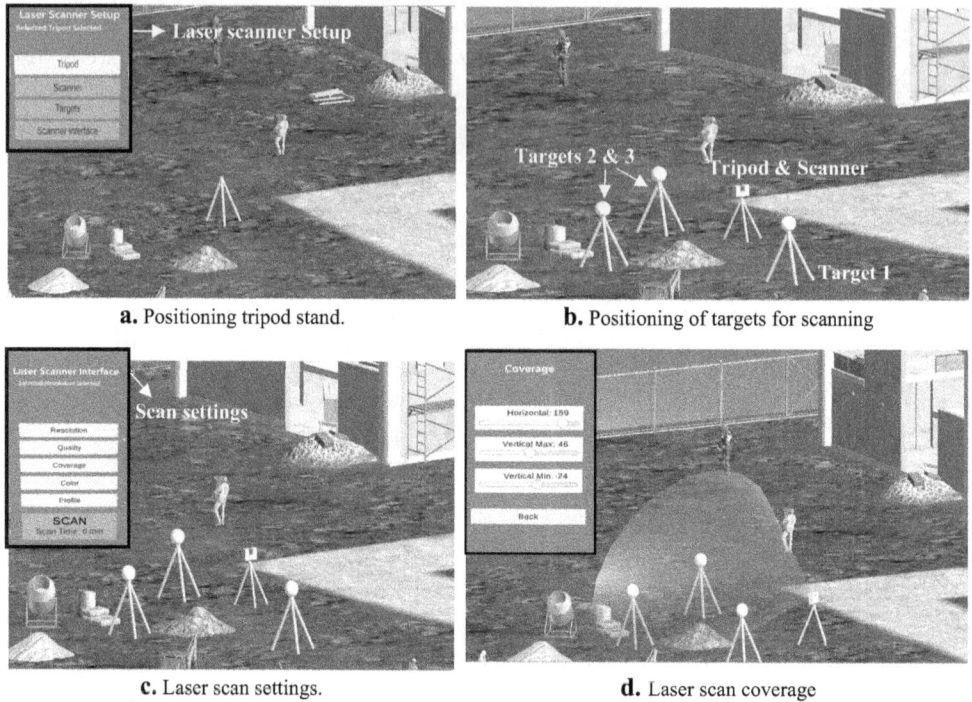

a. Positioning tripod stand.

b. Positioning of targets for scanning

c. Laser scan settings.

d. Laser scan coverage

Figure 2.3 Illustration of laser scanning activity. a. Positioning tripod stand. b. Positioning of targets for scanning stockpile. c. Laser scan settings. d. Laser scan coverage visualization.

2.3.2   Data Collection

During the experimental procedure, eye-tracking data was procured to investigate the cognitive activities while the think-aloud protocol provided immediate feedback on the interaction difficulty during the experiment. To validate the data from the think-aloud protocol, 'HoloLens capture' was employed to video-record the interactions with the learning environment. The data collection procedure is described further in the subsequent sections.

2.3.2.1   Think-Aloud Protocol

The study adopted a think-aloud protocol during the experimental procedure to procure immediate feedback on the interaction difficulty during the laser scanning activities. During the think-aloud protocol, participants were asked if they experienced difficulties while interacting with the environment. This question prompted further responses and comments on their interaction difficulty with the laser scanner components, which were audio recorded. For example, when conducting laser scanning of a stockpile, participants were required to pick up and move laser scanner components like the tripod stand and targets to their desired location. During this process, some participants had trouble interacting with the laser scanner components, while others did not. The responses and comments served as classification labels for the interaction difficulty in the learning environment.

2.3.2.2 Eye-Tracking Data

The eye-tracking data from HoloLens 2 provides eye daze duration, eye origins, hit positions, head origins, and positions at 30 frames per second. Fixation duration for each participant during the 'sensor tutorial' and 'sensor implementation' scenes was extracted from the eye-tracking data based on a minimum duration of 75ms. However, eye-tracking data for a participant was incomplete and hence excluded from the study.

2.3.3 Data Analysis

2.3.3.1 Preprocessing

Think-aloud protocol: The think-aloud protocol was analyzed using thematic coding and validated with video analysis to extract participants who had difficulty interacting with the laser scanner components in the learning environment. Two researchers judged the coding, and an interrater reliability test was performed using Cohen Kappa. The interrater agreement between the judges was good (Cohen's Kappa = 0.7).

Eye-tracking: The total task completion time during both scenes was extracted from the eye-tracking data and statistically analyzed using one-way ANOVA. Based on the significance of the fixations for developing machine learning systems, the study adopted fixation duration, fixation positions, fixation distance, and fixation start time (Table 2.1). While the other data inputs were extracted from the eye gaze data, the fixation distance was computed as the Euclidian distance of the fixation positions (Equation 2.1) (Table 2.2).

$$\text{Euclidian distance} = \sqrt{x^2 + y^2 + z^2} \tag{2.1}$$

2.3.3.2 Data Labeling and Justification

Learning stages: Recall that the sensor tutorial scene was posited as the stage of potential development, and 'help' was provided to scaffold students learning in this scene. The sensor implementation scene on the other hand served as the actual development stage. A prior analysis of the usability data (based on ANOVA) revealed a significant effect (p = 0.002) of the learning scenes on task completion time (Figure 2.4). This result demonstrates that users performed the cognitive tasks significantly faster when engaged in the sensor implementation scene. This was similar to the findings of Toker, Steichen [34] where trial order significantly affected task completion time. Hence, training data labels were provided based on the learning scenes. Eye-tracking data corresponding to sensor tutorial and sensor implementation scene were labeled as potential development (PD) and actual development (AD), respectively.

Interaction difficulty: The input data for the machine learning training were labeled based on the think-aloud protocol data. Table 2.3 details the number of participants

Table 2.2 Data inputs and their description

Data Inputs	Description
Fixation duration	Fixation time measured in milliseconds (ms)
Fixation positions	The coordinates (X, Y, and Z) where the fixations focused
Fixation distance	Euclidian distance of the fixation positions
Fixation Start Time	Fixation start time (ms) measured from the beginning of the experiment

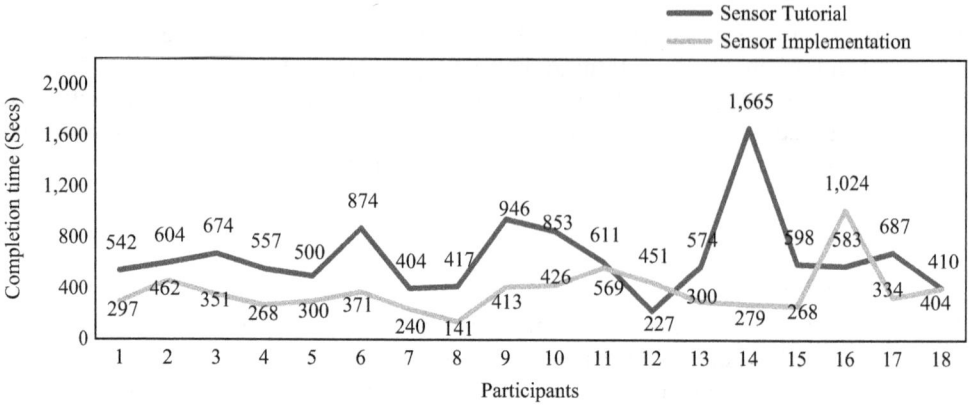

Figure 2.4 Task completion time across both learning scenes (Ogunseiju et al., 2022).

Table 2.3 Difficulty level of participants in both learning scenes

Learning Scenes	Difficulty	No Difficulty	Total
Sensor tutorial	3	14	17
Sensor Implementation	3	14	17
Total	6	28	34

who experienced and did not experience difficulty interacting with the laser scanner components in both learning scenes. By adopting machine learning and eye-tracking, a classification model that automatically identifies interaction difficulty can be developed. The training data can be labeled based on whether difficulty was expressed during the think-aloud protocol as explored in Salminen et al. (2019).

2.3.3.3   Data Classification

The classification model was trained on data from seventeen participants owning to missing fixation position data for a participant. The adopted machine learning classifiers varied for the reviewed papers, thus all the established classifiers for supervised learning were employed for training. The multiple algorithms employed in this study include K-nearest neighbor (KNN), SVM, NN, Ensemble, Naïve Bayes, Decision Tree, LR, and Kernel (Li et al., 2020). It is observed from this study that NN, KNN, SVM, and Ensemble were the top classifiers to understand the learning of the end users. (Witten et al., 2005) provides an in-depth explanation of the employed classifiers. Once trained, fivefold cross-validation was employed for testing the trained model since the sample dataset is small. Cross-validation is an iterative subset splitting that creates data inputs for training and testing exclusively (Mierswa et al., 2006, Bednarik et al., 2012). Hence, the training dataset was divided into five parts. In each cross-validation iteration, some data were used for training, and unseen dataset was used to validate (test the performance) the classification models. For training the model, data features were extracted from the input data (fixation duration, fixation distance, fixation position, and fixation start time). Common statistical data features were chosen to be extracted, which included the mean, median, mode, standard deviation, maximum, minimum, and sum to capture the attributes of the input data (Eivazi and Bednarik, 2011, Conati et al., 2020, Toker et al., 2014, Li et al., 2020). As a result, a total of 42 features were used for training the classifiers.

2.3.3.4 Performance Measures

To evaluate the performance of the classifiers, accuracy, sensitivity (recall), and specificity (precision) (Chatterjee et al., 2011) were employed as performance measures. The sensitivity of the model was measured by examining the rate of true positives, while the specificity revealed the rate of true negatives of the model. The harmonic mean of the model's sensitivity and specificity were evaluated using the F1 score. When the dataset is unbalanced, the accuracy of the classifier may be misleading when adopted as the only performance measure (Egan and Egan, 1975, Bednarik et al., 2012). Hence as adopted by several studies (Egan and Egan, 1975, Bednarik et al., 2012), the Area Under ROC Curve was employed as an additional classifier evaluation in addition to accuracy, specificity, sensitivity, and F1 score. The more the area under the curve, the lesser the impact of the unbalanced distribution of the class labels in the classifier (Bednarik et al., 2012).

2.4 Results

The supervised recognition from the previous section was executed using MATLAB and implemented on a server equipped with a 4.6 GHz Intel[R] Core[TM] i7 9700K CPU, an NVIDIA[R] GeForce RTX[TM] 2080 GPU, and 64 GB RAM. This section focuses on the results of the study, wherein the confusion matrix for the top classifier is explained and the performance of the top three classifiers is compared using the afore-discussed performance measures (Table 2.4).

Table 2.4 Performance measures of classification models

Classifications	Classifier		Accuracy	Specificity	Sensitivity	F1 Score	AUC
Learning Stage	NN	Actual Development	99.90%	1	0.999	0.999	1.00
		Potential Development		0.998	1	0.999	1.00
	KNN	Actual Development	99.70%	1	0.995	0.997	1.00
		Potential Development		0.993	1	0.996	1.00
	SVM	Actual Development	99.70%	1	0.995	0.997	1.00
		Potential Development		0.993	1	0.996	1.00
Difficulty Level	Ensemble	Difficulty	84.60%	0.848	0.530	0.652	0.87
		No Difficulty		0.845	0.965	0.901	0.87
	KNN	Difficulty	83.80%	0.800	0.541	0.645	0.87
		No Difficulty		0.846	0.949	0.895	0.87
	NN	Difficulty	81.40%	0.625	0.684	0.653	0.80
		No Difficulty		0.879	0.863	0.871	0.80

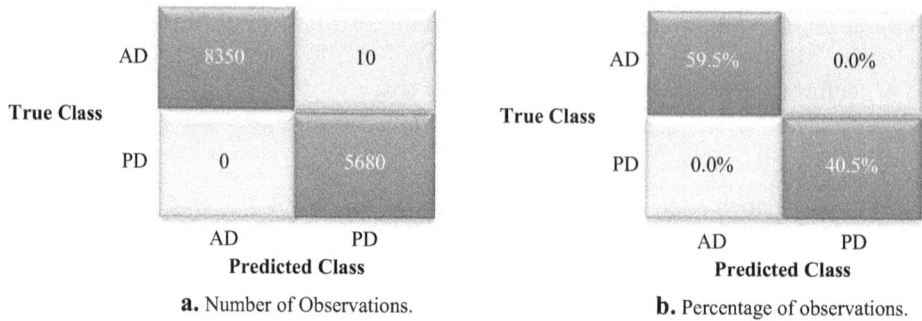

a. Number of Observations.   b. Percentage of observations.

Figure 2.5 Confusion matrixes for learning stages. a. Number of observations. b. Percentage of observations.

2.4.1 Performance Evaluation of Learning Stages Classification Model

The classification algorithm for the development stages performed very well, with the highest accuracy being 99.09% for NN. Whereas the accuracy for models succeeding is 99.70% for KNN and SVM. A total of 14,040 data samples were employed in the study. As shown in Figure 2.5, the confusion matrix for the top classifier successfully classified 8,350 of the sample data as actual development and 5,680 as potential development. Only ten misclassified data samples were wrongly predicted as potential development. The model registered a very high specificity for actual development, which is 1 whereas for potential development (PD) it is 0.998. The model's sensitivity for actual development is 0.999 whereas for potential development it is 1. This is reflected in the F1 score, which is 0.999 for both stages. Also, the area under the curve further supports the high performance of the NN with a high value of 1. The area under the curve is consistent for all the top three classifiers showcasing high performance. The specificity, sensitivity, and F1 score of KNN and SVM for both learning stages are similar but lesser than those of NN.

2.4.2 Performance Evaluation of Difficulty Levels Classification Model

For the difficulty level, the performance of the classification models was high with the highest accuracy of 84.60% for Ensemble classifier which is followed by KNN and NN with 83.80% and 81.40% respectively. The difficulty level training was carried out with 13,560 data samples. The confusion matrix (Figure 2.6) showcases that the Ensemble model correctly classified 1,960 and 9,510 of the data samples as Difficulty (D) and No Difficulty (ND) respectively. In addition, 350 and 1,740 of the data samples were misclassified as D and ND respectively. This could be due to the unbalanced nature of the dataset wherein ND samples were more than D samples, however, the classifier performed well. This is further supported by the high specificity of Ensemble for D (0.848) which is the highest amongst the top three classifiers. Even though the specificity for ND was lowest for Ensemble (0.845), the sensitivity was the highest (0.965) followed by KNN (0.949) and NN (0.863). The F1 score of 0.901 (Ensemble) for ND which is the highest demonstrates a high harmonic mean, whereas, for D, it was 0.652 which was right after NN (0.653). Furthermore, a high value (0.87) of Ensemble for the area under the curve exhibits good performance of the classifier, which is also greater compared to NN (0.80).

a. Number of observations.

b. Percentage of observations.

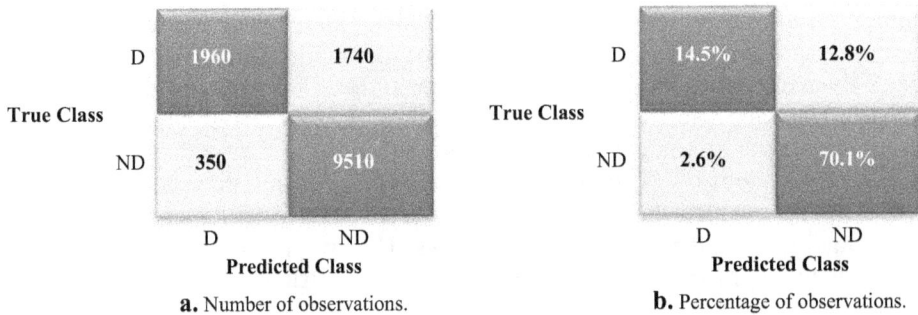

Figure 2.6 Confusion matrixes for difficulty levels. a. Number of observations. b. Percentage of observations.

Overall, the performance of the model was better for ND than D (Figure 2.6b) as also evident from the F1 score of three classifiers, which could also be attributed to the unbalanced data set.

2.5 Discussion

With the promises of MR environments for construction education, there is a need to explore the efficacy of an intelligent MR learning environment that affords students more self-directed and systematic learning. This can be achieved by coupling AI techniques to equip MR learning environments with the ability to control and regulate students' cognitive activities. With the eye being an increasingly important source of cognitive information, this study explores machine learning models that can automatically detect learning stages and human-computer interaction difficulties from eye-tracking data. As a first step, a comprehensive review of literature was conducted to assess the potential of eye-tracking data and machine-learning techniques for designing effective classification models. The review highlights fixations as effective eye movement data for identifying cognitive activities by different machine learning classifiers. Inspired by literature, this study adopts a supervised machine learning approach to detect learning stages and users' interaction difficulty from eye movement data (eye fixations) during laser scanning activities in an MR environment. The study further presented the performance of the top three classifiers for both classification models.

The results revealed that the adopted classification algorithms were effective in detecting learning stages from the learners' eye fixations (Table 2.4). From the adopted machine learning classifiers, the NN classifier displayed superior performance with an accuracy of 99.9%. This was closely followed by KNN and SVM (99.7%), which have been revealed as effective for detecting classes from eye-tracking data (Zhu et al., 2020). Similar to the findings of this study, Carette et al. (2019) also highlighted the ability of NN models to outperform non-NN models during predictions of autism disorder from eye-tracking data. However, the classification model for detecting interaction difficulties performed lesser with the Ensemble classifier revealing the highest accuracy (84.6%). The high performance of Ensemble classifier can be attributed to its structure; the ensemble classifier performs by weighing different separate classifiers and integrating them to obtain one classifier that outperforms others (Saini and Ghosh, 2017). Similar to the study of Salminen et al. (2019) and Eivazi and Bednarik (2011), the results from this study also support the efficacy of think-aloud protocol beyond qualitative data and highlight its effectiveness as data input for detecting interaction difficulties in an MR learning environment. By exploring other performance measures, the results revealed the sensitivity for 'difficulty' class and 'No Difficulty' was 53% and 97%, respectively. This implies that while the ratio of data accurately classified as 'No difficulty' to the total number of 'No difficulty' data was above 97%, only 53% of the total 'Difficulty' data were accurately classified. This suggests that the model was more effective in predicting participants with no interaction difficulties than those with difficulties. The low performance for classifying participants with interaction difficulty can however be attributed to the unbalanced dataset for this model.

The AUC for the classification models for learning stages and difficulty levels were 1.0 and 0.9 respectively. The AUC is a robust and effective measure of the performance of classifier algorithms as it is insensitive to the proportion of true cases. It illustrates the relationship between the true positive and false positive rates along with the change of a threshold parameter (Li et al., 2020). The AUC also reveals the expected performance of each class and has a statistical equivalent to Wilcoxon test of ranks for ranking classification models (Hand, 1997, Batista et al., 2004). As such, a classification model is effective when the AUC is greater than 0.9 (Li et al., 2020). Hence, despite the unbalanced dataset for predicting interaction difficulties, the classification models can be interpreted as effective in classifying the learning stages and difficulty levels. These classification models can be integrated into the learning environment to provide adaptive and seamless learning experiences. This study further provides pathways for creating modern educational systems that integrate virtual learning environment in the traditional teacher-student educational relationship for personalized and self-directed learning in construction education.

2.6   Conclusions and Future Works

Sophisticated ways of improving construction education include the design and implementation of virtual learning environments like MR. Literature has shown that MR environments are currently being applauded for their efficacy in providing interactive and experiential learning experiences in construction education. But to achieve an effective learning environment, it is important to design intelligent systems for seamless learning experiences. Such an intelligent learning environment adapts individual characteristics such as learning difficulties, prior knowledge, learning goals, and objectives to provide improved learning experiences. With the efficacy of eye movement data to provide instantaneous behavioral evidence of users' visual and cognitive processes, this study presents classification models that can reliably predict learning stages and interaction difficulties from users' eye movement data during laser scanning in an MR environment. Such classification models will help detect users who require additional support to acquire the necessary technical skills for deploying laser scanners in the construction industry and inform the specific training needs of users to enhance seamless interaction with the learning environment. The research implication further lies in the potential of an intelligent learning environment for providing personalized learning experiences that often culminate in improved learning outcomes. This study further highlights the potential of such an intelligent learning environment in promoting inclusive learning whereby students with different cognitive capabilities can experience learning tailored to their specific needs irrespective of their individual differences.

Although this study shows the potential of using eye-tracking data for providing intelligent learning experiences, there are some limitations in this study that should be addressed. The dataset for detecting interaction difficulties was unbalanced, and as such impacted the performance of the model. Future works will explore data augmentation methods for providing a robust dataset for the classification model. Likewise, it is important to assess the data inputs with the highest impact on the classification model. For example, exploring the effects of only fixation duration for training the classifiers, and a combined effect of fixation durations and fixation positions. This will inform the choice of training data in the absence of a robust eye-tracking dataset. Furthermore, there are individual features such as prior knowledge of sensing technologies, educational levels, and gender and age differences that can influence learning outcomes in an MR environment. It is therefore beneficial that future studies explore models that automatically adopt these individual features to design a robust and proactive learning environment that provides personalized prompts, feedback, scaffolds self-paced learning, and advances seamless learning experiences.

Acknowledgment

The authors acknowledge National Science Foundation for their support (Grant No. DUE #1916521). Any opinions, findings, conclusions, or recommendations expressed in this material are those of the authors and do not necessarily reflect the views of the National Science Foundation.

References

Andrzejewska, M. & Stolińska, A. 2016. "Comparing the difficulty of tasks using eye tracking combined with subjective and behavioural criteria". *Journal of Eye Movement Research*, 9, 10.16910/jemr.9.3.3.

Batista, G. E., Prati, R. C. & Monard, M. C. 2004. "A study of the behavior of several methods for balancing machine learning training data". *ACM SIGKDD Explorations Newsletter*, 6, 20–29, 10.1145/1007730.1007735.

Bednarik, R., Vrzakova, H. & Hradis, M. 2012. What do you want to do next: A novel approach for intent prediction in gaze-based interaction. Proceedings of the symposium on eye tracking research and applications, 83–90. Available from: 10.1145/2168556.2168569? casa_token=Ml15FRF-FjwAAAAA:UZnkHbHJafkCH3A3mnomAl-AEOK0RhVrDFI8dDW-lYyHtDOmYx9mrtsMGMnMhI0usAZosCOGGQjD, Accessed: July, 2022.

Bosché, F., Abdel-Wahab, M. & Carozza, L. 2016. "Towards a mixed reality system for construction trade training". *Journal of Computing in Civil Engineering*, 30, 04015016, 10.1061/ %28ASCE%29CP.1943-5487.0000479.

Carette, R., Elbattah, M., Cilia, F., Dequen, G., Guerin, J.-L. & Bosche, J. 2019. Learning to Predict Autism Spectrum Disorder based on the Visual Patterns of Eye-tracking Scanpaths. HEALTHINF, 103–112. Available from: https://pdfs.semanticscholar.org/adc5/936edc0f1645 ee05916b5dc784124cbb12ab.pdf, Accessed: July 2022.

Chalhoub, J. & Ayer, S. K. 2018. "Using mixed reality for electrical construction design communication". *Automation in Construction*, 86, 1–10, 10.1016/j.autcon.2017.10.028.

Chatterjee, P., Basu, S., Kundu, M., Nasipuri, M. & Plewczynski, D. 2011. "PPI_SVM: Prediction of protein-protein interactions using machine learning, domain-domain affinities and frequency tables". *Cellular and Molecular Biology Letters*, 16, 264–278, 10.2478/s11658-011-0008-x.

Collins, A., Brown, J. S. & Newman, S. E. 2018. *Cognitive apprenticeship: Teaching the crafts of reading, writing, and mathematics*, Routledge. ISBN: 1315044404.

Conati, C., Lallé, S., Rahman, M. A. & Toker, D. 2020. "Comparing and combining interaction data and eye-tracking data for the real-time prediction of user cognitive abilities in visualization tasks". *ACM Transactions on Interactive Intelligent Systems (TiiS)*, 10, 1–41, 10.1145/3301400? casa_token=7nJFfYZKGdUAAAAA:W9AoFXf7k_KsuWgGpO01DcAGEFZ7eXB2 8KgLqu9O1pjOzw6x2-tXnas8uNEVvqjLIFbAkb9l4Rqj.

Dalrymple, K. A., Jiang, M., Zhao, Q. & Elison, J. T. 2019. "Machine learning accurately classifies age of toddlers based on eye tracking". *Science Reports*, 9, 6255, 10.1038/s41598-019-42764-z.

Egan, J. P. & Egan, J. P. 1975. *Signal detection theory and ROC-analysis*, Academic Press. ISBN: 0122328507.

Eivazi, S. & Bednarik, R. 2011. Predicting problem-solving behavior and performance levels from visual attention data. Proc. Workshop on Eye Gaze in Intelligent Human Machine Interaction at IUI, 9–16. Available from: http://www.cs.joensuu.fi/pages/bednarik/Eivazi_Bednarik_IUI2011_ workshop.pdf, Accessed: July 2022.

Ensafi, M., Thabet, W., Devito, S. & Lewis, A. 2021. "Field testing of mixed reality (MR) technologies for quality control of as-built models at Project handover: A case study". *EPiC Series in Built Environment*, 2, 246–254, https://www.researchgate.net/profile/Mahnaz-Ensafi/publication/ 353418142_Field_Testing_of_Mixed_Reality_MR_Technologies_for_Quality_Control_of_As- Built_Models_at_Project_Handover_A_Case_Study/links/60fb29c1169a1a0103b1e884/.

Goldberg, J. H. & Helfman, J. I. 2010. Scanpath clustering and aggregation. Proceedings of the 2010 symposium on eye-tracking research & applications, 227–234. Available from: 10.1145/ 1743666.1743721, Accessed: July 2022.

Hand, D. J. 1997. *Construction and assessment of classification rules*, John Wiley & Sons. ISBN: 0471965839.

He, J. & Mccarley, J. S. 2010. "Executive working memory load does not compromise perceptual processing during visual search: Evidence from additive factors analysis". *Attention, Perception, & Psychophysics*, 72, 308–316, 10.3758/APP.72.2.308.

Hennessy, S. 1993. "Situated cognition and cognitive apprenticeship: Implications for classroom learning". *Studies in Science Education*, 22, 1–41, 10.1080/03057269308560019.

Hercegfi, K., Komlódi, A., Köles, M. & Tóvölgyi, S. 2019. "Eye-tracking-based wizard-of-oz usability evaluation of an emotional display agent integrated to a virtual environment". *Acta Polytechnica Hungarica*, 16, 145–162, http://epa.niif.hu/02400/02461/00087/pdf/EPA02461_acta_ polytechnica_2019_02_145-162.pdf.

Hwang, W. & Salvendy, G. 2010. "Number of people required for usability evaluation: The 10±2 rule". *Communications of the ACM*, 53, 130–133, 10.1145/1735223.1735255.

Kandemir, M. & Kaski, S. 2012. Learning relevance from natural eye movements in pervasive interfaces. Proceedings of the 14th ACM International Conference on Multimodal Interaction, 85–92. Available from: 10.1145/2388676.2388700.

Lagun, D., Manzanares, C., Zola, S. M., Buffalo, E. A. & Agichtein, E. 2011. "Detecting cognitive impairment by eye movement analysis using automatic classification algorithms". *Journal of Neuroscience Methods*, 201, 196–203, 10.1016/j.jneumeth.2011.06.027.

Lee, S., Hwang, Y., Jin, Y., Ahn, S. & Park, J. 2019. "Effects of individuality, education, and image on visual attention: Analyzing eye-tracking data using machine learning". *Journal of Eye Movement Research*, 12, 10.16910/jemr.12.2.4.

Li, J., Li, H., Umer, W., Wang, H., Xing, X., Zhao, S. & Hou, J. 2020. "Identification and classification of construction equipment operators' mental fatigue using wearable eye-tracking technology". *Automation in Construction*, 109, 103000, 10.1016/j.autcon.2019.103000.

Lunsford, K. J. 2017. Challenges to implementing differentiated instruction in middle school classrooms with mixed skill levels. Available: https://scholarworks.waldenu.edu/dissertations/ 5021, Accessed July 2022.

Mierswa, I., Wurst, M., Klinkenberg, R., Scholz, M. & Euler, T. 2006. Yale: Rapid prototyping for complex data mining tasks. Proceedings of the 12th ACM SIGKDD international conference on

Knowledge discovery and data mining, 935–940. Available from: 10.1145/1150402.1150531, Accessed: July 2022.

Mohammadpour, A., Karan, E., Asadi, S. & Rothrock, L. 2015. Measuring end-user satisfaction in the design of building projects using eye-tracking technology. Computing in Civil Engineering 2015, 564–571. Available from: https://www.researchgate.net/profile/Ebrahim-Karan/publication/282929769_Measuring_End-User_Satisfaction_in_the_Design_of_Building_Projects_Using_Eye-Tracking_Technology/links/57190d4a08aed43f63234fcf/Measuring-End-User-Satisfaction-in-the-Design-of-Building-Projects-Using-Eye-Tracking-Technology.pdf, Accessed: July 2022.

Moore, H. F. & Gheisari, M. 2019. "A review of virtual and mixed reality applications in construction safety literature". *Safety*, 5, 51, 10.3390/safety5030051.

Ogunseiju, O., Akanmu, A. & Bairaktarova, D. 2021a. Sensing Technologies in Construction Engineering and Management Programs: A Comparison of Industry Expectations and Faculty Perceptions. EPiC Series in Built Environment, 505–513. Available from: https://par.nsf.gov/biblio/10232747, Accessed: July 2022.

Ogunseiju, O. O., Akanmu, A. A. & Bairaktarova, D. 2021b. "Mixed reality based environment for learning sensing technology applications in construction". *Journal of Information Technology in Construction*, 26, 863–885, 10.36680/j.itcon.2021.046.

Ogunseiju, O. R., Gonsalves, N., Akanmu, A. A., Bairaktarova, D., Bowman, D. A. & Jazizadeh, F. 2022. "Mixed reality environment for learning sensing technology applications in construction: A usability study". *Advanced Engineering Informatics*, 53, 101637, 10.1016/j.aei.2022.101637.

Ramsier, L. 2019. *Evaluating the usability and user experience of a virtual reality painting application*, Masters Paper. Available: 10.17615/s9z1-m163, Accessed July 2022.

Saini, R. & Ghosh, S. K. 2017. Ensemble classifiers in remote sensing: A review. 2017 International Conference on Computing, Communication and Automation (ICCCA), IEEE, 1148–1152. Available from: 10.1109/CCAA.2017.8229969, Accessed: July 2022.

Salminen, J., Nagpal, M., Kwak, H., An, J., Jung, S.-G. & Jansen, B. J. 2019. Confusion prediction from eye-tracking data: Experiments with machine learning. Proceedings of the 9th International Conference on Information Systems and Technologies, 1–9. Available from: 10.1145/3361570.3361577, Accessed: July 2022.

Satter, K. & Butler, A. 2015. "Competitive usability analysis of immersive virtual environments in engineering design review". *Journal of Computing and Information Science in Engineering*, 15, 10.1115/1.4029750.

Schechter, D., Patel, H., Aiyar, K. & Vasquez, J. 2021. Intelligent tutor assistant: Predicting confusion from pupillometry data with multiple classification models. IIE Annual Conference. Proceedings. Institute of Industrial and Systems Engineers (IISE), 1022–1027. Available from: https://www.proquest.com/docview/2560888544?pq-origsite=gscholar&fromopenview=true, Accessed: July 2022.

Shanbari, H., Blinn, N. & Issa, R. R. 2016. "Using augmented reality video in enhancing masonry and roof component comprehension for construction management students". *Engineering, Construction and Architectural Management*, 23, 765–781, 10.1108/ECAM-01-2016-0028.

Shojaeizadeh, M., Djamasbi, S., Paffenroth, R. C. & Trapp, A. C. 2019. "Detecting task demand via an eye tracking machine learning system". *Decision Support Systems*, 116, 91–101, 10.1016/j.dss.2018.10.012.

Simon, P. 2013. *Too big to ignore: The business case for big data*, John Wiley & Sons. ISBN: 1118638174.

Toker, D., Steichen, B., Gingerich, M., Conati, C. & Carenini, G. 2014. Towards facilitating user skill acquisition: Identifying untrained visualization users through eye tracking. Proceedings of the 19th international conference on Intelligent User Interfaces, 105–114. Available from: 10.1145/2557500.2557524, Accessed: July 2022.

Veresov, N. 2004. "Zone of proximal development (ZPD): The hidden dimension". *Language as Culture – Tensions in Time and Space*, 1, 42–48, 10.1.1.464.5199&rep=rep1&type=pdf.

Vygotsky, L. S. 1980. *Mind in society: The development of higher psychological processes*, Harvard University Press. ISBN: 0674076680.

Wang, T.-K., Huang, J., Liao, P.-C. & Piao, Y. 2018. "Does augmented reality effectively foster visual learning process in construction? An eye-tracking study in steel installation". *Advances in Civil Engineering*, 2018, 10.1155/2018/2472167.

Wang, X. & Dunston, P. S. 2008. "User perspectives on mixed reality tabletop visualization for face-to-face collaborative design review". *Automation in Construction*, 17, 399–412, 10.1016/j.autcon.2007.07.002.

Witten, I. H., Frank, E., Hall, M. A., Pal, C. & Data, M. 2005. *Practical machine learning tools and techniques*. ISBN: 978-0-12-374856-0.

Wu, W., Tesei, A., Ayer, S., London, J., Luo, Y. & Gunji, V. 2018. Closing the skills gap: Construction and engineering education using mixed reality–A case study. 2018 IEEE Frontiers in Education Conference (FIE), IEEE, 1–5. Available from: 10.1109/FIE.2018.8658992, Accessed: July 2022.

Ye, X. & König, M. 2019. Applying eye tracking in virtual construction environments to improve cognitive data collection and human-computer interaction of site hazard identification. ISARC. Proceedings of the International Symposium on Automation and Robotics in Construction. IAARC Publications, 1073–1080. Available from: https://www.proquest.com/docview/2268537650?pq-origsite=gscholar&fromopenview=true, Accessed: July 2022.

Yusoff, R. C. M., Zaman, H. B. & Ahmad, A. 2010. "Design a situated learning environment using mixed reality technology-A case study". *International Journal of Computer, Electrical, Automation, Control and Information Engineering*, 47, 887–892, 10.1.1.1034.1965&rep=rep1&type=pdf.

Zemblys, R., Niehorster, D. C., Komogortsev, O. & Holmqvist, K. 2018. "Using machine learning to detect events in eye-tracking data". *Behavior Research Methods*, 50, 160–181, https://link.springer.com/article/10.3758/s13428-017-0860-3.

Zhu, J., Wang, Z., Gong, T., Zeng, S., Li, X., Hu, B., Li, J., Sun, S. & Zhang, L. 2020. "An improved classification model for depression detection using EEG and eye tracking data". *IEEE Transactions on Nanobioscience*, 19, 527–537, 10.1109/TNB.2020.2990690.

3 Marker-Based Augmented Reality Framework for Checking the Installation Status of Onsite Components

Hung Pham[1], Linh Nguyen[2], Man-Woo Park[2], and Chan-Sik Park[3]

[1]*Eindhoven University of Technology, Netherlands;* [2]*Myongji University, South Korea;* [3]*Chung-Ang University, South Korea*

3.1 Introduction

Construction site inspections are crucial to ensure the project is on schedule and meets all requirements, standards, and regulations (Alzahrani & Emsley, 2013). A construction project requires coordination between multiple teams with various equipment and locations (Vo-Tran & Kanjanabootra, 2013). Construction quality is generally assured via the three-level quality management system with inspections carried out by contractors, sampling inspections and re-inspections by supervisory units, and results supervision by construction authorities (Israngkura Na Ayudhya & Kunishima, 2009; Ling & Bui, 2010). Project managers can make plans based on the inspection results to conduct necessary adjustments to ensure the project's quality and progress. However, the inspector usually has limited time to plan and conduct the inspection, resulting in missing potential defects. Most of these inspections are done manually with pen and paper, protracting the time gap for the inspection results to be delivered to the person in charge, resulting in the overall extended project time (Kang et al., 2011). The whole inspection process can lead to multiple problems in the project, affecting contractors, subcontractors, and related stakeholders (Alzahrani & Emsley, 2013). The inspection process can be streamlined by reducing the work that needs to be done manually by storing inspection results in digital form, which can be directly shared across project participants. The inspector will also have more time for closer inspection of each construction component (Kopsida et al., 2015). Digitalization is expected to be the solution to streamline the inspection process.

The construction industry is known for its slow adaptation to new technologies, despite significant changes compared to the old practice (Zainon et al., 2011). Over the past decade, the rise in labor costs and the drop in technology expenses have promoted the construction industry's digitalization trend. For instance, computer-aided design (CAD) and building information modeling (BIM) are now considered the universal standards in the industry. BIM has changed construction information and management practices by focusing on related information and speeding up workflows (Elghaish et al., 2020). Various project stakeholders, such as engineers, contractors, and architects also use numerous specialized software programs. (Saidi et al., 2003; Tsai et al., 2014). In recent years, the construction industry has started to adopt augmented reality (AR), a cutting-edge technology, to solve these problems in the construction industry. It merges physical and virtual environments by allowing users to access and interact with digital information, including virtual 3-D models and contextual data, which are superimposed on the user's perception (Azuma, 1997). A variety of AR research was conducted to make

DOI: 10.1201/9781003408949-3

smart construction feasible, but from previous studies, most of the AR application was in the stage of development, and only lab-based prototypes were presented.

This chapter introduces a framework to automatically check the installation status of construction components by utilizing head-mounted AR devices and marker-based AR. The 3-D model of the construction is extracted from the BIM model and configured based on the inspection plan. The virtual 3-D model is superimposed on the inspector's view via marker detection. The head-mounted AR with a ToF camera can detect collisions between the virtual 3-D model and the actual construction components to determine the installation status of onsite components. The installation status of each component will be recorded in digital form and can be accessed or shared among stakeholders. The experiments show that the application is feasible for an actual construction site and has high accuracy.

3.2 Literature Review

3.2.1 *AR Principles and Concepts*

Previous studies have explored AR technology for applications in the construction industry. In 2008, Shin and Dunston (2008) proposed 17 tasks in construction that can benefit from using AR: layout, excavation, penetrating, conveying, cutting, positioning, placing, connecting, spreading, finishing, spraying, covering, inspection, coordination, supervision, commenting, and strategizing. AR has become more popular in the construction industry in recent years, both in research and in practical use (Ahmed, 2019; Behzadan & Kamat, 2013; Piroozfar et al., 2018). AR allows for superimposing digital information in a real-life environment, resulting in an interactive interface for working with information. AR can be used to create a collaborative workspace between stakeholders (Lin et al., 2015), but most of its applications in the construction industry mainly focus on using its 3-D model visualization capability. Based on how the 3-D model is instantiated, AR is divided into two main categories: marker-based AR and markerless AR.

3.2.1.1 *Markerless AR*

Markerless AR places virtual 3-D objects based on the surrounding environment features rather than marker detection (Sato et al., 2016). It detects objects and distinctive points in a scene without prior knowledge of the environment (Bosché et al., 2012; Meža et al., 2015). Markerless AR utilizes pattern recognition to calculate the position between the user's camera and the real world. Markerless AR can also use additional input data from the device's inertial measurement unit (IMU) or global positioning system (GPS) to visualize the AR content (Brata et al., 2015). Using markerless AR with mobile devices can simplify the 3-D model instantiate process. However, due to size limitations, mobile devices' sensor accuracy is relatively low, affecting 3-D model placement. The accuracy of the 3-D model placement can be improved by using a high-accuracy GPS device. Companies have applied markerless AR with global navigation satellite system and real-time kinematic positioning to improve accuracy (vGIS, 2022). This system greatly improves position accuracy, but it requires an external backpack to store the antenna and GPS sensors. Equipping the system with a high-precision position sensor increases its weight and cost. This limitation makes the markerless AR system unsuitable for application, requiring high-accuracy 3-D model placement.

3.2.1.2   Marker-Based AR

Marker-based AR requires one or multiple static images, typically in a specific format, as a trigger to be scanned with AR-capable devices. The scan will trigger the content linked with the marker, and marker recognition results in the virtual objects' position in a 3-D coordinate (Fazel & Izadi, 2018). The AR application also uses the device's IMU and camera to determine the spatial movement across six degrees of freedom, allowing the virtual objects to retain their positions when the user moves. Marker-based AR offers higher position accuracy in placing 3-D models suitable for precision use in the construction industry.

One of the most popular markers used in AR applications is the ArUco marker. This fiducial marker is well known for its reliability and fast detection time (Garrido-Jurado et al., 2014). The ArUco marker was designed to promptly provide the 3-D position between the camera and the marker. It has a low false detection rate compared to other popular markers such as ARTag or ARToolkitPlus (Garrido-Jurado et al., 2014). The ArUco marker library has multiple dictionaries, divided based on their bit data, ranging from 4×4 bits to 7×7 bits. Each dictionary contains 1000 marker IDs, each of which can be linked to particular content. The 3-D spatial position of the marker center is determined by using the 2-D image combined with camera optics parameters and distortion matrixes from camera calibration (Hartley & Zisserman, 2003; Marquardt, 1963).

3.2.2   AR Devices

The rising trends of AR have presented various AR mobile and head-mounted devices. Many companies have released their devices with integrated technologies to enhance AR experiences. One of them is the iPad Pro, launched in 2020 and equipped with a LiDAR sensor to support AR applications. The LiDAR scanner can measure the distance of up to 5 meters around the device while maintaining its performance indoors and outdoors (Spreafico et al., 2021). The combination of the surround mapping by the LiDAR scanner and the processor optimized for computer vision algorithms makes the iPad Pro one of the most popular AR mobile devices. However, the iPad requires users to hold it by at least one hand and look at the screen to see AR content, thus making it dangerous to use on construction sites.

Head-mounted AR devices can address the problems of using AR with tablet devices. Head-mounted AR devices are wearable computer-capable devices that register virtual information in the physical world in users' field of view and offer immersive experiences for users (Lutz et al., 2017; Rauschnabel et al., 2015). Several head-mounted AR devices have been released in recent years; however, most of them are not significantly different from mobile devices and are made in limited quota. In 2019, Microsoft introduced and released HoloLens 2, a mixed-reality wearable device for research and development (Ungureanu et al., 2020). HoloLens 2 has improved its attributions to become a more immersive, effective, and user-friendly version compared to its predecessor. It is also commercially available and used in various fields. This device includes a 3:2 light engine 2 K resolution display for each eye with a 52 degrees field of view, a holographic density of greater than 2.5 K radiance, and an optimized display for 3-D eye position, providing better immersive performance for users. The device is equipped with various types of cameras: four visible-light cameras for head tracking, two IR cameras for eye tracking, a 1-MP time-of-flight (TOF) depth sensor, and a camera of 8-MP stills to record 1080p video at 30 frames per second (Microsoft, 2022). Whereas other AR devices rely on IMU to determine the position in 3-D space, four visible-light tracking cameras and the ToF camera allow HoloLens 2 to track its position better in 3-D space, reducing the 3-D model potential offset caused by IMU sensor error. Furthermore, due to built-in cameras continuously scanning the surrounding environment, HoloLens 2 can construct virtual content registering the physical world and spatial mesh as the virtual interpretation of the physical surrounding (Unity, 2017).

3.2.3   AR in Construction

The term AR is increasingly mentioned in the scientific literature of the construction industry. For example, to streamline construction management, Ratajczak et al. (2019) proposed an application that used markerless AR and GPS to support site managers in viewing data related to construction components in their field of view. The data are extracted from the BIM, including data about the progress, performance ability ratio, delay indicator, construction errors, and extra costs. The display information is based on the user's location and images from the camera. The application is expected to provide context-specific information as the user walks through the construction site. AR can also be combined with other cutting-edge technologies to improve construction site management. Zollmann et al. (2014) proposed a framework for combining AR and aerial vehicles to support construction site monitoring. The images from the aerial vehicles are combined with data from the BIM model to create a 3-D model with attached construction site information. Users can superimpose the 3-D model on the construction site via multiple methods such as GPS, IMU, and vision-based. The user can see the construction at multiple stages in the past or present and look through the area covered by obstacles such as walls and columns. Annotations about construction components can be viewed directly on the construction site to help with onsite documentation and monitoring.

Since operation and maintenance is the longest phase in the lifecycle of a building, the number of studies that apply AR in construction maintenance is increasing. In 2009, a framework for using markerless AR was proposed to improve damage prevention and the maintenance of underground infrastructure (Behzadan & Kamat, 2009). The 3-D model of the excavation site was determined via a system of precision GPS and Real-Time Kinematics. The position of the excavator was compared with the 3-D model of the site to ensure the excavation area. Koch et al. (2014) proposed a marker-based framework for indoor navigation and maintenance. The markers in this study are the natural indoor markers in the building such as fire extinguisher signs, device ID tags, or textual information signs; therefore, no additional signs need to be attached. Combining marker detection results with indoor navigation, the AR content is visualized based on user's viewing context, helping the maintenance process.

Some research studies have focused on applying AR in construction inspections. Loporcaro et al. (2019) tested the accuracy of HoloLens as a construction measurement tool. They compared the HoloLens' ToF camera with traditional tape measurement and total station, precise surveying equipment. Chen et al. (2020) proposed a fire safety equipment inspection framework. A cloud-based database was created from BIM, combined with markerless AR, to facilitate the inspection. The inspection report is in digital form and is created based on BIM. However, most research is conceptual or tested in a lab-based environment. Meža et al. (2015) diagnose and classify challenges with AR in construction, namely the misalignment when overlaying the virtual models on the actual components, the limited number of AR devices and their resolution, safety risks from interference, and restriction in the field of view when wearing AR devices. Additionally, AR applications and their popularity still face technological constraints such as GPS accuracy, indoor localization, processing power, or interoperability between multiple platforms.

3.3 Research Objective and Scope

Despite modern construction inspections supported by CAD and BIM, the onsite inspection process is mainly paper-based, time-consuming, and labor-intensive. With the ability to superimpose virtual components in real-world surroundings, AR can be a useful approach for conducting fieldwork inspections. Some previous studies have applied AR in onsite inspections. Many steps were performed manually on a tablet or computer, limiting the AR content to the tablet size and requiring the inspector to hold the device with one hand, potentially hindering the inspector's productivity. The main objective of this study was to propose a feasible, practical AR framework using a head-mounted device to support the construction site inspection process by visualizing the installation status of construction components. The proposed framework uses marker-based AR to superimpose the planned 3-D model to inspector view, then compares the superimposed model with the spatial 3-D scan from the ToF sensor to determine the component's installation status. The framework was tested on an actual construction site with multiple configurations. The performance of the proposed framework was evaluated based on the precision and accuracy of the inspection report compared to actual onsite components.

3.4 Methodology

The framework consists of three stages: (1) preparing the virtual 3-D model, (2) super-imposing the virtual 3-D model to the inspector's view, and (3) comparing the super-imposed 3-D model with the spatial 3-D model of the inspector's surroundings. In the first stage, the 3-D model of actual construction is exported from BIM and extracted with its components of interest. The virtual 3-D model has the same position, rotation, and scale as the actual construction. Multiple anchor points, which are the locations of ArUco markers in actual construction, are specified in the model and instantiated via marker detection and anchor point position adjustment. The virtual 3-D model is superimposed on the construction site based on the instantiated anchor points. Then, the inspector needs to walk through the construction site for the device to scan and form a real-time spatial 3-D model of the inspector's surroundings. As the construction site is on-progress and several components have not been established, some components are included in the BIM model but not in actual construction. These components are denoted as "Not-installed", while the others are "Installed". Lastly, the checking installation step is implemented to classify "Installed" and "Not-installed" components and record their status in the inspection plan. This last step, which is the main contribution of this study, is achieved by using a ToF camera to reconstruct the surrounding environment as a 3-D model, called a spatial 3-D model. The component's installation status is determined as "Not-installed" when it is not constructed on the site, so there is no collision between the superimposed 3-D model and the spatial 3-D model. The status of each component is recorded and visualized in different colors based on their status for the inspector to distinguish.

3.4.1 3-D Model Preparation

In this study, the inspection plan includes the virtual 3-D model of the construction site and information on the inspected component from the BIM model. As it contains 4-D information about the site at various stages, all components of the finished work are involved in the 3-D BIM model, with some not irrelevant to the inspection plan. Thus, only relevant components are configured in the virtual 3-D model (see Figure 3.1), and multiple components can be grouped into one, or one component can be divided into multiple components. Data and information related to the inspection plan are then stored in the head-mounted device before the inspection takes place at the construction site. The origin and rotation are predefined based on the corresponding setup to superimpose the 3-D model (circle in Figure 3.1).

3.4.2 Superimposing Virtual 3-D Model to the Inspector's View

Superimposing the virtual 3-D model on the actual construction requires the transformation's configurations, that is, position and rotation, to match the actual construction. These configurations are then determined based on the anchor points, which are specified in the BIM model in the previous section and located on the actual site via ArUco marker detection. At the actual site, the ArUco markers are attached to a flat surface with a fixed position, and their origin is at the center of the marker. The framework requires two or more anchor points to fix the position and rotation of the virtual 3-D model in global coordinates. Figure 3.2 illustrates the steps to determine the position of one anchor point.

First, the head-mounted device's camera records video frames and detects the marker via a 2-D image of each frame. The results of 2-D image detection are then converted into 3-D camera coordinates $(x_a, y_a, z_a)^c$ using three camera matrices, which include focal distances, intrinsic parameters, and distortion coefficients determined in the camera calibration process. As the inspector moves around the site, the camera coordinate system follows his position and orientation, and the correlation between marker coordinates and camera coordinates is undefined. Thus, converting these coordinates into one unified system is required, and this system is denoted as the global coordinate system. The origin and orientation of the global coordinate system are the camera position and heading direction when the device is switched on. By utilizing the IMU, ToF camera, and visible-light camera, the camera position is tracked in the global coordinate system and returned as rotation and

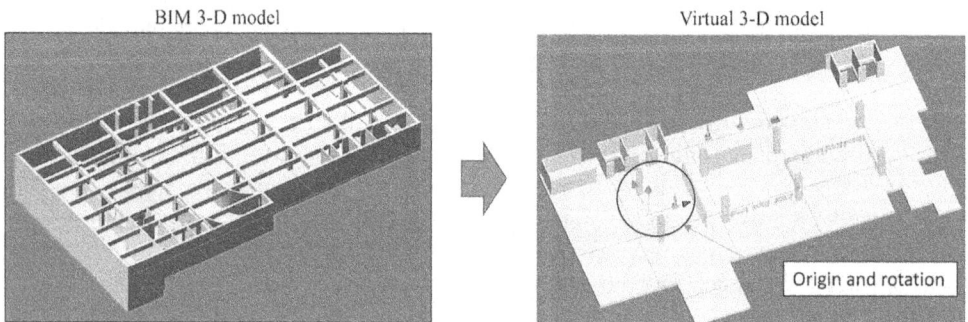

Figure 3.1 Exporting the virtual 3-D model from BIM.

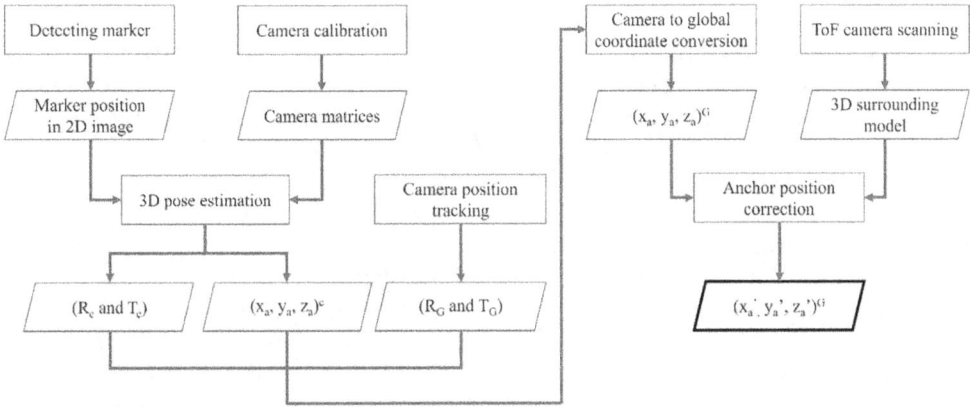

Figure 3.2 Anchor point instantiation procedure.

translation matrices (R_G and T_G) between the global coordinate origin and the position of the camera at the time. Later, the global coordinate of the anchor point $(x_a, y_a, z_a)^G$ is calculated based on two sets of translation and rotation vectors between different coordinate systems: camera-marker (R_c and T_c) and camera-global (R_G and T_G). The above-mentioned three coordinate systems are demonstrated in Figure 3.3. In the marker local coordinate system, the X_m and Y_m axes are horizontal and vertical, respectively, and the Z_m axis is perpendicular to the marker plane with the direction of pointing away from the marker. The camera's local coordinate system has the origin of the head-mounted device's camera, while X_c and Y_c are horizontal and vertical axes, and the Z_c axis points forward.

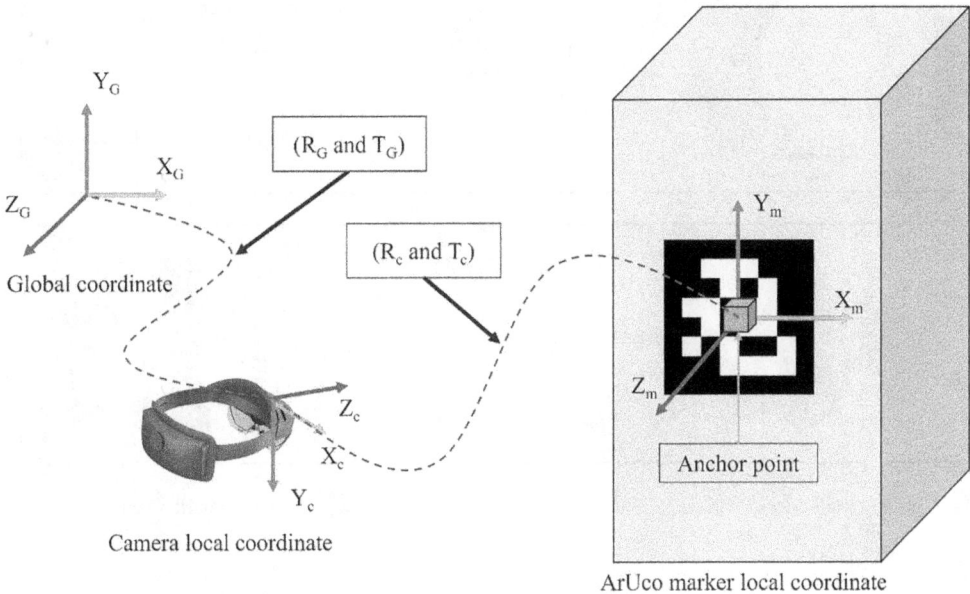

Figure 3.3 Coordinate system transformation.

In the preliminary experiment illustrated in Section 3.5, the 3-D position estimation can be affected by the angle and distance between the camera and marker in the Z axes. An offset between the actual marker center and the anchor point position based on marker detection (A_E) may occur (Figure 3.4), leading to an inaccurate transformation of the superimposed 3-D model. This is a limitation of 3-D position estimation using the ArUco marker; thus, additional steps are performed to adjust the position of the anchor point. First, the angle between Z_c axis and Z_m axis is limited to ensure the 3-D position estimation performance, and values below 15 degrees are indicated to achieve satisfactory performance in instantiating the anchor point. An anchor point position adjustment is then performed to minimize the offset between the anchor point position and the actual marker position. The marker is attached to a surface, and this surface exists in the spatial 3-D model of the construction site, so a virtual ray is cast from the camera toward A_E and hits the spatial 3-D model of the surface. This process is repeated five times and returns five anchor points. The midpoint of these anchor points is the adjusted anchor point (A_D) at the coordinate $(x_a', y_a', z_a')^G$. Using the adjusted anchor point helps increase marker detection confidence and reduce potential errors based on a single detection result.

The origin and rotation of the virtual 3-D model are required for superimposing on the construction site. The model instantiation starts when the number of anchor points requirement is met, and each model requires a specific set of anchor IDs to begin the superimpose process. Due to the offset in angle measurement found in the preliminary experiment in Section 3.5.1, the framework requires at least two markers on the same plane in the set of markers to superimpose the virtual 3-D model. With the multiple-marker setup, the anchor points corresponding to these two markers on the same plane

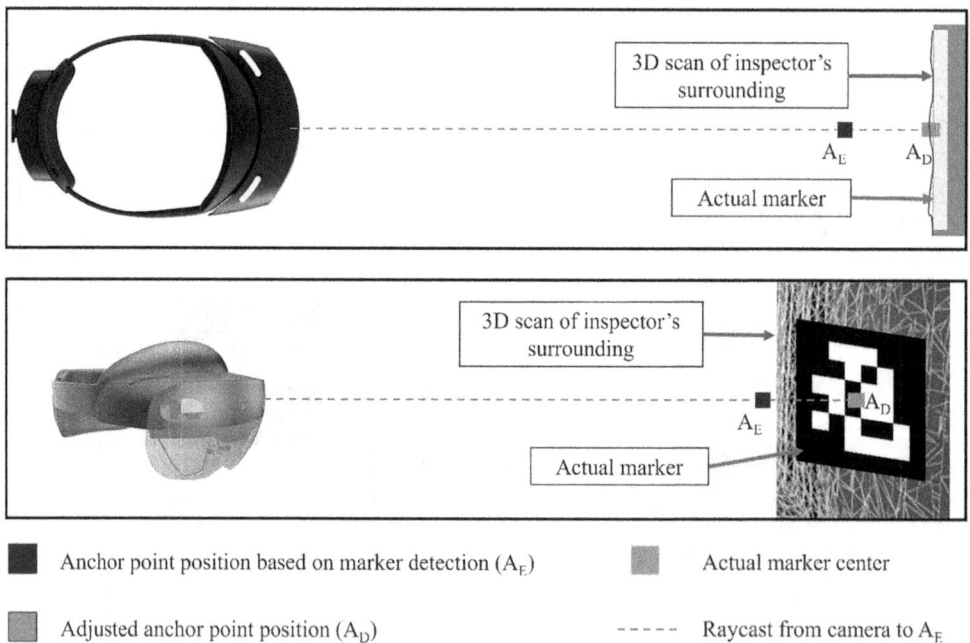

Figure 3.4 Adjusting the anchor point position.

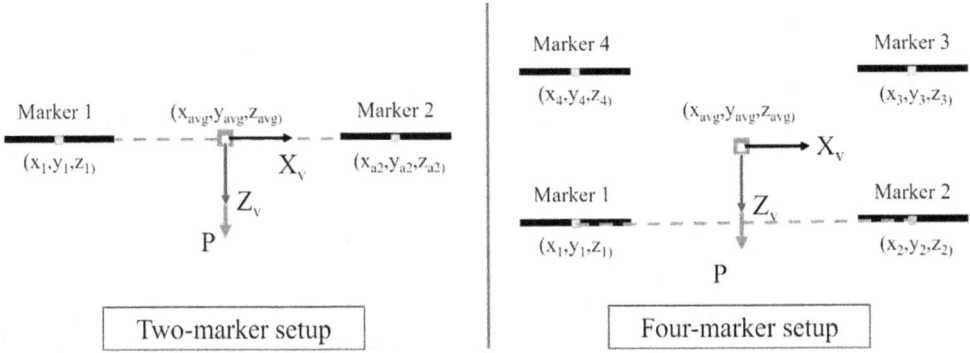

Figure 3.5 Top view of 3-D model transformation.

are instantiated first. In this study, two setups are used: the two-marker setup and the four-marker setup. In both setups, the origin of the virtual 3-D model coordinate is the average of the markers' coordinates. In the two-marker setup, the X_v-axis is on the line connecting between two markers. In the four-marker setup, the X_v-axis is parallel to the line connected between two markers on the same plane. The Y_v-axis is in the same direction as the Y_G-axis, which has an upward direction compared to the earth's surface. The Z_v-axis of the 3-D model is in the same direction as vector P, perpendicular to both X_v and Y_v axes (Figure 3.5). The alignment of these axes fixed the rotation of the superimposed virtual 3-D model in global coordinates. Using multiple anchor points can reduce the errors caused by spatial drifting, since the model position does not depend on only a single spatial point.

3.4.3 Comparing the Superimposed 3-D Model and the Surrounding Spatial 3-D Model

Once the virtual 3-D model is successfully superimposed on actual construction, the inspection stage takes place to distinguish the installation status of the components. The inspector wears the head-mounted AR device to scan the physical surroundings via the ToF camera to create the spatial 3-D model, allowing the device to reconstruct surrounding environments in 3-D form with multiple meshes (Figure 3.6). Then, the status of the components can be determined by detecting collisions between the superimposed 3-D model in the previous section and the spatial 3-D model of reconstructed surroundings. The inspection stage finishes when all interested components are inspected, and the inspection result is saved into the device.

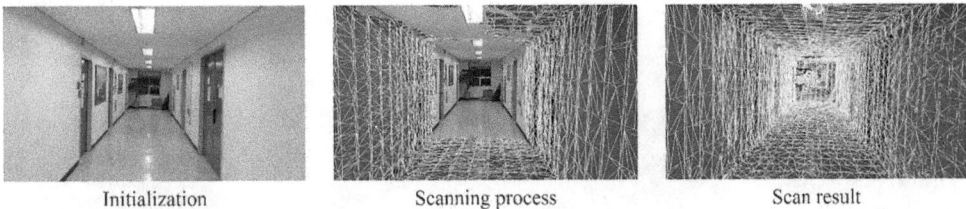

Initialization   Scanning process   Scan result

Figure 3.6 Visualization of spatial mesh during the scanning process.

The device's ToF camera initially scans the environment around the inspector to create a spatial mesh consisting of multiple triangles overlaid on the physical surroundings. These triangles are then converted into a convex hull to form a water-tight surface, reducing physical calculation in the following steps and maintaining the framework's stable performance. Spatial mesh is updated progressively as the ToF camera continuously scans the construction site, and there are several configurations—update interval, level of detail, and display—for this process:

- The spatial observer is the volume of the spatial 3-D model rendered around the inspector. It has the shape of a cube, with its origin fixed at the camera position in global coordinates (Figure 3.7). The size of the cube is defined in the X, Y, and Z directions. This setting is affected by other settings, such as a high level of detail or enabled display of the spatial mesh, requiring more computational power and reducing the amount of spatial mesh rendered in the spatial observer.
- The update interval is the time period between each spatial mesh update, including creating a new scan mesh and deleting outdated meshes. Continuously updating the spatial scan results gives a more accurate view of the surroundings but consumes more computational power.
- The level of detail is the number of triangles for an area and can be configured as fine, medium, and coarse. A finer level of detail can scan small details better by creating more triangles in a mesh; however, this setting requires more computational power and a slower update rate.
- The display option determines whether to display the spatial mesh. In this study, the spatial mesh is not shown to reduce the computational power and avoid obstructing the inspector's view.

In this study, the installation status was defined based on collisions between the superimposed and spatial 3-D models. With the ToF camera, large components, such

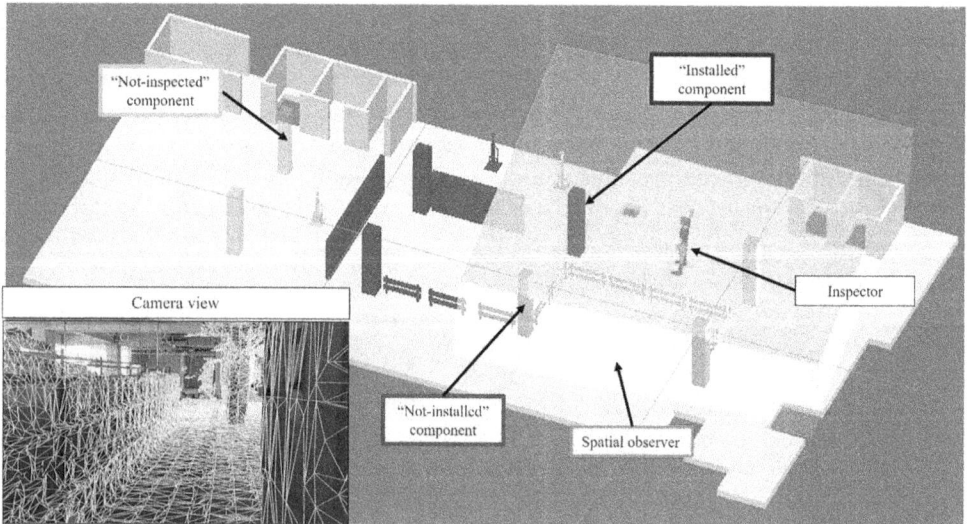

Figure 3.7 Virtual 3-D model simulation.

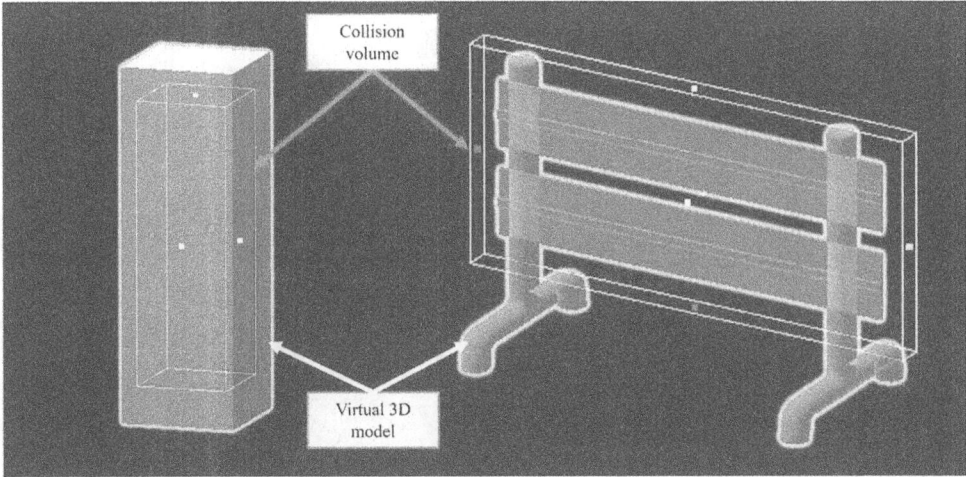

Figure 3.8 Virtual 3-D model collision volume.

as columns or walls, are mapped faster, but the spatial mesh needs time to map the surroundings precisely. There can be small dents or extrusions in the spatial mesh when the spatial 3-D model was first formed due to the limitation of surrounding reconstruction based on the ToF camera. Detecting the collision between this model and the superimposed 3-D model can potentially cause incorrect results due to the collision between the extrusions of the spatial 3-D model with other objects in the virtual 3-D model. By contrast, small, thin objects with voids, such as fences, require more time to scan and form the spatial 3-D model. On the first scan, only parts of these objects are scanned and mapped by the ToF camera, limiting collision detection between the spatial 3-D model and superimposed 3-D model. Each component's interactable volume can be adjusted based on its characteristics to address this problem. This volume can be larger or smaller than the actual size of the component and is called the collision volume. The position of the collision volume was defined by its center position. The size of the collision volume is a factor time of the original size of the model in the X, Y, and Z directions. The collision volume of large components (e.g., columns, walls) can be decreased compared to the component's original size, while the size of small and thin components (e.g., fences) can increase compared to the component's original size (Figure 3.8). Using the collision volume instead of the actual volume of the components helps improve the overall accuracy of the framework and overcomes the limitations of the existing ToF camera.

The framework can automatically detect collisions between the spatial and superimposed 3-D models to determine the status of the components. The inspection components are divided into three categories: "Installed", "Not-installed", and "Not-inspected", which are also visualized in contrasting shades (Figure 3.9). If there is a collision between the spatial 3-D model and the superimposed 3-D model for 10 seconds, the installation status of the corresponding component is set to "Installed" in the inspection report, and the component's model is marked with red color. Since the spatial mesh is only rendered in the spatial observer, components not in the spatial observer range are marked with light color and noted as "Not-inspected" in the report until they

Figure 3.9 Model visualization based on component's installation status.

are in the spatial observer range. These components are in the spatial observer range but have no collision with the spatial mesh marked with a contrasting color to the previous oneand are noted as "Not-installed" in the report (Figure 3.9). The inspector can check the list and status of components to verify that all components have been inspected. The final inspection report is then saved on the device.

3.5   Results and Validation

In this study, two experiments were conducted. The first was a preliminary experiment to estimate distance and angle measurement offset based on marker detection. The second experiment was conducted on an operating construction site to validate the framework's performance and search for possible limitations in a site-based experiment. The device used in both experiments was HoloLens 2 with a calibrated camera.

3.5.1   *Preliminary Experiment on Marker Distance and Angle Measurement*

Preliminary experiments were conducted to evaluate the 3-D pose estimation of the ArUco marker. From the observation and results of these experiments, there were considerable offsets between the estimated anchor point and the actual marker's center, and two main factors, that is, the distance and angle between HoloLens 2 and the ArUco marker, led to these offsets. In these experiments, the offset values were calculated by comparing the ground truth distance, measured by rulers from the camera to the marker center (the ray coming from Zc-axis and noted as "Distance measurement" in Figure 3.10), with the distance measurement returned from marker detection (the ray noted as "Ground truth" in Figure 3.10). Marker detection was performed with a video stream resolution of 1920 × 1080 at 30 frames per second. The markers with different sizes of 0.12 m, 0.15 m, and 0.18 m were placed 2.4 m or 3 m away from the camera, and rotated to create angles of 0, 15, and 30 degrees between the Z_c and Z_m axes (Figure 3.10). It is worth noting that the camera and the marker center had the same x and y co-ordinates in the global coordinate system, and the offset occurred on the Z axis. Table 3.1

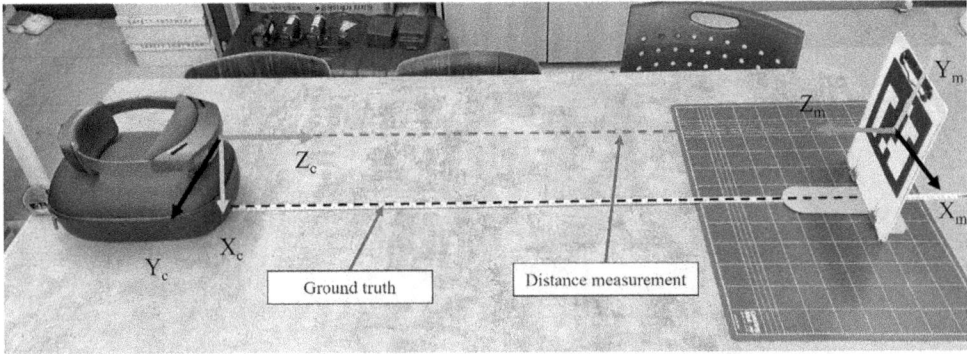

Figure 3.10 Preliminary marker distance and angle measurement experiment.

shows the distance and angle measurements returned from the marker detection and compared with the ground truth data. All of the configurations in the experiment were performed five times, and their average values are recorded in Table 3.1.

Table 3.1 illustrates the differences between the ground truth and the results of measurement from ArUco marker detection. In other words, there are errors in anchor point positions when relying only on ArUco marker detection. From Table 3.1, it is clear that the distance and angle offset increase when the angle between the inspector and the marker is larger, while the difference in marker size does not affect the results. The distance measurement offset was from 0.02 to 0.12 m, and the angle measurement offset was from 1 to 18 degrees. Based on the offset values recorded, the measurement was more

Table 3.1. ArUco marker distance measurements with different angles

Size (m)	Ground Truth		HoloLens 2			
	Distance (m)	Angle (degree)	Distance (m)	Offset (m)	Angle (degree)	Offset (degree)
0.12	2.4	0	2.37	0.03	3	3
		15	2.38	0.02	8	7
		30	2.32	0.08	19	11
	3	0	2.94	0.06	3	3
		15	3.05	0.05	3.5	11.5
		30	3.06	0.06	17	13
0.15	2.4	0	2.35	0.05	2	2
		15	2.37	0.03	2.5	12.5
		30	2.35	0.05	17	13
	3	0	2.95	0.05	2.5	2.5
		15	2.91	0.05	4	11
		30	2.93	0.07	17.5	12.5
0.18	2.4	0	2.37	0.03	1	1
		15	2.36	0.04	7	8
		30	2.3	0.1	13	17
	3	0	3.03	0.03	3	3
		15	2.97	0.03	4.5	10.5
		30	2.88	0.12	12	18

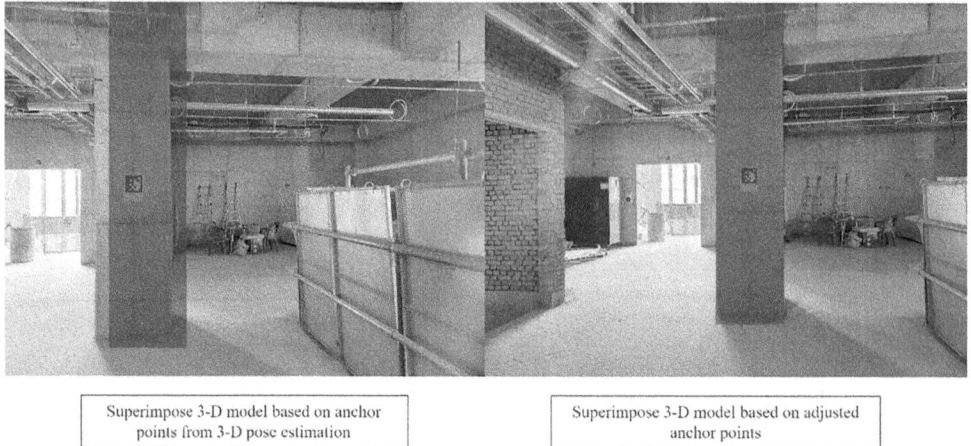

Superimpose 3-D model based on anchor points from 3-D pose estimation	Superimpose 3-D model based on adjusted anchor points

Figure 3.11 Unadjusted and adjusted superimposed 3-D model visualization.

accurate when the offset angle between Z_c axis and Z_m axis was less than 15 degrees. An additional adjustment step was performed to reduce the offset distance in the Z direction to reduce errors in anchor point positions. The method is presented in detail in Section 3.4.2. After the adjustment step was performed, the offset between the anchor point and the ground truth actual marker center was less than 0.01 m. If the virtual 3-D model is superimposed based on unadjusted anchor points, the position and rotation of the model can be incorrect, especially for components far from the model's origin (left side of Figure 3.11), especially components far from the origin of the model. Using the adjusted anchor point helps the framework correctly superimpose the virtual 3-D model to the inspector's field of view.

3.5.2   Onsite Construction Installation Status Experiment

The experiment was performed to validate the framework's performance on an actual construction site. The BIM was on the second floor of the building and had an area of 1300 m². The inspection aimed to determine the installation status of eight columns, three support jacks, eleven fences, and a wall. When the experiment was conducted, all the columns and fences were installed, while three support jacks and the wall were not installed on the construction site. The virtual 3-D model for the experimental area was prepared according to Section 3.4.1, and 10 anchor points were configured in columns marked C, E, F, and G (see Figure 3.12a). Due to safety reasons, markers were not attached to columns A, B, D, and H. Each marker was attached to the column's surface 1.6 m above the ground, and its center was in the middle of the column. Eight experiments with different marker setups were conducted, including two-marker setups and four-marker setups. To examine the influence of marker positions on the framework's performance, the distance between markers was varied due to the selections of two or four-marker setups (see Figure 3.12b). The inspector followed the same inspection route (the path was demonstrated in Figure 3.12) at a moderate walking speed and detected markers to instantiate anchor points based on the chosen setup. Once the required anchor points were determined, the virtual 3-D model was superimposed onto the real site.

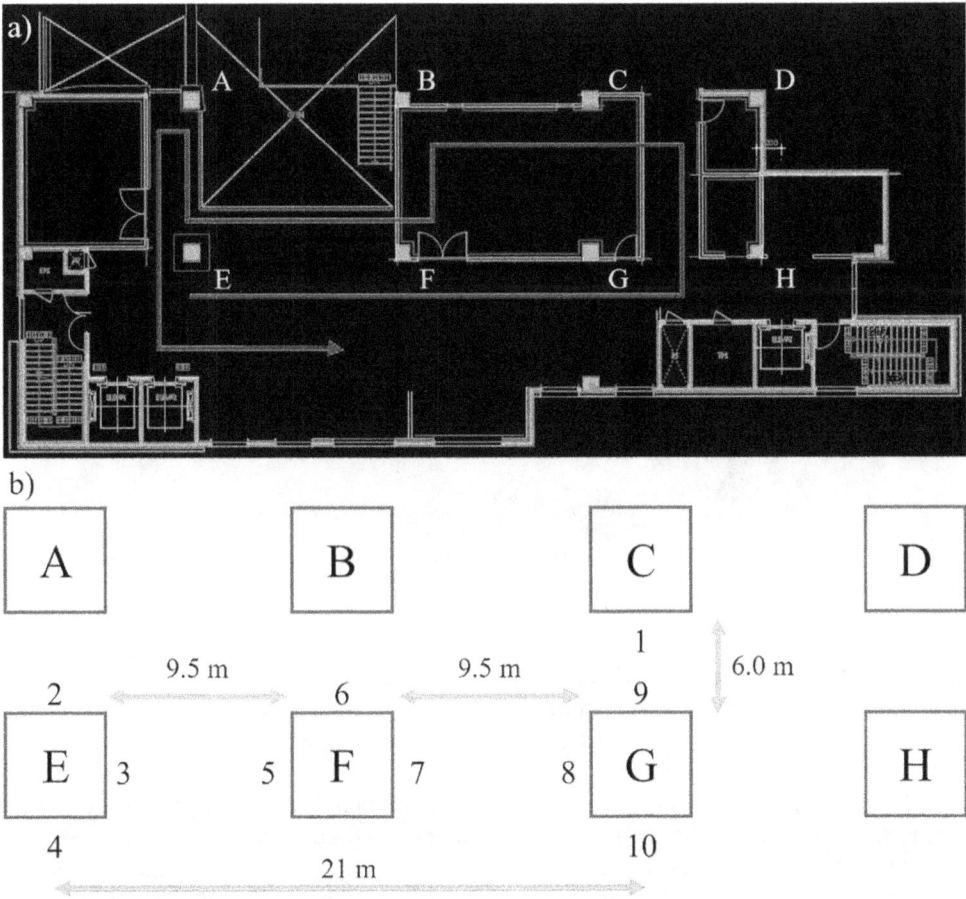

Figure 3.12 Columns and marker positions.

"Installed", "Not-installed", and "Not-inspected" components were visualized in red, blue, and light blue color, respectively, after comparing the superimposed 3-D and spatial 3-D models. The installation status of each component was shown in the inspection list and was automatically saved on the device (see Figure 3.13). The framework's performance was evaluated based on processing time, precision, and accuracy.

The installation status was determined by the collision between the spatial 3-D model and the superimposed 3-D model. The collision volume was 80% of the actual component volume for large components, such as concrete columns, while for thinner components, such as fences, the collision volume increased to 120% of the actual component volume. The spatial mesh settings in most of the experiments were 0.1 s update interval, a spatial observer of 20 × 5 × 20 meters, and the level of detail was coarse. When the update interval was set to 0.1 s, any new scanning result from the ToF camera was added to the spatial observer. The level of detail was set to Coarse to obtain a faster update time and maintain stable performance during the inspection process. These settings allow the head-mounted device to scan the surroundings faster and maintain spatial mesh volume. There were two experiments with different settings,

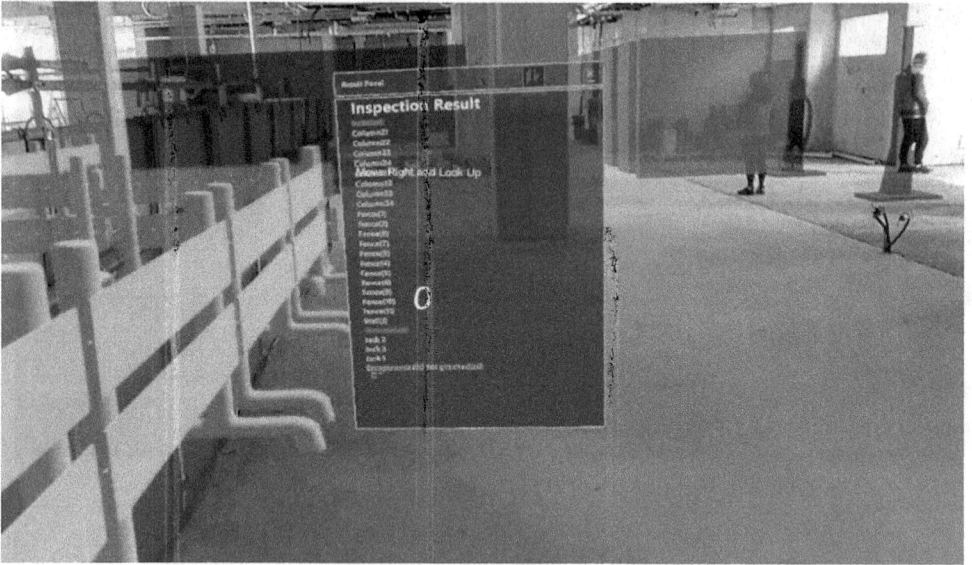

Figure 3.13 Inspection results.

with one having its level of detail as fine and an update interval of 0.1 seconds, and the other having its level of detail as coarse and an update interval of 2.5 s. Experiments with these two setups were conducted to validate whether these settings affect the framework's overall performance.

The processing time to finish the inspection, which includes superimposing the virtual 3-D model, following the inspection route at moderate speed, and surveying components in the inspection list, was examined. In the first stage, performing the marker detection and anchor point position adjustment took 4 to 6 seconds for each anchor position. Once the required anchor points were instantiated, the approximate time to superimpose the virtual 3-D model was 30 seconds when using two anchor points and 50 seconds when using four anchor points. It is worth noting that the inspector's trajectory and walking speed for all experiments remained unchanged. The total time to conduct inspection was around 5 minutes for two-marker setups and 7 minutes for four-marker setups. The inspection is considered to be done when all the components of interest are inspected (Figure 3.13).

The accuracy of this framework was identified based on the correct and mistaken installation statuses of all inspected components. To evaluate the performance of the framework in this experiment, the results of the inspection were presented as precision and accuracy. The two are defined as follows:

$$Precision = TP/(TP + FP)$$
$$Accuracy = (TP + TN)/(TP + FP + TN)$$

where true positive (TP) is the actual component on the construction site determined as "Installed", true negative (TN) is the virtual component not existing on the construction site determined as "Not-installed", and false positive (FP) is the virtual component not

Table 3.2. Results of the experiment on an actual construction site

Marker Setup	Level of Detail	Update Interval	True Positive	True Negative	False Positive	False Negative	Precision (%)	Accuracy (%)
4–10	Coarse	0.1	19	4	0	0	100	100
3–5			19	4	0	0	100	100
1–9			19	4	0	0	100	100
7–8			19	3	1	0	95	95.65
2–4–9–10			19	4	0	0	100	100
3–5–7–8			19	3	1	0	95	95.65
4–10	Fine	0.1	19	3	1	0	95	95.65
4–10	Coarse	2.5	19	4	0	0	100	100
						Average	98.13	98.37

existing on the construction site determined as "Installed". The results of the experiment are summarized in Table 3.2.

The experiments with two markers and four markers performed similarly, whereas the experiments with two markers but with the level of detail set to fine or the update time set to 2.5 seconds showed different user experiences. With the experiment using two markers with a fine detail level and an update interval of 0.1 seconds, the spatial mesh cannot cover the whole volume of the spatial observer due to computational limitations. The number of meshes and triangles was greater than what the device was capable of storing, leading to the reduction of inspection volume simultaneously. With the experiment using two markers with the level of detail set to coarse and the update interval set to 2.5 seconds, the update of new spatial meshes as the inspector walks through the construction site was slower, delaying the process of checking the installation status and display color corresponding to its status.

Table 3.2 shows the excellent performance of the proposed framework in all experiments when the average accuracy and precision were 98.13% and 98.37%, respectively. In the experiments with marker setups 7–8, 3–5–7–8, and 4–10, the false positive resulted from the virtual wall colliding with an electric wire, which was not included in the BIM model (see Figure 3.14). In the preparing model step, this electric wire's position was not considered, leading to unexpected components in the spatial 3-D model. Thus, it is essential to conduct a better survey before conducting an inspection of the actual construction site.

3.5.3 Discussion and Future Research

From the experiment in Section 3.5.1, the error in the position of anchor point instantiation based on the 3-D pose estimation was indicated. This error can lead to an offset in the position of the superimposed 3-D model and affect the performance of the whole framework. By performing an extra step to adjust the anchor's position, the offset was minimized to less than 0.01 m and did not affect the accuracy of superimposing the 3-D model. In the experiment in Section 3.5.2, the inspection result returned the correct installation status for most components in all experiments. In the experiments, the time to perform and the accuracy of each inspection were recorded to determine the performance of each setup. The results of the experiment in Section 3.5.2 demonstrate that the framework is feasible to work on the construction site.

Figure 3.14 The spatial mesh of the component caused false positive results.

Some experiments showed that the superimposed 3-D model drifted from its original position despite the correct final result. The 3-D model was superimposed correctly on the construction site based on the anchor points, but after the inspector moved too far from the position of the first anchor points, the 3-D model started to drift a few centimeters compared to the actual components. This problem occurred due to the error in accumulative error from the IMU sensor and HoloLens 2's spatial observer volume limitation. The spatial mesh of the surrounding can be stored for the area of up to 400 m^2 around the inspector, but in a large and open construction site, such as in this experiment, the amount of accumulative error from the IMU becomes noticeable. However, the 3-D model can quickly recover once it encounters familiar environments to recalculate its position in the space. The familiar environments are the surroundings scanned by HoloLens 2's cameras, which contain distinctive features for the device to recognize. The error can be reduced by dividing the inspection area into multiple parts, with each part having an area equal to the spatial observer to minimize the error from the IMU. This problem can also be addressed by applying the framework to a device with a better IMU sensor and higher computational power than HoloLens 2. With the progression of AR devices and technology in recent years, improved head-mounted AR devices are expected to be released in the near future. To address the false results that occurred by unexpected objects on the construction site, the framework can be improved by determining the installation status of onsite components based on the overlap volume ratio between the superimposed 3-D model and the spatial 3-D model of the inspector's surroundings.

3.6 Conclusion

The rising trend of digitalization in the construction industry is a step forward from traditional analog practices. AR is one of the cutting-edge technologies adopted to address various construction issues. The ability to visualize virtual content in the user's field of view can bring an immersive and intuitive approach to information. Even though there have been some studies using AR in construction inspections, most of them required complex equipment to work and only tested in a lab-based environment. In this study, a marker-based AR framework was introduced to automatically inspect the installation status of components on a construction site. The framework compares the spatial 3-D model of the inspector's surroundings and the superimposed 3-D model of the construction site extracted from the BIM model to find the collisions between the two models. The output of the framework is the list of the inspection components with their installation status in digital format.

An experiment was conducted on an operating construction site to validate the framework's performance. The average accuracy and precision of the framework were 93.13% and 98.37%, respectively. The experimental results proved that the framework has high accuracy and is feasible for streamlining the construction inspection process. However, during the experiment on the operating construction site, the framework faced certain limitations due to the existing hardware. This calls for the continuous development of this framework through the use of the latest head-mounted AR devices and the application of new methods for determining the installation status of onsite components.

Acknowledgments

This work was supported by the National Research Foundation of Korea (NRF) grant funded by the Korean government (MSIT) (No. NRF-2020R1A4A4078916).

References

Ahmed, S. (2019). A review on using opportunities of augmented reality and virtual reality in construction project management. *Organization, Technology and Management in Construction: An International Journal by Sciendo* 11, 1839–1852.

Alzahrani, J. I., & Emsley, M. W. (2013). The impact of contractors' attributes on construction project success: A post construction evaluation. *International Journal of Project Management,* 31(2), 313–322.

Azuma, R. T. (1997). A survey of augmented reality. *Presence: Teleoperators and Virtual Environments,* 6(4), Article 4.

Behzadan, A. H., & Kamat, V. R. (2009). Interactive augmented reality visualization for improved damage prevention and maintenance of underground infrastructure. *Construction Research Congress,* 2009, 1214–1222.

Behzadan, A. H., & Kamat, V. R. (2013). Enabling discovery-based learning in construction using telepresent augmented reality. *Automation in Construction,* 33, 3–10. 10.1016/j.autcon. 2012.09.003

Bosché, F., Tingdahl, D., Carozza, L., & Van Gool, L. (2012). Markerless vision-based augmented reality for enhanced project visualization. 2012 Proceedings of the 29th ISARC, 11.

Brata, K., Liang, D., & Hadi Pramono, S. (2015). Location-based augmented reality information for bus route planning system. *International Journal of Electrical and Computer Engineering,* 5(1), 142–149.

Chen, Y.-J., Lai, Y.-S., & Lin, Y.-H. (2020). BIM-based augmented reality inspection and maintenance of fire safety equipment. *Automation in Construction*, 110, 103041.

Elghaish, F., Matarneh, S., Talebi, S., Kagioglou, M., Hosseini, M. R., & Abrishami, S. (2020). Toward digitalization in the construction industry with immersive and drones technologies: A critical literature review. *Smart and Sustainable Built Environment*, 10(3), 345–363.

Fazel, A., & Izadi, A. (2018). An interactive augmented reality tool for constructing free-form modular surfaces. *Automation in Construction*, 85, 135–145. 10.1016/j.autcon.2017.10.015

Garrido-Jurado, S., Muñoz-Salinas, R., Madrid-Cuevas, F. J., & Marín-Jiménez, M. J. (2014). Automatic generation and detection of highly reliable fiducial markers under occlusion. *Pattern Recognition*, 47(6), 2280–2292.

Hartley, R., & Zisserman, A. (2003). *Multiple View Geometry in Computer Vision*. Cambridge University Press.

Israngkura Na Ayudhya, B., & Kunishima, P. (2009). The performance of disbursement procedures in Highway Public Works in Thailand. *International Journal of Engineering Science and Technology*, 61.

Kang, S.-C., Lin, K.-Y., & Tsai, M.-H. (2011, August 1). The Development of an Innovative Tool Kit to Support Construction Safety Audits. The Development of an Innovative Tool Kit to Support Construction Safety Audits. The Development of an Innovative Tool Kit to Support Construction Safety Audits, Washington, D.C., USA.

Koch, C., Neges, M., König, M., & Abramovici, M. (2014). Natural markers for augmented reality-based indoor navigation and facility maintenance. *Automation in Construction*, 48, 18–30.

Kopsida, M., Brilakis, I., & Vela, P. A. (2015). A review of automated construction progress monitoring and inspection methods. *Proceedings of the 32nd CIB W78 Conference on Construction IT*. 12.

Lin, T.-H., Liu, C.-H., Tsai, M.-H., & Kang, S.-C. (2015). Using augmented reality in a multiscreen environment for construction discussion. *Journal of Computing in Civil Engineering*, 29(6), 04014088.

Ling, F. Y. Y., & Bui, T. T. D. (2010). Factors affecting construction project outcomes: Case study of Vietnam. *Journal of Professional Issues in Engineering Education and Practice*, 136(3), 148–155.

Loporcaro, G., L, B., P, M., & H, R. (2019). *Evaluation of Microsoft HoloLens Augmented Reality Technology as a construction checking tool*. University of Canterbury.

Lutz, O., Burmeister, C., Ferreira dos Santos, L., Morkisch, N., Dohle, C., & Krüger, J. (2017). Application of head-mounted devices with eye-tracking in virtual reality therapy. *Current Directions in Biomedical Engineering*, 3, 53–56.

Marquardt, D. W. (1963). An algorithm for least-squares estimation of nonlinear parameters. *Journal of the Society for Industrial and Applied Mathematics*, 11(2), 431–441. 10.1137/0111030 of a mobile BIM-based augmented reality system. Automation in Construction, 42, 1–12.

Meža, S., Turk, Ž., & Dolenc, M. (2015). Measuring the potential of augmented reality in civil engineering. *Advances in Engineering Software*, 90, 1–10.

Microsoft. (2022). HoloLens 2—Overview, Features, and Specs | Microsoft HoloLens.

Piroozfar, P., Essa, A., Boseley, S., Farr, E., & Jin, R. (2018, July 2). Augmented Reality (AR) and Virtual Reality (VR) in construction industry: An experiential development workflow. *The Tenth International Conference on Construction in the 21st Century* (CITC-10).

Ratajczak, J., Riedl, M., & Matt, D. (2019). BIM-based and AR application combined with location-based management system for the improvement of the construction performance. *Buildings*, 9, 118.

Rauschnabel, P., Brem, A., & Ro, Y. (2015). *Augmented Reality Smart Glasses: Definition, Conceptual Insights, and Managerial Importance*. Working Paper, The University of Michigan-Dearborn.

Saidi, K. S., Lytle, A. M., & Stone, W. C. (2003). Report of the NIST Workshop on Data Exchange Standards at the Construction Job Site. *ISARC Proceedings*, 617–622.

Sato, Y., Fukuda, T., Yabuki, N., Michikawa, T., & Motamedi, A. (2016). A Marker-Less Augmented Reality System Using Image Processing Techniques for Architecture and Urban Environment. *The 21st International Conference on Computer-Aided Architectural Design Research in Asia* (CAADRIA 2016), 713–722.

Shin, D. H., & Dunston, P. S. (2008). Identification of application areas for augmented reality in industrial construction based on technology suitability. *Automation in Construction*, 17(7), 882–894.

Spreafico, A., Chiabrando, F., Losè, L. T., & Tonolo, F. G. (2021). The iPad Pro built-in lidar sensor: 3D rapid mapping tests and quality assessment. *The International Archives of Photogrammetry, Remote Sensing and Spatial Information Sciences*, XLIII-B1-2021, 63–69.

Tsai, Y.-H., Hsieh, S.-H., & Kang, S.-C. (2014). A BIM-enabled approach for construction inspection. *Computing in Civil and Building Engineering*, 2014, 721–728.

Ungureanu, D., Bogo, F., Galliani, S., Sama, P., Duan, X., Meekhof, C., Stühmer, J., Cashman, T. J., Tekin, B., Schönberger, J. L., Olszta, P., & Pollefeys, M. (2020). HoloLens 2 research mode as a tool for computer vision research. ArXiv:2008.11239 [Cs].

Unity. (2017). Unity – Manual: HoloLens Spatial Mapping. Unity in the Unity User Manual, 2019(1).

vGIS. (2022). What is the best augmented reality AR system for BIM and Esri GIS? VGIS – Leading Augmented Reality Solutions for BIM, GIS and 3D Scans.

Vo-Tran, H., & Kanjanabootra, S. (2013, May 5). *Information Sharing Problems Among Stakeholders in the Construction Industry at the Inspection Stage: A Case Study*. CIB World Building Congress 2013. CIB World Building Congress 2013, Brisbane, Australia.

Zainon, N., Rahim, F. A., & Salleh, H. (2011). The information technology application change trend: its implications for the construction industry. *Journal of Surveying, Construction and Property*, 2(2), Article 2.

Zollmann, S., Hoppe, C., Kluckner, S., Poglitsch, C., Bischof, H., & Reitmayr, G. (2014). Augmented reality for construction site monitoring and documentation. *Proceedings of the IEEE*, 102(2), 137–154.

4 Schedule-Driven BIM Model Breakdown Framework for Construction Monitoring with Augmented Reality

Thai-Hoa Le, Yun Ting Shih, and Jacob J. Lin

Department of Civil Engineering, National Taiwan University, Taiwan

4.1 Introduction

Recently, augmented reality (AR) appears to be a potential method for construction monitoring since it can save time, reduce manual work, and ineffective exertion on construction sites. According to Rahimian et al. (2020), AR has attracted more attention since it can provide an experience that seamlessly integrates virtual content into real-world scenes. A number of AR applications have been developed to support the architecture, engineering, and construction (AEC) industry in different use cases illustrated by Delgado et al. (2020). AR can generally support visualization with real-time project information interaction by combining documentation and digital information. In the planning and design phase, AR supports the customers to better understand the end product without specialized knowledge of construction (Mutis & Ambekar, 2020). By having a visualization of the construction product at an early stage, project teams can be sure of high performance of the project's presentation, constantly working on problem-solving without physical contact with the object and probably, other team members. These lead to an improved teamwork capability, more safety, and a decrease in cost and downtime wasted for rework or overlaps (Ahmed et al., 2017; Forcada Matheu et al., 2017). Although applicable in most fields of the AEC, the applicability of AR is at the early testing level, and most applications have been made with AR in the design phase. Delgado et al. (2020) reported that only design-related use cases achieved higher levels of AR adoption, from 2.1 to 2.47 out of 5.0 in the UK construction industry. In addition, dynamic conditions of outdoor site environments and the technical limitation of current AR systems seem to thwart the technology from embracement.

Among various digital solutions, BIM designs are currently recommended as the initiation for AR implementations. Integrating BIM and AR can address practical issues in construction projects such as slow information retrieving, error committing in assembly, and low efficiency of communication and problem-solving (Wang et al., 2014). Additionally, BIM knowledge is highly suggested for solving multi-faced problems, including the multi-organizational and multi-disciplinary issues in the AEC industry (Li et al., 2017). 4D BIM models can be visualized in AR devices to compare the as-planned models to practical progress of the project (Han & Golparvar-Fard, 2015). Nevertheless, there is no particular method for the incompleteness and complication of BIM-to-AR process. Han and Leite (2022) introduced a BIM-to-XR workflow that can automatically update the information and changes to an AR-ready model; however, the conversion is challenging if the original model size is large.

DOI: 10.1201/9781003408949-4

This chapter presents a framework that can support AR-based construction monitoring by exporting semantic information from the BIM model considering thresholds in AR devices. The approach investigates three factors to create the BIM model ready for AR-based monitoring: work packages, schedule, and working areas for the model breakdown process. The work packages are the first factor to be considered to target relevant BIM components that have engineering relations. Next, scheduling information is examined to provide reasonable sequence of construction areas. The final filter for BIM model segmentation is working areas exported based on the work breakdown structure (WBS). The following section discusses the state-of-the-art research in AR applications in construction and explains the findings from the literature about AR application requirements in Section 4.3. The next sections introduce the proposed framework followed by the case studies and results.

4.2 Related Work

4.2.1 Construction Monitoring Activities with AR Support

Being mentioned as a pillar of Industry 4.0 (Nassereddine et al., 2021), AR technology has proved its potential when being used to monitor construction practices. An AR prototype with the combination of a head-mounted device (HMD) and a computer screen was set up to perform orientation tasks such as measurement, examination, or inaccuracy detection (Wang & Dunston, 2006). Golparvar-Fard et al. (2009) established D^4 AR approach that can collect data and superimpose reconstructed scenes to the real site. The method supports the engineer in comparing real-time progress to the approved schedule. Another AR prototype system, the ARCam, was introduced to execute steel column inspection (Dunston, 2010). Zhou et al. (2017) presented an AR method to inspect segment displacements when constructing a tunnel. Lee and Akin (2011) elicited an AR-based equipment operation and maintenance (O&M) fieldwork support to observe and facilitate on-site activities. Validation experiments showed that the application obtained sensor-based information faster than the traditional method and saved 51% of the time spent locating work areas. Fazel and Izadi (2018) proposed a framework using an AR tool to support complex masonry. The system can monitor and provide virtual guidance to users, leading to fewer construction errors. Additionally, various applications of AR use for assembly system were systematically reviewed by Wang et al. (2016). Although having initial achievements in applying AR in monitoring process, challenges associated with the AR implementation are identified under the modeling and alignment barriers, hardware limitation, and data management (Noghabaei et al., 2020).

4.2.2 BIM and AR Integration in Construction

Engineers and managers can use AR to visualize BIM data on-site. Integrating AR and BIM can improve performance in different fields such as architectural visualization, progress monitoring, dynamic site visualization, defect detection, construction inspection, training, etc. (May et al., 2022; Sidani et al., 2021). AR application can overlay BIM information straightly on the site scene; thus, it can support detecting conflicts during the coordination of construction parties and visualize work sequences to optimize the site logistics. Chu et al. (2018) integrated mobile BIM into AR system to evaluate the technical performance of workers. Experimental validation of the research showed that the mobile BIM-AR system increased the productivity of participants and reduced committed errors. The combination of BIM and AR also provides an intelligent solution to monitor complex plumbing facilities (Diao & Shih, 2019). This BIM-based AR maintenance system (BARMS) can be used in smartphones to consider subjects, paths, and actions following related knowledge guidance and management protocols. In the research of Mutis and Ambekar (2020), the i-Tracker was developed in four phases, in which the first phase is rendering BIM model to a 3D game object and initialize the AR visualization features. Subsequently, it can be seen that integrating BIM and AR is one of the most promising concepts to improve the applicability of BIM in fieldwork. Yet, compared to VR technology, the number of AR applications integrated with BIM is still restricted because of the limited AR devices available on the market (Alizadehsalehi et al., 2020).

Considering current trends in integrating BIM and AR, it is significant to have a framework for data sharing and information collaboration between the two technologies. Chi et al. (2013) mentioned four phases necessary to develop an AR application for AEC utilization, in which BIM is required to create a cloud computing environment. Such BIM support is known as "CloudBIM" which can enhance stand-alone applications with network and dynamic features such as shared environments and allocated tasks (Chi et al., 2013; Redmond et al., 2012). Sharing the same cloud-based functions with BIM, AR applications can be updated with huge data and allow users to simultaneously work with massive amounts of information and multi-stakeholder communication. According to Banfi et al. (2019), cloud-based system such as Autodesk BIM360 is the most supportive and popular for sharing BIM/XR (extended reality) on the market. Alizadehsalehi et al. (2020) introduced a comprehensive BIM-to-XR toolsets workflow to give scientists and engineers a general view of supported software. Nevertheless, the challenge of completeness and smoothness of file transfer still exists. Amin et al. (2023) made a systematic review and identified six key functions for BIM-based AR platform evaluation called PIVCAT. Those functions include: positioning (P)_to evaluate the accuracy of BIM model localization and alignment, interaction (I)_to evaluate the user interface efficiency, visualization (V)_to evaluate the system's visibility and graphic fidelity, collaboration (C)_to evaluate the communication and information sharing capabilities, automation (A)_to evaluate the ability to work without interruption, and integration (T)_to evaluate the extension with additional technologies and services. Yet the research did not specify BIM requirements for AR-based applications. In most cases, the researcher must convert BIM data to AR development platform through a compli-cated process. Following this, it takes time to receive the result; therefore, it cannot afford the requirement of immediate feedback on construction field.

4.3   Requirements of BIM-Based Data Used for the Input of AR Applications

As previously stated, BIM functions as a digital platform that facilitates the integration of voluminous and disparate data into a unified model. This information must be meticulously delineated in an AR environment to enable effective monitoring. Nevertheless, the efficacy of AR visualization is likely to be contingent on the detail present in the BIM model, as depicted by the Level of Development (LOD). Currently, "Level of Detail" and "Level of Development" are mixed in research. This chapter clarifies that LOD stands for Level of Development, which covers geometry and non-geometry information in the 4D model (Boton et al., 2015). Besides, LOD specifications must be compatible with usage requirements through the project stages. There are five levels of development, from LOD 100 to LOD 500 (AIA, 2013). It is suggested that LOD 100 to 300 for design, LOD 300 and 400 for construction, and LOD 500 for turnover (Sacks et al., 2018). Overall, it is evident that there is a lack of LOD base for AR-based monitoring methods in research. The following table shows some BIM-supported information and recommendations of LOD required in developing AR applications in construction. The categories are aggregated based on the functional similarity of the information used.

Table 4.1 summarizes results of requirements from previous studies, in which each AR task had been assigned with different required data. In their works, features like geometry, object ID, hierarchy, 2D images can be extracted from BIM model. Additional information such as schedule data, quality parameters, locations, etc. need to be obtained from different software or tools, for instance, Microsoft Project, point cloud, etc. Visual style describes how the model is presented for employment. Most cases only require virtual visualization for deviation inspection and comparison. A noticeable concern is that reports rarely state BIM LODs for AR applications. Among reviewed papers, only El Ammari and Hammad (2019) referred to LOD 400 as an essential input for the inspection and maintenance of fire safety systems with AR technology. The use of LOD 400 is reasonable since the document from AIA has defined LOD 400 for the model as specific assemblies necessary for complete fabrication, including the MEP systems installation. The detailed information includes element quantity, location, orientation, size, and shape in precision. On the other hand, the suggested LOD 300 are resulted by combining LOD descriptions and extracted BIM information mentioned in the other references. The multi-user collaborative BIM-AR system of Garbett et al. (2021) allows real-time sharing of 2D and 3D geometric BIM data in simulation. For monitoring tasks such as construction progress monitoring or the structural components inspection, geometric features are required.

Table 4.1 Extraction of information for AR tasks

AR Application	BIM Information	Other Information	Visual Style	Reference	Suggested LOD
Multi-user collaborative AR system	Geometry	N/A	Model display in mobile for interaction	Garbett et al. (2021)	300
Progress monitoring	Geometry	Task ID, schedule	As-built data detection and deviation	Kopsida and Brilakis (2020); Wang and Chen (2020)	300
Structural elements inspection	Geometry	Quality parameters, schedule	Structural component visuals	Mirshokraei et al. (2019)	300
Safety inspection and maintenance	As-build document	Cost, technical data, operational manual	Virtual model in iPad	Chen et al. (2020)	400
Collaboration in facility management	Geometry, object ID, hierarchy	Task ID, images, schedule, assigned materials, and location	AR navigation	El Ammari and Hammad (2019)	400

Wang and Chen (2020) then exported schedule information from BIM to align with spatial information in their AR system for real-time feedback. Whilst, Kopsida and Brilakis (2020) claimed 3D elements from as-planned BIM model to augment real-time comparison with captured data of building progress using HoloLens. For BIM-AR quality control, Mirshokraei et al. (2019) stated that three types of data that must be collected and synchronized to prepare the 4D BIM quality model are model of physical object containing geometric data, schedule data, and quality control data. Subsequently, those BIM information requirements are relevant to LOD 300. Therefore, it is probably to recommend LOD 300 as the lowest level of development for BIM-to-AR development.

4.4   The Framework to Create an AR-Ready BIM

As illustrated in Figure 4.1, the schedule-driven framework of creating an AR-ready BIM for construction monitoring tasks includes three parts: (1) AR Tasks Model Requirements, (2) Model preparation, and (3) On-site Implementation.

4.4.1   AR Tasks Model Requirements

The first phase includes identifying the monitoring tasks that can be performed by AR and determining model requirements for AR. In order to ascertain the requisite information for monitoring performance, it is imperative to define the "AR-based monitoring task". As different construction projects involve distinct resources, diverse model information becomes necessary. Of all the components to be constructed and completed, structures and MEP systems are likely to be the most pivotal parts of AR application development. Consequently, this research endeavors to primarily investigate the structural components and MEP systems to perform monitoring tasks. With respect to the selected monitoring task, BIM-supported information can be defined, and the data prerequisites of the BIM model can be stipulated. The suggested LODs are demonstrated in Table 4.1. The results are based on literature and current required documents in the construction industry. "Model requirements for AR use" is the next necessary step since the entire model shall not be converted to AR device. To implement monitoring activities in AR environment, five factors are recommended, including (1) BIM LOD, (2) Model elements, (3) Schedule level, (4) Visual style (or rendering type), and (5) Visualization criteria. According to the literature review, those five are typical results that can be extracted from BIM model. Detailed requirements are illustrated in Table 4.2.

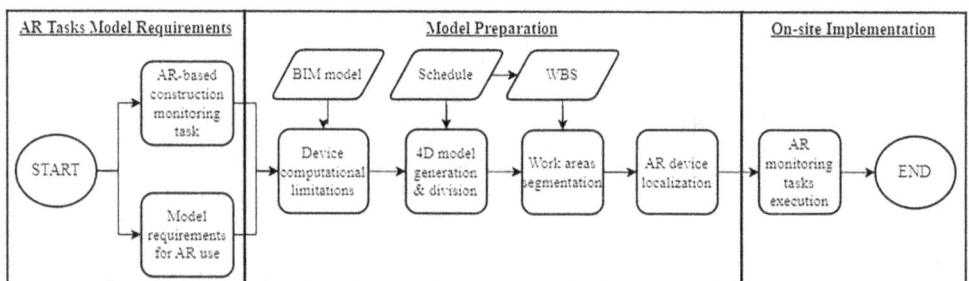

Figure 4.1 The proposed framework of BIM to AR monitoring tasks.

Table 4.2 Model requirements for AR-based monitoring tasks

Task	BIM LOD	Selected Element	Schedule Level	Visual Style	Visualization Criteria
Structural monitoring	300	• Structural component (compulsory) • Openings (optional)	Production level (weekly work plan)	Color-coded (delayed/in progress)	4D model
MEP installation monitoring	400	• Wall, slab • MEP systems	Make-ready planning (4–6 weeks lookahead)	Color-coded (correct/ incorrect position)	4D model

To conduct structural monitoring, BIM model having LOD 300 is satisfied since it can specify vital 3D object geometry, including dimensions, capacities, and connections. The structural elements are compulsory, and openings are optional. The model needs to be updated weekly for inspection as weekly coordination meetings are often organized for observation and evaluation of working performance. On the other hand, MEP installation monitoring requires higher **BIM LOD (LOD 400)**, which must include details such as fabrication and installation of the components. In order to show the installation of MEP system, MEP elements and other structural objects like walls and slabs are required. Apparently, the attached MEP will be affected if there are any mistakes in wall and slab construction. A four- to six-week lookahead schedule is recommended for this monitoring task. This requirement comes from experts' feedback. To visualize the results in AR devices, 4D model is suggested, user can see 3D model overlapped to relevant positions in the real world with adjusted milestones. Checking elements are presented in different colors to show the current progress.

4.4.2 AR Model Preparation

The second module, "Model Preparation", has four steps to load the appropriate model to the AR device: device computational limitations, 4D model generation & division, work areas segmentation, and AR device localization. "Device computation limitation" is necessary to test the limit of the AR device while working with a large number of model elements. Schedule and WBS are basis for model segmentation in accordance with time and engineering logic. Final step is to create a QR code for AR implementation. This phase aims to establish a separate model that contains adequate monitoring information regarding building sequence, engineering properties, and performs well in the AR device. In addition, this phase is necessary to avoid unexpected crashes when using the entire BIM model to develop the AR application because of its large and complicated file size.

Device computational limitations: At the start, the original 3D BIM model must be verified regarding its size and the AR device's technical specifications. The purpose of this step is to limit the crash issues that might come due to the insufficient computational power of the AR device. To examine the limit of model element loading, we made numerous trials of dividing the model and export them to HoloLens. In each trial, we changed the size of the model by cutting at different gridlines and counted the number of model elements. Structural models and MEP models were tested separately.

Corresponding areas and time for copying models to HoloLens were recorded for comparison. It is recognized that within the same area, an MEP model has more elements than a structural model, and it crashes more often. The following thresholds are then set for model breaking: 5,000 square meters (sqm) of area, 70 meters maximum for gridlines, 30,000 elements, and rectangle shape. The results showed that number of elements should be considered as the prior threshold in order to ensure there is no crash when releasing the application to HoloLens.

4D model generation and division: After that, schedule information is imported to the 3D model to generate the 4D simulation model. In the proposed framework, the schedule data needs to be reviewed before using the WBS to separate the model since it is necessary to identify the tasks' sequence and their completion milestones of inspection. Theoretically, schedule and WBS are closely associated because of technical requirements. A schedule outlines the duration and sequence of tasks required to complete a project, while a WBS is a hierarchical decomposition of the project into smaller, more manageable components and work packages (Sutrisna et al., 2018). Activities and tasks outlined in the schedule are based on the components identified in the WBS. Without a well-structured WBS, it is difficult to develop an accurate schedule. Conversely, without a meaningful schedule, it is challenging to monitor and manage the progress of the entire project and ensure that it is completed on time and within budget. In the schedule, there are tasks that are critical to the timeline of the project and must be completed on the fixed time. When creating the schedule, constructor has to rely on the supplier's ability to provide necessary materials, along with the working capacity of the construction unit. Therefore, the schedule generally illustrates segmented construction zones and construction tasks, determined by appropriate resource allocation, and construction means and methods. The segmented construction zones can be found in construction drawings. In this research, the schedule is the base for generating the workspace to avoid conflict during segmentation. By tracking the approved schedule information and examining the dependencies between tasks, we can clarify the required space and sequence of construction for all activities executed at a particular time. Upon engineering-related tasks, it is feasible to designate their respective locations and align them with the relevant sections of the model. For instance, when conducting concrete framing construction, formwork, reinforcement work, and concrete placement are associated. Those are works required to complete a specified construction segment in a required time. The relevant BIM elements then should be structural components belonging to such segment's position. After clarifying the associated tasks, the models' elements related to such activities can be assigned to relevant milestones. Figure 4.2 shows an example of potential segmenting components in a building level based on the schedule, which has been embedded with the 3D model in Synchro 4D. The purple elements are highlighted when selecting a period in schedule. In this case, it is the fourth floor of the building, that includes the elements of beams, columns, and walls. Other activities and their workspaces can also be visualized in sequence according to the designed timeline. While monitoring activity in an AR environment, the related details should appear correctly with relevant WBS codes in a working area being constructed at that time. Hence, the working areas are now fixed to schedule. In the next phase, only selected areas with schedule-driven information are extracted when converting the segmented model to AR.

Work areas segmentation: The third step is separating the model into working areas based on the WBS. In construction, the manager needs a clear and comprehensive WBS

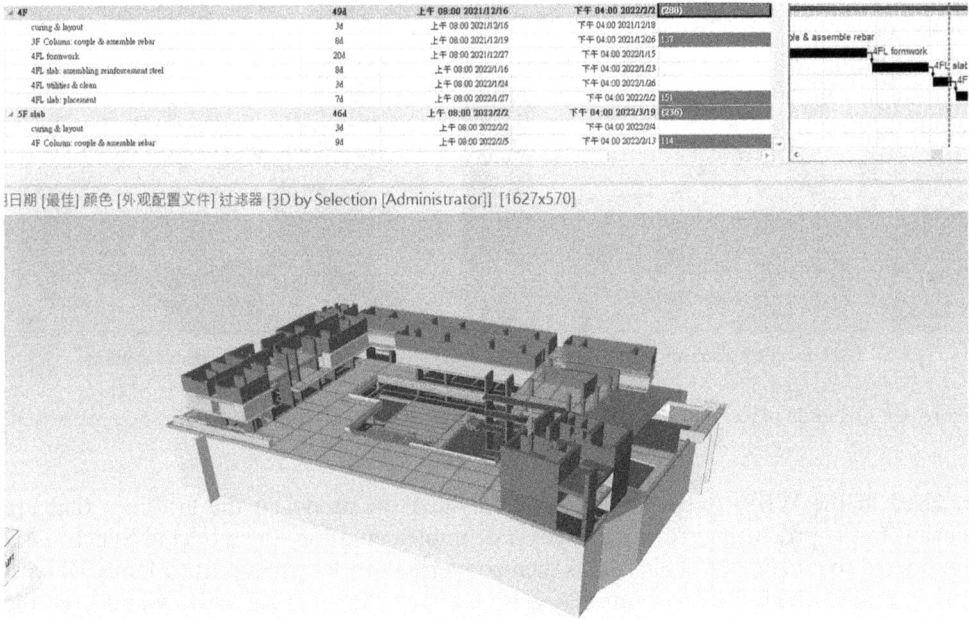

4F	49d	上午 08:00 2021/12/16	下午 04:00 2022/2/2 (288)	
curing & layout	3d	上午 08:00 2021/12/16	下午 04:00 2021/12/18	
3F Column: couple & assemble rebar	8d	上午 08:00 2021/12/19	下午 04:00 2021/12/26	ble & assemble rebar
4FL formwork	20d	上午 08:00 2021/12/27	下午 04:00 2022/1/15	4FL formwork
4FL slab: assembling reinforcement steel	8d	上午 08:00 2022/1/16	下午 04:00 2022/1/23	4FL slab
4FL utilities & clean	3d	上午 08:00 2022/1/24	下午 04:00 2022/1/26	4F
4FL slab: placement	7d	上午 08:00 2022/1/27	下午 04:00 2022/2/2 (15)	
5F slab	46d	上午 08:00 2022/2/2	下午 04:00 2022/3/19 (236)	
curing & layout	3d	上午 08:00 2022/2/2	下午 04:00 2022/2/4	
4F Column: couple & assemble rebar	9d	上午 08:00 2022/2/5	下午 04:00 2022/2/13 (14)	

日期 [最佳] 颜色 [外观配置文件] 过滤器 [3D by Selection [Administrator]] [1627x570]

Figure 4.2 Level-based components are selected to segment the model.

to identify the project's deliverables and tasks to accurately reflect the technical requirements of the project. Details of activities in the WBS can be found in the WBS dictionary. They are essential factors in determining the critical path and project schedule. After reviewing schedule in the previous step, major deliverables, which are the top-level of the WBS, have been determined. Work packages, which are the lowest level of the WBS, are typically defined by project's scope, duration, and required resources. They are directly used to organize and manage construction methods or to establish a framework for project monitoring and control. In general, work packages can provide a detailed plan of each individual task or activity, including its duration, resources' quantity and allocation, and dependencies. Typically, a specific team or individual usually is responsible for such assigned works. According to work packages, relevant elements of the BIM model, which have been determined in the first phase ("AR Tasks Model Requirements"), must be selected. The result of this step should be a definite model for a specified monitoring task. By referring to the WBS, the model can be subdivided into less complex and easily manageable parts. Therefore, this is a reasonable basis for creating work segments before importing them into AR devices. In this research, the building models are split into level-based components since constructing a building by level is a standard method. After that, in case of a larger layout and a more complicated structure, architectural/structural gridlines are suggested to split up the model. Gridlines provide a visual reference point to accurately position objects or elements within the model and multiple drawings, ensuring that all components are located in the correct place. Additionally, they help to maintain consistency and accuracy throughout the design and construction process. Communication among stakeholders can also be facilitated with ease by identifying the appropriate gridlines system. Generally, gridlines area can align with the work areas that

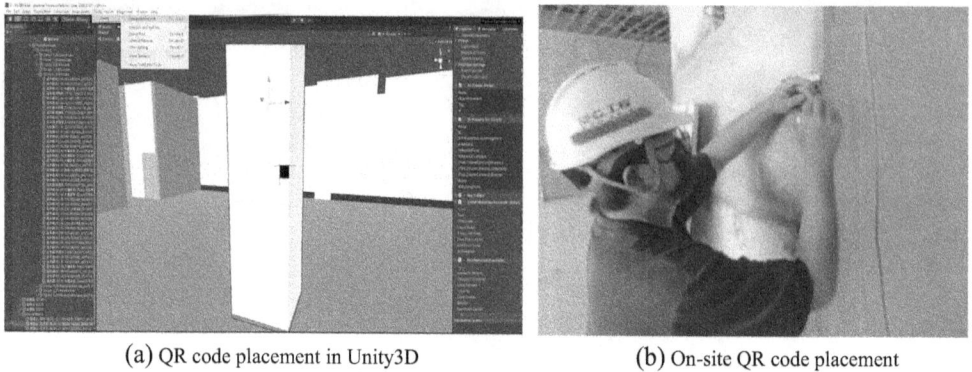

(a) QR code placement in Unity3D   (b) On-site QR code placement

Figure 4.3 QR code placement. (a) QR code placement in Unity3D. (b) On-site QR code placement.

are used in the WBS. Accordingly, we subdivided the model in the gridlines that are relevant to the processing work packages. For implementation, the model in Synchro 4D is extracted to an IFC file. This IFC is then converted to FBX to import to Unity3D with Tridify Tools, which provides automated tools to cut the model effectively and keep the assigned materials correctly.

AR device localization: The final step of model preparation is creating a QR code so that the separated model can be localized in the AR environment. The QR code is generated in Unity (Figure 4.3a). Each segmented model needs a unique QR code. Although placing the QR code on the floor can provide more accurate positioning, it sometimes cannot be scanned. Microsoft suggests a minimum detection distance for QR code is from 0.15 to 0.5 meters with a range of angle less than 45 degrees looking straight. As a result, the marker is attached at 1.75 meters high on the column to ensure the relevantly perpendicular view from the AR device's camera, which seems to give the best performance.

4.4.3 Onsite Implementation

After preparation, the user needs to place the QR code in the relevant position on the field. The position should be determined and well-protected during the work. Figure 4.3b shows the attachment of a QR code to a column, which will be inspected for geometric features. In order to initiate monitoring, it is necessary to align and maintain the HoloLens 2 camera directly toward the QR code. As soon as the QR code is scanned, the relevant model will materialize and overlap onto the surrounding area. It is advisable for inspectors to move slowly around different objects for inspection, and mark pertinent colors to the objects once they have been examined.

4.5 Experimental Validation

The proposed framework has been implemented with Microsoft HoloLens 2. This device has a second-generation custom-built holographic processing unit, a memory with 4-GB LPDDR4x system DRAM, and storage of 64 GB. Model preparation was executed in an advanced computer with CPU 12th Gen Intel® Core™ i7–12700K, 64 GB RAM, and GPU NVIDIA RTX 3070.

4.5.1 Experiment 1

The first experiment is an educational building (building A) with an area of 3,521 sqm. The test for this experiment is limited to structural components. Therefore, only structural elements of the building model are selected. This structural Revit model is exported to IFC format while schedule information in Microsoft Project is saved as an XML file. The schedule includes various tasks generated in the WBS and assigned time milestones. Those two data types are then imported into Synchro 4D to link relevant objects with the prepared schedule. The segmentation of the building layout can be seen in Figure 4.4. The building in experiment 1 has similar structures in all floors, therefore, a typical floor has been selected to visualize the segmentation. The three-segmented blocks are numbered according to the construction sequence as established in the imported schedule. After being established, the structural model has 3,305 elements, which is good enough to be copied to the HoloLens. By splitting the model into three working areas, the device shall load the model faster and run smoothly. The number of structural elements can be seen in Table 4.3. The three segmented areas are divided in Synchro 4D with assigned WBS code for related activities and schedule-driven data. Each separate model shall be loaded to HoloLens for on-site practical monitoring tests.

4.5.2 Experiment 2

The second experiment is a multifunctional building (building B) that has an area of 19,176 sqm, including four basements and eight floors.

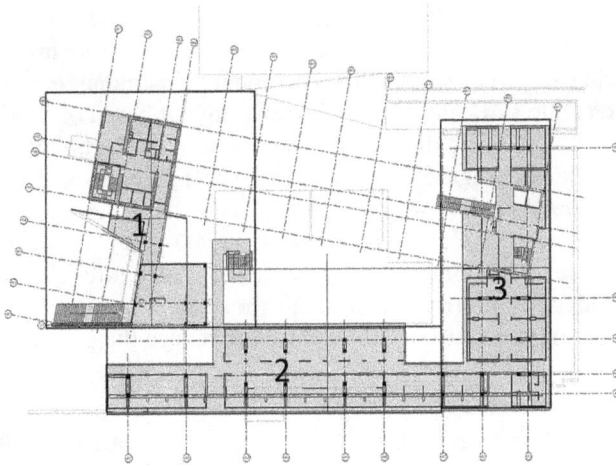

Figure 4.4 Breaking the model by working areas in a floor layout (building A).

Table 4.3 Detail of the working areas layout on the fourth floor of building A

No.	Semantic Segment	Number of Structural Elements
1	4F-WA1	511
2	4F-WA2	687
3	4F-WA3	304

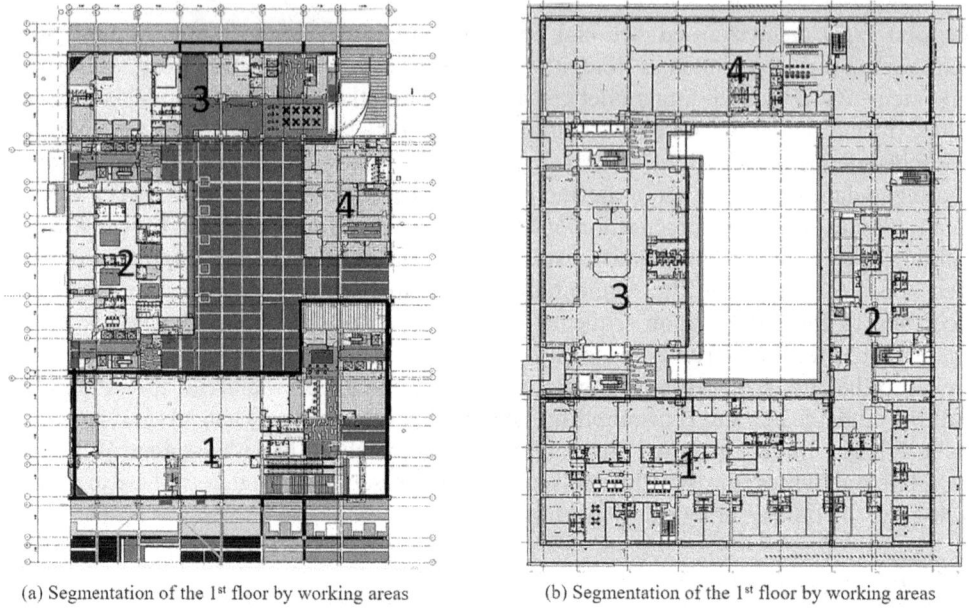

(a) Segmentation of the 1ˢᵗ floor by working areas   (b) Segmentation of the 1ˢᵗ floor by working areas

Figure 4.5 Breaking the model by working areas in two different floors (building B). (a) Segmentation of the first floor by working areas. (b) Segmentation of the second floor by working areas.

In this case, the original model is separated into structural and MEP models according to the proposed thresholds. In practice, the structural part is constructed before installing the MEP systems. The following steps are similar to experiment 1, including exporting relevant data to Synchro 4D to create a 4D model and divide the layout based on working areas. In this case, the layouts are segmented differently by floor because the building has different functions on different levels (Figure 4.5). Segmentation is based on the construction sequence of the MEP elements following the actual schedule. In the first tests, because the number of MEP systems' elements exceeded 30,000, they crashed the transmission. The number of elements is summarized in Table 4.4. The figures show that the MEP model contains considerably more elements than the structural model, which may cause slower data transfer from computer to AR device and reduce the performance of the HoloLens in practice.

Table 4.4 Detail of the working areas layout on the first floor and second floor of building B

No.	Semantic Segment	Number of Structural Elements	Number of MEP Elements
1	1F-WA1	437	8917
2	1F-WA2	511	6199
3	1F-WA3	356	6120
4	1F-WA4	142	3500
5	2F-WA1	885	9450
6	2F-WA2	835	5558
7	2F-WA3	790	6399
8	2F-WA4	629	6880

4.6 Onsite Result and Discussion

The results of QR placement and site experience are shown in the illustrations below. The building models have been successfully overlaid on the construction site following the proposed framework. The segmented models are updated with the latest schedule before loading into the AR device. Users can use HoloLens to look into the models while moving in the relevant zones. Four experts were invited to try the device and implement progress monitoring tasks such as checking deviation between design and the real-time status. Some screenshots of the HoloLens screen are presented in Figure 4.6.

In this illustration, color-code illustration can be applied to visualize a correctly constructed structure. For example, the column in Figure 4.6(a) perfectly overlaps the real element and is often marked green. In contrast, the missing pipelines presented in Figure 4.6(b) can be recognized without physical objects. They should be highlighted (commonly red in the application) as a warning that they have not been installed as scheduled. Participants can interact with models through the user interface designed with specified functions for the AR monitoring tasks mentioned in the previous section.

After testing, participants shared positive feedback on the proposed framework. The issue of complexity in the 4D model has probably been solved. Since the AR experiments were run with the segmented models, they brought a decent performance for the user while checking the quality and progress of the construction project. Participants agreed that segmenting the models provided better visualization and more precise inspection than using the original models, which comprise almost all types of designed elements. Consequently, showing more objects on the screen may cause occlusion and negatively affect the inspector's sight. However, there is feedback mentioned that visualizing a mixed model could expose conflicts when all components are completed. Those conflicts may not be detected if only one segmented model is applied. To argue, the proposed framework states that segmented models are regularly exported from the 4D models, which are schedule driven. A mandatory condition of application is using the latest updated model for inspection. Additionally, only one construction activity can be implemented in a space in a moment. A working crew must finish their specialized tasks before the others come and start the next item. Therefore, inspectors can compare and evaluate one's performance before giving permission for the next tasks.

(a) AR superimposed column in building A   (b) AR superimposed MEP system in building B

Figure 4.6 Experimental test run on-site. (a) AR superimposed column in building A. (b) AR superimposed MEP system in building B.

In general, qualified performance has been achieved for AR-based monitoring tasks with the proposed framework. Nonetheless, there seem to be technological limitations, such as light issues, output recording, and localization. Besides, safety should be another concern, as there is no research on the impact of AR HMD to the usual movement of the wearer. A suggestion is to create a specific route for AR users to follow. Even so, this moving zone may keep them from potential hazards, but it also prevents users from active inspection and causes problems in viewing.

4.7   Conclusion

This chapter first reviewed the literature to define the required LOD for BIM model before converting it to an AR-based model. After that, a novel framework to create an AR-ready BIM for construction monitoring activities with schedule-driven data has been proposed. The process succeeds in comparing separate semantic models with the construction progress. In addition, by separating the original model, segmented models provide a smooth performance in AR devices when executing monitoring activities on-site. Furthermore, the practical applications of the workflow have been demonstrated in monitoring the construction of structural components and MEP systems in two different construction sites, and positive feedback from construction experts was received. In the future, the research will focus on navigation in the AR environment which can support the supervisor in locating necessary components and items during the monitoring work.

References

Ahmed, S., Hossain, M. M., & Hoque, M. I. (2017). A brief discussion on augmented reality and virtual reality in construction industry. *Journal of System and Management Sciences, 7*(3), 1–33.

AIA. (2013). Project building information modeling protocol form. *AIA Document G202-2013.*

Alizadehsalehi, S., Hadavi, A., & Huang, J. C. (2020). From BIM to extended reality in AEC industry. *Automation in Construction, 116,* 103254.

Amin, K., Mills, G., & Wilson, D. (2023). Key functions in BIM-based AR platforms. *Automation in Construction, 150,* 104816.

Banfi, F., Brumana, R., & Stanga, C. (2019). Extended reality and informative models for the architectural heritage: From scan-to-BIM process to virtual and augmented reality. *Virtual Archaeology Review, 10*(21), 14–30.

Boton, C., Kubicki, S., & Halin, G. (2015). The challenge of level of development in 4D/BIM simulation across AEC project lifecyle. A case study. *Procedia Engineering, 123,* 59–67.

Chen, Y.-J., Lai, Y.-S., & Lin, Y.-H. (2020). BIM-based augmented reality inspection and maintenance of fire safety equipment. *Automation in Construction, 110,* 103041.

Chi, H.-L., Kang, S.-C., & Wang, X. (2013). Research trends and opportunities of augmented reality applications in architecture, engineering, and construction. *Automation in Construction, 33,* 116–122.

Chu, M., Matthews, J., & Love, P. E. (2018). Integrating mobile building information modelling and augmented reality systems: An experimental study. *Automation in Construction, 85,* 305–316.

Delgado, J. M. D., Oyedele, L., Demian, P., & Beach, T. (2020). A research agenda for augmented and virtual reality in architecture, engineering and construction. *Advanced Engineering Informatics, 45,* 101122.

Diao, P.-H., & Shih, N.-J. (2019). BIM-based AR maintenance system (BARMS) as an intelligent instruction platform for complex plumbing facilities. *Applied Sciences, 9*(8), 1592.

Dunston, P. S. (2010). Technology development needs for advancing augmented reality-based inspection. *Automation in Construction, 19*(2), 169–182.

El Ammari, K., & Hammad, A. (2019). Remote interactive collaboration in facilities management using BIM-based mixed reality. *Automation in Construction, 107*, 102940.

Fazel, A., & Izadi, A. (2018). An interactive augmented reality tool for constructing free-form modular surfaces. *Automation in Construction, 85*, 135–145.

Forcada Matheu, N., Gangolells Solanellas, M., Casals Casanova, M., & Macarulla Martí, M. (2017). Factors affecting rework costs in construction. *Journal of Construction Engineering and Management (ASCE), 143*(8), 04017032/04017010–04017032/04017039.

Garbett, J., Hartley, T., & Heesom, D. (2021). A multi-user collaborative BIM-AR system to support design and construction. *Automation in Construction, 122*, 103487.

Golparvar-Fard, M., Peña-Mora, F., & Savarese, S. (2009). D4AR–A 4-dimensional augmented reality model for automating construction progress monitoring data collection, processing and communication. *Journal of Information Technology in Construction, 14*(13), 129–153.

Han, B., & Leite, F. (2022). Generic extended reality and integrated development for visualization applications in architecture, engineering, and construction. *Automation in Construction, 140*, 104329.

Han, K. K., & Golparvar-Fard, M. (2015). Appearance-based material classification for monitoring of operation-level construction progress using 4D BIM and site photologs. *Automation in Construction, 53*, 44–57.

Kopsida, M., & Brilakis, I. (2020). Real-time volume-to-plane comparison for mixed reality-based progress monitoring. *Journal of Computing in Civil Engineering, 34*(4), 04020016.

Lee, S., & Akin, Ö. (2011). Augmented reality-based computational fieldwork support for equipment operations and maintenance. *Automation in Construction, 20*(4), 338–352.

Li, X., Wu, P., Shen, G. Q., Wang, X., & Teng, Y. (2017). Mapping the knowledge domains of Building Information Modeling (BIM): A bibliometric approach. *Automation in Construction, 84*, 195–206.

May, K. W., KC, C., Ochoa, J. J., Gu, N., Walsh, J., Smith, R. T., & Thomas, B. H. (2022). The identification, development, and evaluation of BIM-ARDM: A BIM-based AR defect management system for construction inspections. *Buildings, 12*(2), 140.

Mirshokraei, M., De Gaetani, C. I., & Migliaccio, F. (2019). A web-based BIM–AR quality management system for structural elements. *Applied Sciences, 9*(19), 3984.

Mutis, I., & Ambekar, A. (2020). Challenges and enablers of augmented reality technology for in situ walkthrough applications. *Journal of Information Technology in Construction, 25*, 55–71.

Nassereddine, H., Veeramani, A., & Veeramani, D. (2021). Exploring the current and future states of augmented reality in the construction industry. In *Collaboration and integration in construction, engineering, management and technology* (pp. 185–189). Springer.

Noghabaei, M., Heydarian, A., Balali, V., & Han, K. (2020). Trend analysis on adoption of virtual and augmented reality in the architecture, engineering, and construction industry. *Data, 5*(1), 26.

Rahimian, F. P., Seyedzadeh, S., Oliver, S., Rodriguez, S., & Dawood, N. (2020). On-demand monitoring of construction projects through a game-like hybrid application of BIM and machine learning. *Automation in Construction, 110*, 103012.

Redmond, A., Hore, A., Alshawi, M., & West, R. (2012). Exploring how information exchanges can be enhanced through Cloud BIM. *Automation in Construction, 24*, 175–183.

Sacks, R., Eastman, C., Lee, G., & Teicholz, P. (2018). *BIM handbook: A guide to building information modeling for owners, designers, engineers, contractors, and facility managers.* John Wiley & Sons.

Sidani, A., Dinis, F. M., Duarte, J., Sanhudo, L., Calvetti, D., Baptista, J. S., Martins, J. P., & Soeiro, A. (2021). Recent tools and techniques of BIM-based augmented reality: A systematic review. *Journal of Building Engineering, 42*, 102500.

Sutrisna, M., Ramanayaka, C. D., & Goulding, J. S. (2018). Developing work breakdown structure matrix for managing offsite construction projects. *Architectural Engineering and Design Management, 14*(5), 381–397.

Wang, S.-K., & Chen, H.-M. (2020). A construction progress on-site monitoring and presentation system based on the integration of augmented reality and BIM. ISARC. Proceedings of the International Symposium on Automation and Robotics in Construction,

Wang, X., & Dunston, P. S. (2006). Compatibility issues in Augmented Reality systems for AEC: An experimental prototype study. *Automation in Construction, 15*(3), 314–326.

Wang, X., Truijens, M., Hou, L., Wang, Y., & Zhou, Y. (2014). Integrating augmented reality with building information modeling: Onsite construction process controlling for liquefied natural gas industry. *Automation in Construction, 40,* 96–105.

Wang, X., Ong, S. K., & Nee, A. Y. (2016). A comprehensive survey of augmented reality assembly research. *Advances in Manufacturing, 4*(1), 1–22.

Zhou, Y., Luo, H., & Yang, Y. (2017). Implementation of augmented reality for segment displacement inspection during tunneling construction. *Automation in Construction, 82,* 112–121.

5 Building Information Model Visualisation in Augmented Reality

Vishak Dudhee and Vladimir Vukovic

School of Computing, Engineering and Digital Technologies, Teesside
University, Middlesbrough, UK

5.1 Introduction

During construction, "omission errors" are often committed due to the lack of information available at hand (Love et al., 2009). Team members do not consistently and
enthusiastically spend time and effort to get access to the required information, which is
stored remotely or distantly and would require them to be physically away from the task
that they are performing. Wearing a head-mounted augmented reality (AR) device that
includes data from a building information modelling (BIM) model can mitigate this issue
as the worker would not have to be detached from their task and the information can be
made directly relevant to the task. The integration of BIM and AR can also enhance the
visualisation of different phases of a project lifecycle and eventually improve the final
product (Wang et al., 2014). The use of a three-dimensional (3D) BIM model in architecture, engineering and construction (AEC) has already shown its numerous benefits. It
allows designers to virtually test the proposed alternative designs and identify any possible problems at an early stage of the construction process. Using AR, BIM information
can be visualised in real-time and in the actual physical setting, particularly during the
construction, maintenance and operational stages of a building (Yan, Culp and Graf,
2011). This can improve decision-making, eventually allowing tasks to be efficiently and
effectively executed.

To fully benefit from the integration of BIM and AR, the 3D model of the building
needs to be precisely superimposed to the actual physical environment in AR. At the
moment, the superimposing techniques that are used to overlay the 3D BIM model on
the actual physical environment mainly use reference points (Mahmood, Han and Lee,
2020). The accuracy of the reference point techniques depends on various factors such
as the user's knowledge of the pre-set reference point and the ability to set the reference
points at the exact location in the physical AR environment. Although the integration
of BIM and AR has been researched (Alizadehsalehi, Hadavi and Huang, 2020), the
superimposing techniques for the practical application of the integration processes have
not been studied sufficiently. This research aims to explore the application of BIM and
AR for the effective visualisation of building information models in an AR environment and evaluate some of the currently available AR tools. The objectives include the
following:

- Analysing the compatibility of BIM models to AR tools.
- Demonstrating the BIM-AR integration and visualisation process.
- Evaluating the superimposing techniques.

DOI: 10.1201/9781003408949-5

This chapter describes and analyses the various superimposing techniques that are being used in current BIM-AR applications. The limitations of the superimposing techniques and access to real-time information are discussed, followed by the proposal of an automated superimposing approach.

5.2   Background

5.2.1   Building Information

BIM brings together independent components such as structural, architectural, mechanical, electrical and plumbing (MEP) by connecting the project stakeholders through technology (Dossick and Neff, 2010; Dossick, Osburn and Neff, 2019). BIM models are usually 3D models representing the different building components, but they can also be in 4D with the integration of time and 5D with the combination of time-based cost (Taylor and Bernstein, 2009; Goulding, Rahimian and Wang, 2014). There is a difference between a 3D BIM Model and a typical 3D Model. 3D models consist of only geometrical representations such as shape and dimensions, whereas a 3D BIM model contains a building's architectural, construction and installation information (Tang et al., 2010). Unlike computer-aided design (CAD) models, BIM models are generated using intelligent building elements which are comprised of related data and parameters (Ding, Zhou and Akinci, 2014). A 3D BIM model is comprised of component layers that define and represent the actual material of the building. Each individual property of the elements is flexible, making them "parametric" (Lee, Sacks and Eastman, 2006; Rausch and Haas, 2021). A 3D BIM model contains detailed information about the building, meeting the client's expectations and the designers as they expect to physically replicate it as accurately as possible (Adami, Scala and Spezzoni, 2017).

To facilitate such accuracy in the operational phase, internet of things (IoT) sensors installed in buildings continuously monitor, collect and store the physical environment data that can be used for near real-time analyses (Bashir and Gill, Dec 2016; Bebelaar et al., 2018). IoT data enrich building information models by providing real-time information of actual operations of the building (Tang et al., 2019; Brynskov et al., 2018). The type of information captured from the sensors varies and includes both the positioning information and physical measurements. The data captured by individual sensors are usually organised in time series. The advancements in IoT technologies have facilitated the introduction of the Digital Twin concept in construction. A digital twin is defined as "a realistic digital representation of assets, processes or systems in the built or natural environment". The digital twin concept emerged from the manufacturing, production and operations domain over the past decade and is rapidly gaining popularity in the AEC industry (Sacks et al., 2020; Boje et al., 2020). Although digital twins can enhance all different phases of a building lifecycle, its application is mainly limited to the operational and maintenance phase. The digital twin-enabling technologies include IoT, Cloud, Extended Reality (XR), Big Data and Machine Learning, where XR has an essential role in allowing full fidelity of the digital twin concept (Aheleroff et al., 2021). The scope of digital twins is advancing beyond the BIM concept to enable asset-centric establishments to congregate engineering, operational, and information technologies into immersive experiences.

5.2.2   Spatial Mapping and Tracking

Spatial mapping is the production of 3D maps of physical environments that allow AR devices to understand and interact with the actual world (Park et al., 2020). Spatial mapping is useful in motion planning, preventing collision and blending the virtual and physical worlds in a realistic way. "Spatial Surface Observer" and "Spatial Surface" are the two main kinds of objects used for spatial mapping (Zeller and Coulter, 2018). The application gives one or more bounding volumes to the Spatial Surface Observer to mark the regions of interest for collection of spatial mapping data. A series of spatial surfaces are associated with each of the bounding volumes. The boundary volume may be either pinned in the actual physical environment or attached to the AR device. When attached to the device, it only moves as the device moves without rotating. A small area of the physical environment is represented by a spatial surface as a triangle mesh attached to the spatial coordinate system locked in position. Spatial surfaces are illustrated as numerous triangle meshes (Wang, Gao and Wu, 2019). A large amount of computational and storage resources is needed to store, render, and process such meshes. Therefore, to minimise the processing and storage resources used for the mesh, all software must have a suitable mesh-caching scheme. The scheme should dictate which meshes are preserved or eliminated and when to revise each spatial surface mesh. Simultaneous Localisation And Mapping (SLAM) are methods employed to simultaneously calculate the position and reconstruction problems in 3D when the AR device is moving in a real environment (Taketomi, Uchiyama and Ikeda, 2017). The SLAM system can quickly and precisely track camera visuals of both handheld and wearable AR devices in an unknown environment.

XR equipment, such as head-mounted displays (HMD), uses tracking techniques to determine the position of an object in the virtual or augmented real-world environment (Sherman and Craig, 2019). The difference between the two tracking methods, outside-in tracking and inside-out tracking, lies in the location of the cameras and sensors that define the object's position in space (Ribo, Pinz, and Fuhrmann, 2001). For inside-out tracking, the camera and sensors are located on the HMD, whereas for outside-in tracking, a stationary tracking device is mounted to a fixed location in the scene. While using the inside-out tracking method, the camera incorporated in the HMD determines the changes of the equipment's position in relation to the external environment. To create an illusion of walking through the virtual environment in real-time, sensors adjust the user's spatial position when the headset moves (Keil, Edler, and Dickmann, 2019). Such tracking method can be used both for marker-based tracking, where markers are used as reference points, and markerless inside-out tracking, which uses natural features to determine positions. Markerless inside-out tracking technique identifies specific shapes and images that are already present in the physical environment to calculate and determine the position and orientation of an AR device in physical space. Such positional tracking system uses data from accelerometers and gyroscopes to enhance precision. In marker-based tracking, fiducial markers, positional markers or infrared markers are used to allow the tracking system to easily detect and position the HMD in the specific environment (Borrego et al., 2016). The fiducial markers are in the form of primitive shapes such as points, circles, or squares. Quick Response (QR) codes are commonly used positional markers placed in the physical environment as reference points. Infra-red (IR) markers and cameras sensitive to different types of light can also be used. The marker-based tracking is effective only if the markers are within the field of view of the HMD and can be clearly detected.

5.2.3 *Integration of BIM and AR*

The integration of BIM and AR has been researched in various areas of the construction industry (Alizadehsalehi, Hadavi and Huang, 2020; Rahimian et al., 2020;

Dudhee and Vukovic, 2021). The areas of research include construction safety (Li et al., 2018; Moore and Gheisari, 2019), design visualisation (Wang et al., 2014; Sampaio, 2018), project management (Ahmed, 2018), construction progress evaluation (Ratajczak, Riedl and Matt, 2019) and construction education and training (Wang et al., 2018). To provide a realistic and accurate AR experience, digital information must be aligned with the elements of the real physical space (Montero et al., 2019). Digital models can be superimposed in an AR space in two ways: (1) through tapping the digital model into the physical environment's surface, or (2) by snapping the model to a QR code, which physically represents the origin of the model, or a reference point (Hashimoto, 2020). Measuring placement of the QR code can facilitate accurate positioning of the model in the physical environment. Alternatively, model placement and adjustment in the AR environment is possible using the length, width and height planes if high accuracy is not required. Snapping digital objects in an AR setting is inherently more challenging than snapping in a virtual reality because of physical constraints such as sensor noise, environmental complexity and hardware limitations (Nuernberger et al., 2016). In an AR environment, the likelihood of extracting the wrong constraints is higher and can negatively affect the snapping performance. An ideal AR system creates an illusion that the real and virtual components coexist in the space (Montero et al., 2019). Although AR technology has advanced, most software applications still cannot imitate the ideal experience of virtual objects being part of the real physical setting. The research question originating from the identified gaps in knowledge is:

• How can BIM models be accurately superimposed within AR?

The presented research on integrating BIM and AR in various aspects of construction mainly focuses on technical integration and system architecture development. The practical application of using AR software and devices has not been sufficiently explored. This research expands on previous studies on the integration process (DaValle and Azhar, 2020; Alizadehsalehi, Hadavi and Huang, 2020; Prabhakaran et al., 2020) and AR application evaluation (Huang, Shakya and Odeleye, 2019, Huang, 2020) with an in-depth study on superimposing techniques. Although much prior research has established the process for integrating BIM and AR, this research addresses the need for more precise superimposing and examines the integration of BIM and AR from a practical point of view.

5.3   Methods

To analyse the different techniques used to superimpose a 3D BIM model in AR environment and evaluate the methods' effectiveness, a BIM model of an office room located in the Phoenix Building at Teesside University's Middlesbrough campus was created. The two pieces of AR equipment used in this experimentation were Microsoft HoloLens 1 Development Edition, which is the first mixed reality head-mounted headset developed and manufactured by Microsoft Corporation, and Daqri Smart Glasses, which was created for use in a variety of areas, including technical visualisation with BIM (Engineering.com, 2019). The Microsoft HoloLens was selected as it runs Windows mixed reality platform under Windows 10, which is the most popular operating system and can include third-party applications (Borycki, 2018). The Daqri Smart Glasses were selected as they provide a bespoke solution tailored specifically towards the AEC industry. Daqri Smart Glasses consists of a wired control box that provides the required computing power, while the device operates on an internally built Visual Operating System (VOS). Testing examined four different BIM-AR applications: 3D Viewer Beta (Microsoft, 2016), BIM Holoview (Revit-Holoview, 2017), HoloLive (VisualLive, 2018) and Model-BIM (Daqri, 2018). HoloLens has a wider range of software available that allows visualisation of digital models in AR compared to Daqri which is limited to its inbuilt application Model-BIM. The applications chosen for Microsoft HoloLens were based on compatibility with BIM models and availability.

5.3.1   Building Information Model

The Phoenix Building is a four-storey multipurpose building that accommodates the Centre for Construction Innovation and Research (CCIR). The Room P1.09 selected to conduct the experiment is an office on the first floor of the building. The room was selected as it was accessible and available. A model of the room was generated in Revit (Autodesk, 2019) version 2019.2 based on available information and observation. Available 3D CAD models and 2D plans allowed positioning of the MEP components in the room model. Being an office, the room consisted of electrical and mechanical (heating, ventilation and air-conditioning) system components. As the exact designs and information on some of the room elements, such as the doors, windows and electrical trays were not available, similar components' designs were used keeping the size and properties as close as possible to reality.

The room's BIM model is shown in Figure 5.1. Due to the limited textures available as default in Revit, assigning the desired colours to the different components was challenging. The heating/ventilation system of the room is highlighted in yellow (lighter shade of grey in figure), the electrical cable trays in brown, the walls in white and the doors and windows in a darker brown wooden texture (darker shade of grey in figure) to differentiate different elements in the room model.

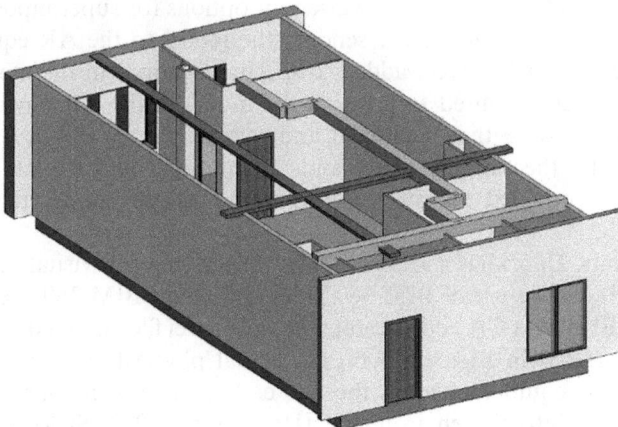

Figure 5.1 Created BIM model in Revit file format (RVT).

5.3.2   *Superimposing Model*

The superimposing test used the 3D viewer software 3D Viewer Beta, BIM Holoview and HoloLive on Microsoft HoloLens and Model-BIM on DAQRI Smart Glasses. The 3D viewers descaled the model to a reasonable size fitting the HoloLens display so that positioning, moving, and rotating the model were easier. To superimpose the model into the actual building, the model was first placed at a correct position and then scaled accordingly (Huang, 2020). After an inbuilt room scan, the software added a grid on the bottom level to guide and increase the precision while positioning the model on the floor or table. The software scaled the model in terms of percentages rather than using ratios. Therefore, the scaling process relied on trial and error, personal judgement, and consideration of reference points such as walls, roof and other various components. As the model scaled up, minor adjustments using rotation and moving options were frequently made to secure the model in the correct position. The scaling, rotating, and moving process repeated until the model was superimposed on the actual building.

The procedure of integrating, processing and visualising BIM models using different AR devices is shown in Figure 5.2. The developed process map shows the method of visualising BIM models on Microsoft HoloLens and DAQRI Smart Glasses.

A federated BIM model generated by the inputs of AEC professionals must be converted into a suitable format before it can be transferred to the AR device (Alizadehsalehi, Hadavi and Huang, 2020). Having the BIM model file in the desired format, the different device-specific approaches can be used to transfer files, introduce reference points or landmarks to the model, send the model to AR devices and access them through the devices. BIM-related software on both Microsoft HoloLens and DAQRI Smart Glasses uses cloud-based file storage systems to facilitate file access from both the computers and the AR devices. Models on which reference points, such as fiducial markers or landmarks, have been created can be superimposed using those reference points. However, models without reference points must be superimposed manually by trial and error. Once superimposed, the model can be visualised, and the system can be used for inspection, decision-making, locating hidden elements and other activities.

The Holoview and HoloLive software included a variety of options for superimposing the BIM model in a physical environment. Before sending the model to the AR equipment, fiducial markers as reference points were added to specific locations in the model. Two or more reference points were required to position the model in the AR environment. The same reference points were then identified and selected in the physical environment to fit the model to the correct place and orientation. In case of any misalignment, the application provided various options to adjust the model to the physical environment.

Similar to the reference points, Daqri has a set procedure of accessing and visualising a BIM model in AR, which includes the use of BIM 360 Docs and Daqri BIM 360 Viewer (Negrete et al., 2018). Model-BIM uses QR codes to retrieve the model file and landmarks set on the model and as illustrated in Figure 5.3 on the actual physical environment. Using this system, once the AR equipment scans the QR code and the landmark, the model is automatically imposed into the environment (Hashimoto, 2020; Sydora and Stroulia, 2018).

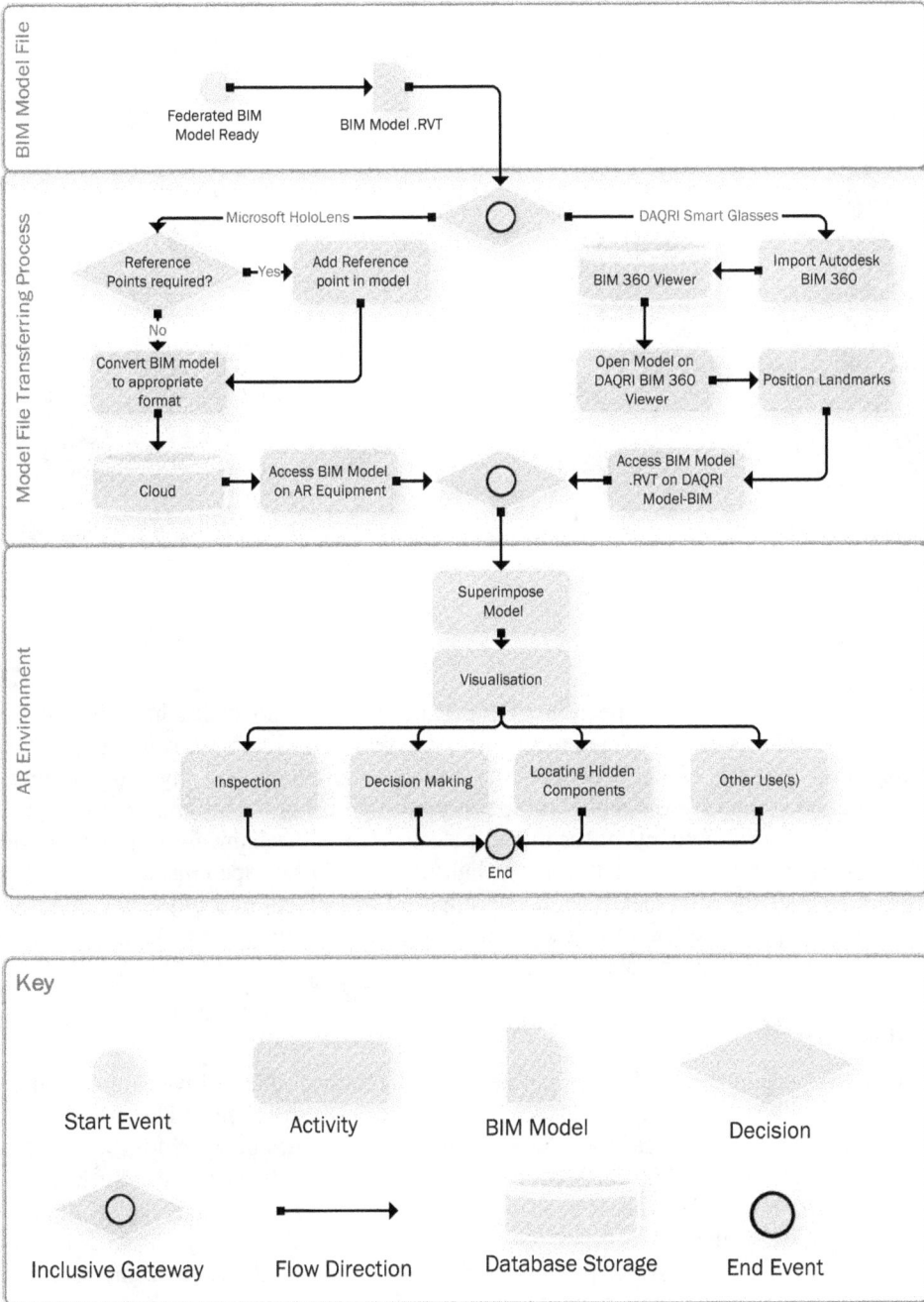

Figure 5.2 BIM visualisation in AR process.

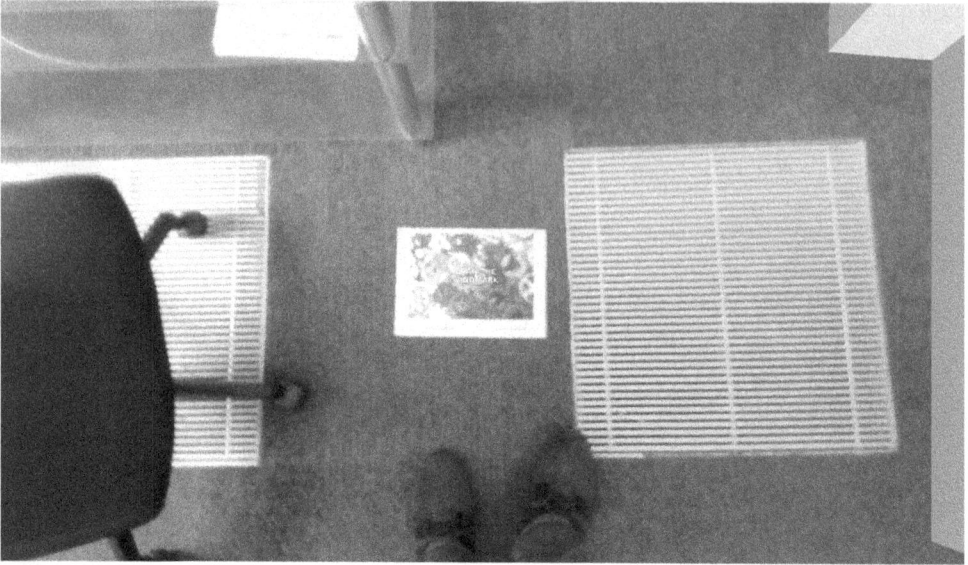

Figure 5.3 DAQRI's landmark.

5.3.3 *Analysis and Evaluation*

Validation included visual inspection of the superimposed model in the actual office room. The model was first validated by ensuring the reference points were at the predicted locations and that the model's layout was perfectly matching the physical building's layout. The time taken for the model to be superimposed using the different methods was recorded. Then, the alignments of the model's electrical and mechanical components were checked in comparison to the actual room lighting, ventilation openings and plumbing. Further review of the alignment accuracy compared the virtual and physical cable tray. The cable tray spans the length of the room so that misalignment at different distances from reference points can be assessed.

5.4   Results

The model superimposed in 3D viewer Beta, BIM Holoview, HoloLive and DAQRI's Model-BIM is illustrated in Figures 5.4, 5.5, 5.6 and 5.7, respectively. The models were imposed on the actual physical room using different techniques described in Section 5.3.

Using the 3D Viewer Beta AR software, the model was superimposed by trial and error. The superimposing process itself was time-consuming and took several attempts before it was closely aligned to the physical room. The model needed to be scaled to the right size, rotated to the correct position and then aligned to the room using reference points set on the model. During the scaling and aligning process, often the model maintained its original placement and created large misalignments. Such issues would result in restarting the entire superimposing process.

HoloLive allowed the option to use either fiducial markers as reference points or QR codes to superimpose the model into the physical environment, whereas in BIM Holoview only the reference point method was available. The reference point system available on

Figure 5.4 Room P1.09 lighting fixture superimposed on 3D Viewer Beta.

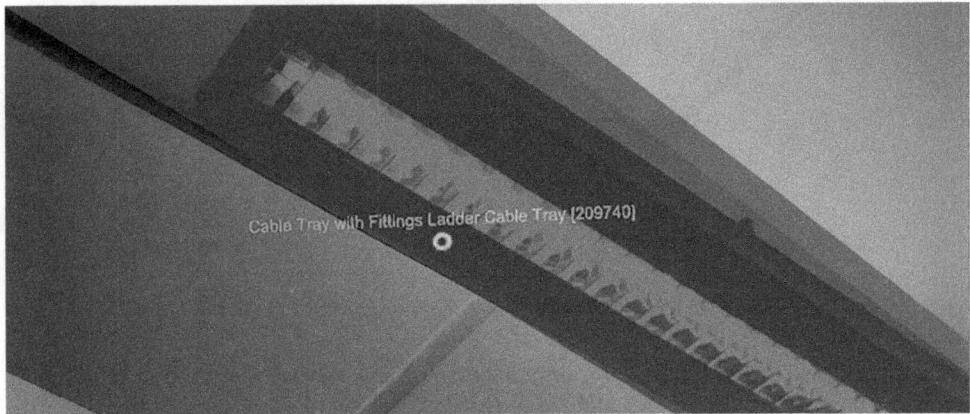

Figure 5.5 Room P1.09 lighting fixture superimposed on BIM Holoview.

HoloLive and BIM Holoview was developed for superimposing specifically BIM models in AR. Therefore, such a system is more accurate than 3D Viewer Beta and relatively easier to use and align the model in AR. Using the reference point system, the BIM model is set in the correct orientation. Model alignment can be finetuned by moving the model along the horizontal and vertical planes.

Daqri Smart Glasses and Model BIM also take into consideration the integration of BIM in AR. Therefore, the superimposing procedure set by Daqri and the available options allow more accurate model alignment to the physical room compared to the other tested software. Using the QR code, the model can be directly accessed and, using the

Figure 5.6 Room P1.09 section superimposed on HoloLive.

Figure 5.7 Room P1.09 cable trays and ductwork superimposed on DAQRI – Model BIM.

landmark, aligned to the room. After imposing the model, the software allows refining the alignment by moving the model along the three-dimensional Cartesian coordinate axes with a high level of precision.

The different software used on Microsoft HoloLens and DAQRI Smart Glasses for visualising BIM models in AR has been evaluated and compared, as illustrated in Table 5.1.

The software has been evaluated based on different criteria relevant to the integration of BIM and AR and rated through product comparison. In terms of the time taken to superimpose the model in the actual physical room, attempting through trial and error took around 10–20 minutes, using the reference points it took between 3–5 mins and using the landmark method it took around 2 mins. The trial and error method was the slowest with the landmark positioning system being around 5–10 times faster. The use of reference points was relatively faster than the trial and error method but slightly slower than the landmark positioning system.

Table 5.1 AR software comparison summary

Device		Microsoft HoloLens		DAQRI
Software	3D Viewer Beta	BIM Holoview	HoloLive	Model BIM
File transfer method	Manual Upload	Cloud	Cloud	Cloud
Revit Plugin	No	Yes	Yes	Yes
Ease of use	★★	★★★	★★★	★★★
Positioning accuracy	★★	★★★	★★★	★★★
Superimposing Method	No set method	Reference points	Reference points and QR	Using a set landmark
Superimposing accuracy	★	★★	★★	★★★
Key				
★	Poor			
★★	Average			
	Good			

5.5   Discussion

5.5.1   *Model Superimposing and Visualisation*

Models which consist of reference points or landmarks can be superimposed using those reference points. However, models without reference points need to be superimposed manually by trial and error. There is a high probability that the manual technique will take a significant amount of time and effort to even come close to superimposing the model accurately. Furthermore, such a technique is not always feasible and can introduce misalignment. The rotation and movement of the model in the AR environment is done by using the scaling system rather than a ratio system, creating inaccuracies.

DAQRI provides a free integrated application for visualising BIM models, however, Autodesk BIM 360 is required to transfer the model to the DAQRI application. 3D Viewer Beta, being a generic application, does not contain BIM-related options whereas the other analysed software do have such options. Most of the BIM-specific applications use cloud systems for transferring BIM models to the devices. The selected way to superimpose the model has an impact on the positioning accuracy. The use of the DAQRI's landmark system has proven to be more accurate than reference points when superimposing the model and results in fewer misalignments. The overall performance of BIM Holoview, HoloLive and Model-BIM is satisfactory, but DAQRI's Model-BIM surpasses them due to the wide variety of options available. The software evaluation was subjective based on the limited case study experience presented in the paper and would require further studies to verify the findings in more general applications.

Meshes are used to scan and place the model in real surroundings. Microsoft HoloLens had a default mesh density (illustrated in Figure 5.8) which could not be reprogrammed in the device.

Therefore, corners, edges and other small elements of a building cannot be precisely scanned. The use of corners and building edges as reference points to superimpose a building model could sometimes cause misalignment. A millimetre difference at a reference point can cause the other end of the model to be drastically misaligned as illustrated in Figure 5.9.

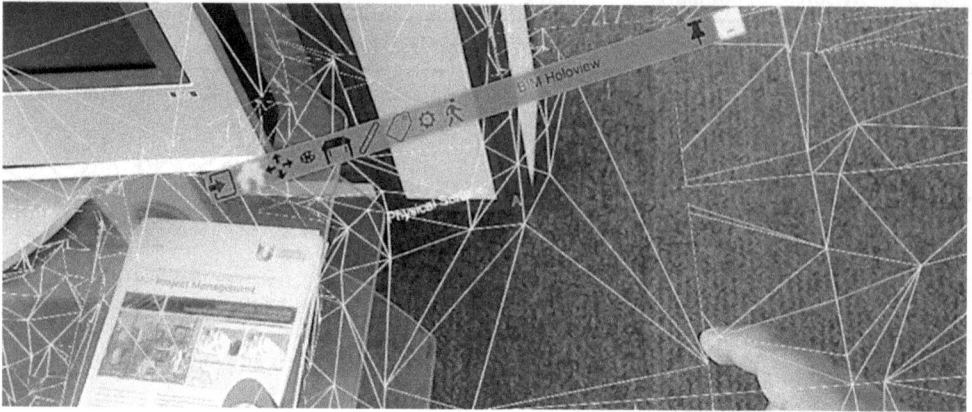

Figure 5.8 Microsoft HoloLens mesh density.

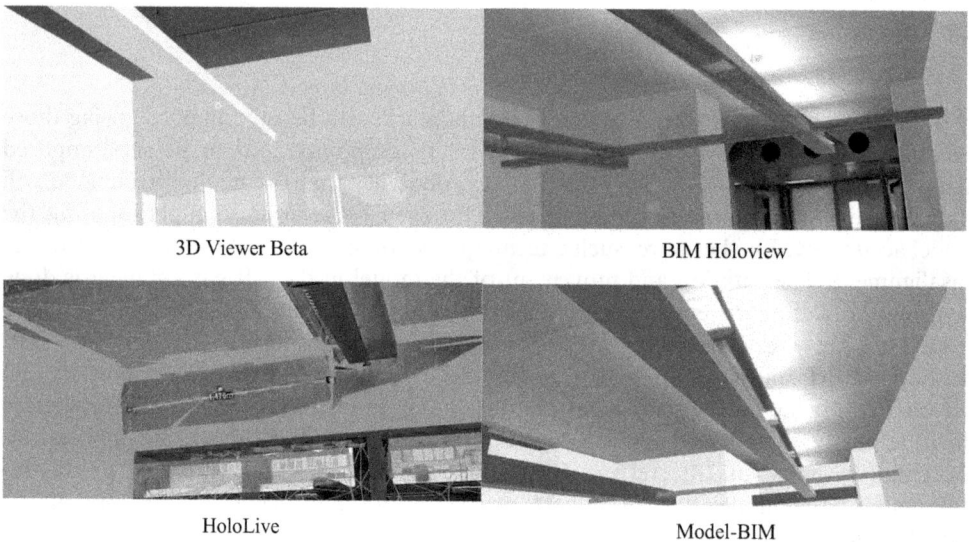

3D Viewer Beta

BIM Holoview

HoloLive

Model-BIM

Figure 5.9 Room P1.09 superimposed model comparison. 3D Viewer Beta. BIM Holoview. HoloLive. Model-BIM.

Previous studies also showed that, although Microsoft HoloLens had spatial mapping capabilities, the mesh it created lacked accuracy for precise model imposition (Evans et al., 2017). Inaccuracies while superimposing BIM models in AR can emerge due to various factors such as the tracking technology, mesh density or the technique and process used. The accuracy of optical see-through systems that head-mounted AR devices use to impose computer-generated information on the real physical environment relies on calibration (Grubert et al., 2018). The complex and unreliable calibration procedures to align the digital and real physical space limit the AR devices' utilisation for highly precise activity (Qian et al., 2017; Cutolo et al., 2020).

Hence, the current experiment compared several approaches to provide best practice guidelines to improve alignment accuracy while superimposing the P1.09 office room model. Some of the practices which improved the alignment and accuracy were (1) Avoiding the positioning of reference markers on corners and edges and, instead, positioning the reference markers on a horizontal or vertical open flat surface that has been measured and marked both in the model and in physical space. Scanning elements such as corners and edges can be imprecise compared to open flat surfaces due to the tracking technology and mesh density; therefore, using an open flat surface, the reference points can be accurately placed in the AR environment. (2) While using a fiducial marker, the use of circles or ring shapes can improve the alignment process and act as a guide while finetuning the model position. The model superimposing techniques largely depend on the AR application's options; however, the aforementioned recommended practice can increase the superimposing accuracy for a better visualising experience.

5.5.2 Implementation Approaches

Superimposing a BIM model in an actual physical building allows the user to visualise the model in relation to the building. Using this technique, hidden components of a building such as the MEP and heating, ventilation and air conditioning (HVAC) system positions can be easily identified. Using BIM and AR, the user does not have to search for the plans of the building or different systems, try to understand the information and locations of the systems in the physical environment. Thus, the system has the potential to improve building construction, maintenance and refurbishment.

Suppose a room is in use and a mechanical component needs to be replaced. Firstly, using the BIM-AR method, the hidden components can be identified and the exact details of the part, such as the model number, size and date installed or maintained, can be easily retrieved without accessing the cavity. The exact component can then be brought on-site and replaced, the room only being disturbed during the replacement process rather than during the identifying and information-gathering processes. During its lifetime, the building undergoes numerous upgrades and modifications. The paperless approach integrating BIM in AR can allow storage and access to the electronic building and maintenance records in one place rather than having high volumes of paper that can get lost or be damaged while used.

While application of appropriate alignment techniques can enhance superimposing accuracy, further innovation of AR technology is needed for higher precision. Preference should be given to AR software designed for the user's intended application, compatible with the AR device and the BIM model file formats. Future apps should include the possibility of customising mesh density, allowing the usage of third-party mesh development software, and allowing the integration of open-source and third-party plug-ins.

Current BIM models lack real-time data to reflect changes in building environmental conditions (Dudhee and Vukovic, 2021). Therefore, the integration of real-time sensory data can make the visualisation process more accurate, detailed, and informative. Environmental monitoring provides stakeholders with valuable information about building performance. Knowledge gained from monitoring data analytics can be used to maintain and continuously improve building performance and operation. A comprehensive environmental monitoring system thus has the potential to align stakeholders around the challenge of ensuring energy efficiency and reducing emissions.

Although the integration and visualisation of BIM and sensory information in the AR environment have been investigated, further research is still needed on their practical implementation and application (Natephra and Motamedi, 2019). The current AR-based visualisation method limits the user to static building information. The integration of real-time sensory information in the current visualisation method can enrich the building model information.

5.6   Conclusion

The research investigates the application of BIM in AR and evaluates some of the currently available AR tools. The presented experimentation shows that pre-set reference points and landmarks both result in slight misalignments of the model to the real environment. The presented outcome is not exclusively linked to using a particular technology but comparatively analyses and evaluates state-of-the-art features and implementation procedures for BIM and AR. Based on the findings, recommendations on the best practice for superimposing building information models in AR has been provided. The AR devices used in this study represent the first generation of head-mounted headsets. AR technology is rapidly evolving, and some improvement has already been made to the newer version of Microsoft HoloLens, whereas the production of the Daqri Smart Glasses has been discontinued. Nevertheless, new approaches that do not require manual model adjustments should be further developed to improve the accuracy of superimposing models into buildings. At this stage, the functions of the AR devices and commercially available BIM-AR applications are limited and still under development. The current scanning technology is insufficiently adequate in terms of efficiency and accuracy. By following the recommended processes, the inaccuracies and the time needed for superimposing BIM models in AR can be reduced by approximately 50%, but the invention of more robust scanning technology may further improve the imposing and superimposing accuracy of BIM models in AR. Integrating real-time sensory information in the current visualisation process will enhance the model and the user's experience. The integration of BIM, AR and IoT Sensors will allow for the visualisation of both static and dynamic building information and enable realisation of digital construction twins.

Acknowledgements

This chapter is an extended version of the paper entitled "Superimposing Building Information Models in Augmented Reality" presented at 20th International Conference on Construction Applications of Virtual Reality (CONVR 2020). The authors would like to thank Professor Nashwan Dawood and Dr Farzad Rahimian for their editorial contributions at CONVR2020.

References

Adami, A., Scala, B. and Spezzoni, A. (2017), "Modelling and accuracy in a BIM environment for planned conservation: the apartment of Troia of Giulio Romano", *Copernicus GmbH, Gottingen*, pp. 17.
Aheleroff, S., Xu, X., Zhong, R.Y. and Lu, Y. (2021), "Digital twin as a service (DTAAs) in industry 4.0: an architecture reference model", *Advanced engineering informatics*, Vol. 47 available at: 10.1016/j.aei.2020.101225.

Ahmed, S. (2018), "A review on using opportunities of augmented reality and virtual reality in construction project management", *Organization, technology & management in construction*, Vol. 10, No. 1, pp. 1839–1852. available at: http://www.degruyter.com/doi/10.2478/otmcj-2018-0012.

Alizadehsalehi, S., Hadavi, A. and Huang, J.C. (2020), "From BIM to extended reality in AEC industry", *Automation in construction*, Vol. 116, pp. 103254. available at: 10.1016/j.autcon.2020.1 03254.

Autodesk. (2019). Revit (2019.2) [Computer Software], Autodesk Inc. Retrieved 31 January 2019, from https://www.autodesk.co.uk/products/revit/.

Bashir, M.R. and Gill, A.Q. (2016). "Towards an IoT big data analytics framework: Smart buildings systems", 18th International Conference on High Performance Computing and Communications; IEEE 14th International Conference on Smart City, IEEE, pp. 1325–1332.

Bebelaar, N., Braggaar, R.C., Kleijwegt, C.M., Meulmeester, R.W.E., Michailidou, G., Salheb, N., van der Spek, S., Vaissier, N. and Verbree, E. (2018), "Monitoring urban environmental phenomena through a wireless distributed sensor network", *Smart and sustainable built environment*, Vol. 7, No. 1, pp. 68–79. available at: https://www.emerald.com/insight/content/doi/10.1108/SASBE-10-2017-0046/full/html.

Boje, C., Guerriero, A., Kubicki, S. and Rezgui, Y. (2020), "Towards a semantic construction digital twin: directions for future research", *Automation in construction*, Vol. 114, pp. 103179. available at: 10.1016/j.autcon.2020.103179.

Borrego, A., Latorre, J., Llorens, R., Alcañiz, M. and Noé, E. (2016), "Feasibility of a walking virtual reality system for rehabilitation: objective and subjective parameters", *Journal of neuroengineering and rehabilitation*, Vol. 13, No. 1, pp. 68. available at: https://search.datacite.org/works/10.1186/s12984-016-0174-1.

Borycki, D. (2018), *Programming for mixed reality with Windows 10, Unity, Vuforia, and UrhoSharp* translated by Anonymous Microsoft Press.

Brynskov, M., Heijnen, A., Balestrini, M. and Raetzsch, C. (2018), "Experimentation at scale: challenges for making urban informatics work", *Smart and sustainable built environment*, Vol. 7, No. 1, pp. 150–163. available at: https://www.emerald.com/insight/content/doi/10.1108/SASBE-10-2017-0054/full/html.

Cutolo, F., Cattari, N., Fontana, U. and Ferrari, V. (2020), "Optical see-through head-mounted displays with short focal distance: conditions for mitigating parallax-related registration error", *Frontiers in robotics and AI*, Vol. 7, pp. 572001. available at: https://www.ncbi.nlm.nih.gov/pubmed/33501331.

Daqri (2018), *Model-BIM* [AR App]. Retrieved 6 August 2019.

DaValle, A. and Azhar, S. (2020), "An investigation of mixed reality technology for onsite construction assembly", *MATEC web of conferences: 9th International Conference on Engineering, Project, and Production Management (EPPM2018)*, Vol. 312, pp. 6001–6010. available at: 10.1 051/matecconf/202031206001.

Ding, L., Zhou, Y. and Akinci, B. (2014), "Building information modeling (BIM) application framework: the process of expanding from 3D to computable nD", *Automation in construction*, Vol. 46, pp. 82–93. available at: 10.1016/j.autcon.2014.04.009.

Dossick, C., Osburn, L. and Neff, G. (2019), "Innovation through practice: the messy work of making technology useful for architecture, engineering and construction teams", *Engineering, construction, and architectural management*.

Dossick, C.S. and Neff, G. (2010), "Organisational divisions in BIM-enabled commercial construction", *Journal of construction engineering and management*, Vol. 136, No. 4, pp. 459–467. available at: http://ascelibrary.org/doi/abs/10.1061/(ASCE)CO.1943-7862.0000109.

Dudhee, V. and Vukovic, V. (2021), "Integration of building information modelling and augmented reality for building energy systems visualisation", in I. Mporas, P. Kourtessis, A. Al-Habaibeh, A. Asthana, V. Vukovic & J. Senior (Eds.), *Energy and sustainable futures: Proceedings of 2nd*

ICESF 2020, Springer International Publishing, Cham, Switzerland, pp. 7. available at: 10.1007/ 978-3-030-63916-7_11.

Engineering.com (2019), "Research report: augmented reality for maintenance, repair and overhaul (MRO)", ProjectBoard, Ontario, Canada. available at: https://www.ptc.com/en/resources/ augmented-reality/report/ar-for-maintenance-repair-overhaul.

Evans, G., Miller, J., Pena, M.I., MacAllister, A. and Winer, E.H. (2017), "Evaluating the Microsoft HoloLens through an augmented reality assembly application", *Evaluating the Microsoft HoloLens through an augmented reality assembly application*, Iowa State University, Anaheim, California.

Goulding, J., Rahimian, F. and Wang, X. (2014), "Virtual reality-based cloud BIM platform for integrated AEC projects", *Journal of information technology in construction*, Vol. 19, No. Special Issue: BIM Cloud-Based Technology in the AEC Sector: Present Status and Future Trends, pp. 308–325. available at: https://www.itcon.org/2014/18.

Grubert, J., Itoh, Y., Moser, K. and Swan, J.E. (2018), "A survey of calibration methods for optical see-through head-mounted displays", *IEEE transactions on visualisation and computer graphics*, Vol. 24, No. 9, pp. 2649–2662. available at: https://ieeexplore.ieee.org/document/8052554.

Hashimoto, J. (2020), *Capturing tacit knowledge through smart device augmented reality (SDAR)*, University of Hawaii.

Huang, Y. (2020), "Evaluating mixed reality technology for architectural design and construction layout", *Journal of civil engineering and construction technology*, Vol. 11, No. 1, pp. 1–12.

Huang, Y., Shakya, S. and Odeleye, T. (2019), "Comparing the functionality between virtual reality and mixed reality for architecture and construction uses", *Tu mu gong cheng yu jian zhu*, Vol. 13, No. 7, pp. 409–414.

Keil, J., Edler, D. and Dickmann, F. (2019), "Preparing the HoloLens for user studies: an augmented reality interface for the spatial adjustment of holographic objects in 3D indoor environments", *KN – Journal of cartography and geographic information*, Vol. 69, No. 3, pp. 205–215. available at: https://search.datacite.org/works/10.1007/s42489-019-00025-z.

Lee, G., Sacks, R. and Eastman, C.M. (2006), "Specifying parametric building object behavior (BOB) for a building information modeling system", *Automation in construction*, Vol. 15, No. 6, pp. 758–776. available at: 10.1016/j.autcon.2005.09.009.

Li, X., Yi, W., Chi, H., Wang, X. and Chan, A.P.C. (2018), "A critical review of virtual and augmented reality (VR/AR) applications in construction safety", *Automation in construction*, Vol. 86, pp. 150–162. available at: 10.1016/j.autcon.2017.11.003.

Love, P.E.D., Edwards, D.J., Irani, Z. and Walker, D.H.T. (2009), "Project pathogens: the anatomy of omission errors in construction and resource engineering project", *IEEE transactions on engineering management*, Vol. 56, No. 3, pp. 425–435. available at: https://ieeexplore.ieee.org/ document/5166564.

Mahmood, B., Han, S. and Lee, D. (2020), "BIM-based registration and localisation of 3D point clouds of indoor scenes using geometric features for augmented reality", *Remote sensing (Basel, Switzerland)*, Vol. 12, No. 14, pp. 2302. available at: https://search.proquest.com/docview/ 2425911685.

Microsoft Corporation (2016), *3D Viewer Beta* [HoloLens], Microsoft Corporation. Retrieved 12 February 2019, from https://www.microsoft.com/en-us/p/3d-viewer-beta/9nblggh5pm4z? activetab=pivot:overviewtab.

Montero, A., Montero, A., Zarraonandia, T., Zarraonandia, T., Diaz, P., Diaz, P., Aedo, I. and Aedo, I. (2019), "Designing and implementing interactive and realistic augmented reality experiences", *Universal access in the information society*, Vol. 18, No. 1, pp. 49–61. available at: https://search.proquest.com/docview/2191955235.

Moore, H.F. and Gheisari, M. (2019), "A review of virtual and mixed reality applications in construction safety literature", *Safety (Basel)*, Vol. 5, No. 3, pp. 51. available at: https://explore. openaire.eu/search/publication?articleId=doajarticles::c115640683c97b4925f49c5bf9badc7a.

Natephra, W. and Motamedi, A. (2019), *Live data visualisation of IoT sensors using augmented reality (AR) and BIM* (pp. 632). IAARC Publications, Waterloo.

Negrete, J., Moffett, J., Tyner, D. and Chen, P. (2018), *Mixing realities: a McCarthy, autodesk, DAQRI partnership*. Autodesk, Las Vegas, NV, USA.

Nuernberger, B., Ofek, E., Benko, H. and Wilson, A.D. (2016), "SnapToReality: aligning augmented reality to the real world", Proceedings of the 2016 CHI Conference on Human Factors in Computing Systems, Association for Computing Machinery, New York, United States pp. 1233–1244.

Park, K., Choi, S.H., Kim, M. and Lee, J.Y. (2020), "Deep learning-based mobile augmented reality for task assistance using 3D spatial mapping and snapshot-based RGB-D data", *Computers & industrial engineering*, Vol. 146, pp. 106585. available at: 10.1016/j.cie.2020. 106585.

Prabhakaran, A., Mahamadu, A., Mahdjoubi, L. and Manu, P. (2020), "An approach for integrating mixed reality into BIM for early stage design coordination", *MATEC web of conferences: 9th International conference on engineering, project, and production management*, Vol. 312, pp. 1–10. available at: 10.1051/matecconf/202031204001.

Qian, L., Azimi, E., Kazanzides, P. and Navab, N. (2017), "Comprehensive tracker based display calibration for holographic optical see-through head-mounted display", *arXiv preprint arXiv:1703.05834.*

Rahimian, F., Seyedzadeh, S., Oliver, S., Rodriguez, S. and Dawood, N. (2020), "On-demand monitoring of construction projects through a game-like hybrid application of BIM and machine learning", *Automation in construction*, Vol. 110, pp. 103012. available at: 10.1016/j.autcon.2019.1 03012.

Ratajczak, J., Riedl, M. and Matt, D.T. (2019), "BIM-based and AR application combined with location-based management system for the improvement of the construction performance", *Buildings (Basel)*, Vol. 9, No. 5, pp. 118. available at: https://explore.openaire.eu/search/publication?articleId=dedup_wf_001::a795b1dc6609580db89a1c54dbe42266.

Rausch, C. and Haas, C. (2021), "Automated shape and pose updating of building information model elements from 3D point clouds", *Automation in construction*, Vol. 124, pp. 103561. available at: 10.1016/j.autcon.2021.103561.

Revit Holoview Ltd (2017), *BIM Holoview* [HoloLens], Revit-Holoview Ltd. Retrieved 29 July 2019, from http://www.bimholoview.com/Features.

Ribo, M., Pinz, A. and Fuhrmann, A.L. (2001), "A new optical tracking system for virtual and augmented reality applications, IMTC 2001. Proceedings of the 18th IEEE Instrumentation and Measurement Technology Conference. Rediscovering Measurement in the Age of Informatics, IEEE, pp. 1932–1936.

Sacks, R., Brilakis, I., Pikas, E., Xie, H.S. and Girolami, M. (2020), "Construction with digital twin information systems" available at: https://www.repository.cam.ac.uk/handle/1810/313570.

Sampaio, A.Z. (2018), *Enhancing BIM methodology with VR technology*. IntechOpen.

Sherman, W.R. and Craig, A.B. (2019), *Understanding virtual reality*, Second edition ed., translated by Anonymous MK, Morgan Kaufmann Publishers, Cambridge, MA.

Sydora, C. and Stroulia, E. (2018), "Augmented reality on building information models", 9th International Conference on Information, Intelligence, Systems and Applications (IISA), IEEE, pp. 1–4.

Taketomi, T., Uchiyama, H. and Ikeda, S. (2017), "Visual SLAM algorithms: a survey from 2010 to 2016", *IPSJ transactions on computer vision and applications*, Vol. 9, No. 1, pp. 1–11. available at: https://explore.openaire.eu/search/publication?articleId=doajarticles::79f79d75bd 1bd93a87f41c0755a6b19e.

Tang, P., Huber, D., Akinci, B., Lipman, R. and Lytle, A. (2010), "Automatic reconstruction of as-built building information models from laser-scanned point clouds: a review of related techniques", *Automation in construction*, Vol. 19, No. 7, pp. 829–843. available at: 10.1016/j.autcon. 2010.06.007.

92   *Vishak Dudhee and Vladimir Vukovic*

Tang, S., Shelden, D.R., Eastman, C.M., Pishdad-Bozorgi, P. and Gao, X. (2019), "A review of building information modeling (BIM) and the internet of things (IoT) devices integration: present status and future trends", *Automation in construction*, Vol. 101, pp. 127–139. available at: 10.1016/j.autcon.2019.01.020.

Taylor, J.E. and Bernstein, P.G. (2009), "Paradigm trajectories of building information modeling practice in project networks", *Journal of management in engineering*, Vol. 25, No. 2, pp. 69–76. available at: http://ascelibrary.org/doi/abs/10.1061/(ASCE)0742-597X(2009)25:2(69).

Visual Live, 3.L. (2018), *HoloLive* [HoloLens], Visual Live 3D LLC. Retrieved 15 July 2019, from https://visuallive.com/hololive-ar-for-construction/.

Wang, J., Wang, X., Shou, W. and Xu, B. (2014), "Integrating BIM and augmented reality for interactive architectural visualisation", *Construction innovation*, Vol. 14, No. 4, pp. 453–476. available at: https://search.proquest.com/docview/2083745274.

Wang, P., Wu, P., Wang, J., Chi, H. and Wang, X. (2018), "A critical review of the use of virtual reality in construction engineering education and training", *International journal of environmental research and public health*, Vol. 15, No. 6, pp. 1204. available at: https://www.ncbi.nlm.nih.gov/pubmed/29890627.

Wang, Q., Gao, B. and Wu, H. (2019), "Triangular mesh generation on free-form surfaces based on bubble dynamics simulation", *Engineering computations*, Vol. 36, No. 2, pp. 646–663. available at: https://www.emerald.com/insight/content/doi/10.1108/EC-09-2017-0352/full/html.

Yan, W., Culp, C. and Graf, R. (2011), "Integrating BIM and gaming for real-time interactive architectural visualisation", *Automation in construction*, Vol. 20, No. 4, pp. 446–458. available at: https://search.datacite.org/works/10.1016/j.autcon.2010.11.013.

Zeller, M. and Coulter, D. (2018), Mar 21, last update, *Spatial mapping*. Available: https://docs.microsoft.com/en-us/windows/mixed-reality/design/spatial-mapping [17 December 2020].

6 Conceptual Framework for Safety Training for Migrant Construction Workers Using Virtual Reality Techniques

Rahat Hussain, Akeem Pedro, Syed Farhan Alam Zaidi, Muhammad Sibtain Abbas, Mehrtash Soltani, and Chansik Park

Department of Architecture Engineering, Chung Ang University, Seoul, South Korea

6.1 Introduction

Ensuring the safety of workers remains a significant challenge for the construction industry. Construction practitioners tend to prioritize productivity and efficiency to meet project deadlines and stay within budget, often leading to safety taking a backseat. Unfortunately, this approach makes construction job sites high-risk environments, particularly for migrant workers, who are more susceptible to accidents and injuries (Peiró et al., 2020; Thamrin, 2018). Challenges such as poor communication (Yap & Lee, 2020), cultural differences (Trax et al., 2015), low literacy rate (Lin et al., 2018), language barrier (Flynn, 2014), and family disruption make workers more vulnerable to risks at construction job sites. The accident rate of migrants was recorded as 1.3 times higher compared to local construction workers in the United States (Dong et al., 2009), and a German study showed that falling object injuries for migrant workers are 40% higher than for non-nationals (Arndt et al., 2004). Due to the increasing trend of statistics about migrants, the Assistant Secretary of Labor for Employment and Training (U.S.) Emily Stover DeRocco predicted that "Migrants make up one out of every three new workers in the country, and by 2025, this proportion will be one out of every two new workers" (Life, 2012).

Limited attention has been given to developing and implementing training tailored to the specific challenges and vulnerabilities encountered by migrant workers in the construction industry (Peiró et al., 2020). Furthermore, safety training may not be undertaken at all; studies report that around two-thirds of migrant workers sampled had not received the same safety training method as local workers (Trajkovski & Loosemore, 2006). A study (Morando & Brullo, 2022) mentioned that complete transferability is hard to achieve from safety training if the training supporting factors are not included. These specific challenges and risks associated with migrant workers generate a gap in the literature to identify the status of current approaches and produce a pressing need to develop specific training by incorporating advanced technologies that can enhance their knowledge acquisition abilities. Over the last decade, virtual reality (VR) technology has matured enough to be used for industry and services. It has grown into a simulation technology for users to interact with self-motivated, three-dimensional virtual content and into practice for fundamental research in design, visualization, training, and evaluation (Cosmar et al., 2013). Despite the potential benefits, there is a gap in the literature and practice regarding VR-based safety training methods for migrant workers.

DOI: 10.1201/9781003408949-6

This research aims to develop a conceptual framework for migrant workers' safety training (MWST) using VR techniques with the following three principles. First, developing scenarios in a virtual environment using Building Information Modeling and Unity game engine support. These scenarios should incorporate safety regulations from governmental/local/state authorities and construction best practices from the contractor's historical knowledge and experience. A language selection module is also included in the scenario development stage so that workers get training in an immersive environment by selecting their preferred language for the relevant trade. For the maximum acquisition of knowledge during the training program, external support, including peer support in terms of incentives/rewards, motivation, and soft skills, including cross-cultural values (Clarke, 2006), workers' self-efficacy, and cognitive ability, have been considered as the second principle for the proposed framework. Literature shows that incorporating these external supportive factors can enhance the workers' ability to gain maximum knowledge during training and transfer the gained knowledge to the actual job site (Hussain et al., 2017). Third, a participatory approach to improving problem-solving and innovative skills is incorporated in the framework (Ahonen et al., 2014). This is the approach of learning in which a worker can participate within a group as a member/leader and within the team as a role-playing character. The evaluation of workers' performance is done by senior health and safety manager. Following this introduction, the next section discusses the challenges faced by traditional training methods for migrant workers and explores the potential of incorporating VR for training. Section 6.3 then presents the proposed conceptual framework and its constituent principles to meet migrant workers' educational needs. Finally, the conclusion outlines the limitations and suggests future research directions.

6.2   Background

6.2.1   *Traditional Methods for Migrant Workers' Safety Training (MWST)*

Construction safety training is a crucial tool for maintaining a safe job site and reducing the rate of accidents (Clarke, 2013; Zhou et al., 2015). Literature suggests that encouraging behavior change and enhancing the safety climate are external factors that can improve workers' safety performance (Cunningham et al., 2018). Similarly, a meta-analysis has shown that safety training can strongly impact construction workers' knowledge and behavior. The ultimate goal of any training method is to deliver maximum knowledge and enhance workers' skills to transfer that knowledge to the job site. Therefore, it is essential to evaluate the effectiveness of a training program before implementation. A study using Kirkpatrick's four-level training evaluation model found that most traditional approaches only evaluate the first level (Asari & Leman, 2015).

It is unrealistic to expect every skill learned during training to be used on the job, especially for migrant workers. Zhao argues that conventional training methods do not maximize knowledge transfer, resulting in losses (Zhao & Lucas, 2014). The shortcomings of traditional safety training modes and learning styles lead to information transfer losses during training. Passive training methods, such as handing out text-heavy safety materials to workers who require high reading proficiency, create a gap between the knowledge that should be conveyed and the knowledge that is actually transferred. Effective knowledge acquisition during training may depend on the training materials and delivery strategies used by instructors. Additionally, organizational support and employee motivation can facilitate the application of information in the workplace.

In summary, most studies on MWST have focused on addressing cultural diversity and lack of safety regulation awareness. While tailored safety training has been used, these studies may be limited to improving migrant workers' language abilities, mainly through translation and audio-visual material. The potential benefits of virtual environment-based learning for migrant workers have not been widely considered in safety training research.

6.2.2 *Empowering VR Tools for Safety Training*

By integrating advanced input/output devices with 3D computer-generated immersive graphics, VR represents a significant step forward in visual technology, creating a synthetic environment that provides a sense of reality and the feeling of "being there" for users (Jayaram et al., 1997). As VR technology continues to mature, it has been successfully applied across various industries, including aviation for operational training (Biggs et al., 2018), mining for safe work performance (Lucas & Thabet, 2008), and medical and dental training (Kaluschke et al., 2022; Lewis et al., 2011). In the construction industry specifically, VR technology has been widely used for concrete structural analysis (Setareh et al., 2005), electrical design and installation (Sulbaran & Shiratuddin, 2006), civil and architecture design (Jin et al., 2007), and prototyping (Huang et al., 2007). Furthermore, VR-based safety training programs can offer an engaging and interactive tool for training construction workers to increase their ability to identify hazards (Cha et al., 2012). Chandra & Leenders, (2012) focused on incorporating a communication and collaborative environment within the virtual world, while Le et al. (2015) developed a semantic safety platform for construction scenarios to communicate and analyze various hazard cases properly.

Brown & others (2000) suggested that trainers can improve their training methods by integrating technology resources to motivate workers to learn the language. The integration of VR technology in the curriculum aligns with the four components of Keller's ARCS Model, which was developed to enhance motivation (Cunningham et al., 2018) to achieve the training objectives. These components direct workers' attention toward training and create connections between the VR scenarios and recognizable objects at the job site. Additionally, by interacting with multiple avatars, workers may feel a sense of achievement and self-worth, accomplishing the safety training objective. This study aims to create practical and dynamic training scenarios by incorporating interventions for migrant workers to improve their safety performance.

6.3 The Conceptual Framework for MWST

The conceptual framework for MWST is designed to provide a comprehensive and innovative approach to safety training in the construction industry. The framework consists of three core principles: immersive scenario development, enhanced learning support, and participatory approach for training as illustrated in Figure 6.1.

The first principle of immersive scenario development encourages the utilization of cutting-edge technologies such as VR and building information modeling (BIM) to create realistic and interactive training scenarios. These scenarios must incorporate safety regulations and standards, as well as best practices from the contractor's historical knowledge and experience. The language selection module allows workers to receive training in their preferred language, ensuring that they have practical skills and knowledge that they can apply to real-world situations. Enhanced learning support, as a second principle, involves using managerial support to create personalized and engaging learning experiences.

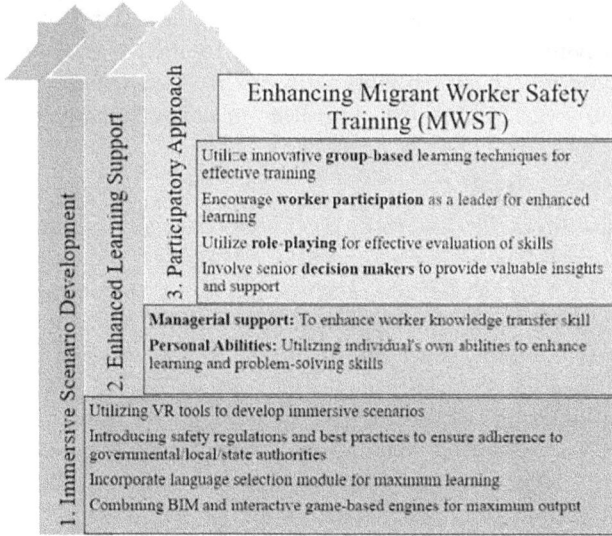

Figure 6.1 Core principles of migrant workers' safety training (MWST).

External supportive factors such as peer incentives and rewards, motivation, and soft skills training are also incorporated to enhance workers' ability to acquire and transfer knowledge gained during training to the actual job site. This principle focused on utilizing the workers' own abilities to ensure workers are motivated to learn and have the necessary skills and knowledge to perform their jobs safely and effectively. The last principle of the proposed framework promotes a participatory approach to training by fostering a culture of innovation and teamwork through collaborative problem-solving. This involves utilizing collaborative virtual environments, role-playing characters, and cross-functional team problem-solving to encourage workers to actively participate in identifying and solving safety issues through innovative solutions. Real-time data analytics provide workers with feedback and insights to improve their performance. This principle enables workers to work together effectively to achieve common safety goals, improving safety outcomes and enhancing the overall productivity of the construction industry.

The MWST framework is presented in further detail in the subsequent sections. The three core principles of immersive scenario development, enhanced learning support, and participatory approach can provide a constructive loop to create an environment that is engaging, practical, and effective, ensuring that workers have the skills and knowledge they need to work safely and productively on the job site.

6.3.1 *Immersive Scenario Development*

Due to the diverse nature of the construction industry, site layout planning, trades, and safety management techniques can vary greatly from project to project. To create effective virtual training scenarios, it is important to include project-specific information that reflects the unique characteristics of each job site. This includes geometrical information, interactive objects, and hazardous zones, which are essential components for tailoring safety training scenarios to the needs of workers.

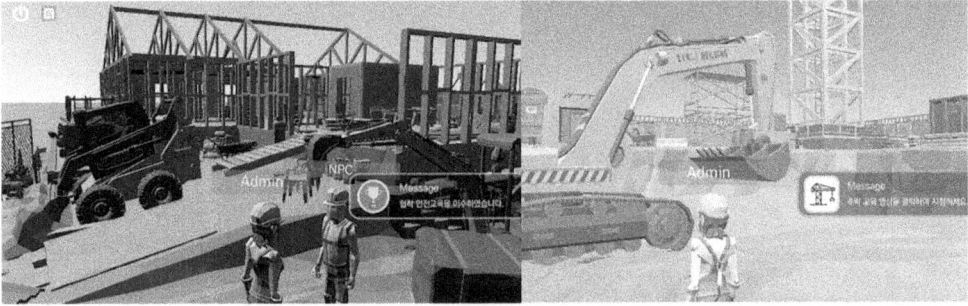

Figure 6.2 An example of training scenario with 3D model and environment.

One effective way to incorporate project information is through the use of BIM tools, which allow for the creation of detailed, three-dimensional models of the construction site. By including design and planning information, as well as information about specific trades and construction methods, BIM models can provide a comprehensive representation of the job site. To further enhance the engagement and effectiveness of VR training, BIM models can be incorporated into the Unity game engine. This approach can create a dynamic and interactive training environment that allows workers to explore and learn in a more immersive and engaging way (Golovina et al., 2019). An example of a training scenario created using the BIM model and Unity-based environment is shown in Figure 6.2.

Once the initial setup of the virtual training environment and model is generated, it is crucial to incorporate relevant safety regulations, rules from local governmental authorities, and best practices from contractors to ensure comprehensive training. However, it is important to note that best practices may vary between contractors and may change over time. Therefore, regularly updating the training material with the latest best practices is recommended. In addition, the natural language processing (NLP) module can be a valuable addition to increase engagement levels for migrant workers and minimize the language barrier problem. The text-to-speech function can cater to the illiterate population of migrants who cannot read even in their preferred language, while the voice-over function can translate the language of the training material into the desired language of the worker.

Furthermore, the NLP module can help to enhance the effectiveness of the training program by providing real-time feedback to workers. For example, NLP-based chatbots can be incorporated into the virtual environment, which can answer workers' queries and provide additional information. This can increase workers' engagement with the training material and improve their understanding of the content. Additionally, the NLP module can also be used to analyze workers' responses and identify areas where they may need additional support or clarification. This can help trainers to tailor the training content to the specific needs of each worker and ensure that they have a clear understanding of the material. The NLP module can be a powerful tool for improving the effectiveness of the virtual training environment and ensuring that migrant workers have access to comprehensive and accessible safety training.

Regarding hardware, workers can access the virtual training scenarios using different devices, but head-mounted display devices are recommended for maximum knowledge

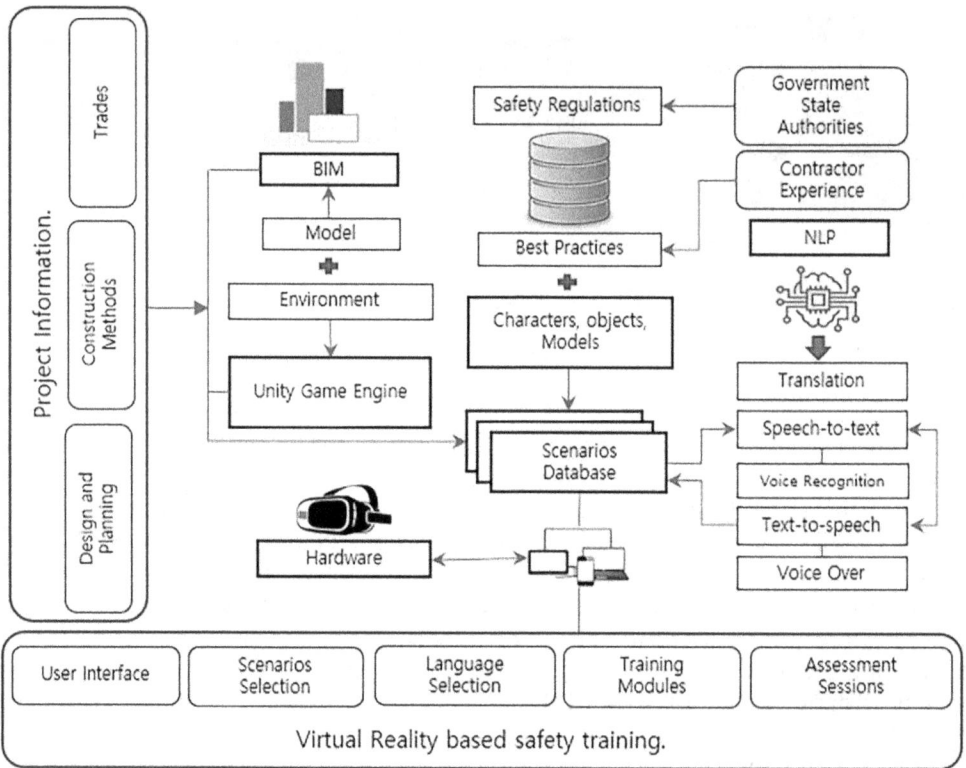

Figure 6.3 System framework for tailoring VR-based scenario development.

acquisition. It is also important to have a user-friendly interface that enables workers to easily select the desired language for their training scenarios. Assessment modules, such as self-assessment and supervisor assessment stages, are also necessary for each trade. These modules provide workers with feedback on their progress and can help identify areas that require further attention. The systematic flow of each step and process involved in creating VR-based scenarios can be seen in Figure 6.3, and it is important to follow this process to ensure the effectiveness of the training program.

In addition to the hardware requirements, software tools play an equally important role in VR-based safety training. Advanced VR software platforms offer interactive and immersive learning experiences that enhance the learning outcomes of workers. These platforms can simulate a range of hazardous scenarios and provide workers with real-time feedback on their performance, allowing them to learn from their mistakes and improve their skills. Moreover, VR simulations enable workers to practice responding to emergencies in a safe and controlled environment, without risking their lives or damaging equipment. VR training scenarios can also be customized based on the type of construction project, trade, and the workers' level of experience. For example, VR scenarios for a demolition project may differ significantly from those for a renovation project. Similarly, the training requirements for a novice worker may differ from those for an experienced worker. Customized scenarios enable workers to learn skills that are relevant to their job roles and the construction projects they are working on. This leads to a more efficient and effective learning experience, with higher knowledge retention and improved safety performance. By following a systematic flow process for creating VR-based scenarios and utilizing advanced VR software platforms and hardware devices, construction companies can provide effective safety training to their migrant workers and reduce the risk of accidents and injuries on job sites.

6.3.2 *Enhanced Learning Support*

To enhance worker knowledge transfer skills and problem-solving abilities, it is crucial to provide enhanced learning support to migrant workers. Furthermore, managerial support is one way to enhance worker knowledge transfer skills. This can involve peer and supervisor support, level assessments with incentives and rewards, and a training process that enriches knowledge acquisition. Additionally, personal abilities should be utilized to enhance learning and problem-solving skills. To ensure that workers can effectively apply their knowledge and skills in practical work settings, it is important to provide training that is tailored to their individual needs and abilities. For instance, workers' literacy and ability to understand local language. Despite being skilled, migrant workers often exhibit poor safety performance in multi-lingual work groups due to gaps in knowledge acquisition and transmission in various working environments. Therefore, to minimize occupational health disparities, it is essential to include external supportive functions that enhance the safety performance of migrants (Hussain et al., 2017).

In addition to providing external support, it is also important to encourage workers to utilize their own personal abilities to enhance their learning and problem-solving skills. This can include self-reflection, self-directed learning, and seeking feedback from supervisors and peers. Motivation is also a critical component of training, which needs to be supported by managers as it is not typically innate. The ARCS Model of Motivational Design by Keller highlights the importance of attention, relevance, confidence, and satisfaction, which must be fostered and maintained throughout the training process. To ensure that workers remain motivated and engaged throughout the training process, it is important to provide ongoing feedback and support. This can include regular check-ins with supervisors and peers, as well as opportunities for workers to apply their newly acquired knowledge and skills in real-world situations.

When designing training programs for migrant workers, it is also important to consider the cultural differences that may affect their learning and transfer of knowledge. For example, workers from collectivistic cultures may place a greater emphasis on teamwork and collaboration, whereas workers from individualistic cultures may prioritize personal achievement and recognition. Cross-cultural values, self-efficacy, and cognitive ability of workers must be considered during training sessions to understand how culture influences learning and transfer.

By integrating the knowledge and cultural values of trainees into the training program, it is possible to create a more inclusive and effective learning environment that meets the needs of all workers. This can be achieved through the use of culturally relevant examples, case studies, and role-playing scenarios. In collectivistic cultures, motivation to get training will be higher when training benefits the workgroup. On the other hand, promoting the individual benefits of migrants will also enhance their motivation. Furthermore, in high uncertainty avoidance cultures, emphasizing well-established practices can be effective in individual self-efficacy. Ultimately, by providing enhanced learning support that is tailored to the needs and abilities of migrant workers, it is possible to improve their safety performance and minimize occupational health disparities in multi-lingual work groups.

6.3.3 *Participatory Approach*

The participatory training program is an integral part of the framework designed to provide practical and easy-to-implement safety measures for workers in diverse construction workgroups. This approach not only addresses individual workers' capabilities but also improves safety and health outcomes through practical means. It is essential to actively engage workers in the group and evaluate their abilities through self-assessment and comparison with other group members. Taking on responsibilities and playing a role within the group can also foster team building, increasing workers' confidence in applying their newly acquired knowledge to the construction job site.

During training, it is recommended to maintain performance records that include data on workers' task execution, completion, and unsafe behavior in terms of safety violations, as illustrated in Figure 6.3. After completing the training, workers can see their auto-assessment, while their immediate supervisor can measure group/team-based performance. This data can be used by decision-makers and instructors to provide feedback to workers, enabling them to improve their safety performance. Trained migrant workers exhibit a higher level of site safety awareness than those trained in a different work environment with no feedback. With the aid of self and performance assessment, each migrant worker can identify and mitigate safety hazards, leading to improved safety outcomes.

Supervisory feedback is a critical component of performance assessment in the construction industry. The immediate supervisor plays a crucial role in the feedback process, providing regular feedback to workers regarding their performance, identifying areas for improvement, and recognizing achievements. Supervisors can use the data collected from the performance records to identify patterns in behavior and address specific issues with workers during training. In addition, the use of performance records can help supervisors identify workers who need additional training or support to ensure that they are performing their tasks safely. The use of performance records can also assist in measuring the effectiveness of the training program. By comparing pre- and post-training performance records, instructors can determine the effectiveness of the training program and identify areas that require improvement. This can lead to a continuous improvement process that ensures that the training program remains up-to-date and relevant to the needs of the workers. Hence, the use of performance records and supervisory feedback is critical in ensuring that workers receive the maximum benefit from the training program. This can lead to improved safety outcomes on the job site, benefiting both workers and the organization (Figure 6.4).

To make the participatory training program effective, innovative group learning techniques can be utilized to enhance the learning process. Encouraging worker participation as a leader can also contribute to effective training. Role-playing can be used for the evaluation of skills, and senior decision-makers can be involved to provide valuable insights and support. During the training process, it is essential to encourage active participation by the workers to ensure that they grasp the necessary knowledge and skills. Utilizing role-playing activities can be an effective way of evaluating their skills and identifying areas that require improvement. Senior decision-makers can provide valuable feedback and support to workers during the training process. By involving senior

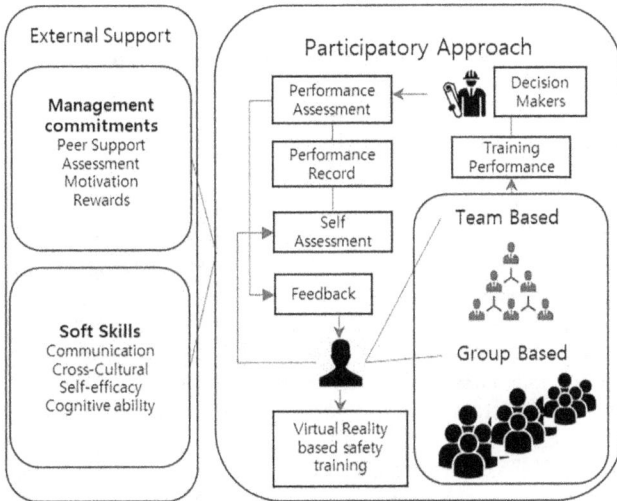

Figure 6.4 Relation between external supporting factors and participatory approach.

decision-makers, the workers can learn about the importance of safety regulations and understand the company's commitment to ensuring the safety and well-being of its employees. Overall, incorporating these techniques can enhance the participatory training program's effectiveness and ensure that workers acquire the necessary knowledge and skills to promote their safety and well-being in the construction industry.

6.4   Conclusion

Migrant workers in the construction industry often experience a higher risk of work-related injuries than their local counterparts. Despite various training approaches, migrants still have unacceptably high accident rates due to insufficient knowledge acquisition during training programs. Even trained workers may not be able to perform safely in multilingual and multicultural environments. To address this issue, a conceptual framework for VR-based safety training with external supporting interventions has been developed for migrant workers. The framework includes a flow process to create VR-based content that incorporates project information for construction safety training. It also integrates safety regulations mandated by local authorities and provides linguistic aid through NLP to maximize knowledge acquisition. Additionally, external support in the form of managerial assistance and soft skills development is provided to promote high-level knowledge development.

To ensure the effectiveness of the training program, migrant workers should be involved in a participative approach during the assessment process, utilizing group-based procedures and teamwork within a virtual environment. Incorporating external peer support and developing soft skills would also assist workers in applying their newly acquired information on the job. One of the primary challenges in adopting the proposed framework is incorporating real project information to build training scenarios. This is due to the diverse nature of the construction industry. However, an assessment database can be generated within the virtual environment for record-keeping purposes.

Future work will include the implementation and validation of the proposed framework by automating the assessment process. Through this framework, VR-based safety training can provide a cost-effective and efficient solution for improving safety and reducing accidents among migrant workers in the construction industry.

References

Ahonen, E. Q., Zanoni, J., Forst, L., Ochsner, M., Kimmel, L., Martino, C., Ringholm, E., Rodríguez, E., Kader, A., & Sokas, R. (2014). Evaluating goals in worker health protection using a participatory design and an evaluation checklist. *New Solutions: A Journal of Environmental and Occupational Health Policy, 23*(4), 537–560.

Arndt, V., Rothenbacher, D., Daniel, U., Zschenderlein, B., Schuberth, S., & Brenner, H. (2004). All-cause and cause specific mortality in a cohort of 20 000 construction workers; results from a 10 year follow up. *Occupational and Environmental Medicine, 61*(5), 419–425.

Asari, M. S., & Leman, A. M. (2015). Safety training evaluation: Approaches and practices. *Journal of Occupational Safety and Health, 12*(2), 23–30.

Biggs, A. T., Geyer, D. J., Schroeder, V. M., Robinson, F. E., & Bradley, J. L. (2018). *Adapting virtual reality and augmented reality systems for naval aviation training*. Naval Medical Research Unit Dayton Wright-Patterson AFB, United States.

Brown, H. D., & others. (2000). *Principles of language learning and teaching* (Vol. 4). Longman, New York.

Cha, M., Han, S., Lee, J., & Choi, B. (2012). A virtual reality based fire training simulator integrated with fire dynamics data. *Fire Safety Journal, 50*, 12–24.

Chandra, Y., & Leenders, M. A. A. M. (2012). User innovation and entrepreneurship in the virtual world: A study of Second Life residents. *Technovation, 32*(7–8), 464–476.

Clarke, S. (2006). The relationship between safety climate and safety performance: A meta-analytic review. *Journal of Occupational Health Psychology, 11*(4), 315.

Clarke, S. (2013). Safety leadership: A meta-analytic review of transformational and transactional leadership styles as antecedents of safety behaviours. *Journal of Occupational and Organizational Psychology, 86*(1), 22–49.

Cosmar, M., Nickel, P., Schulz, R., & Zieschang, H. (2013). Human machine interface. https://oshwiki.osha.europa.eu/en/themes/human-machine-interface

Cunningham, T. R., Guerin, R. J., Keller, B. M., Flynn, M. A., Salgado, C., & Hudson, D. (2018). Differences in safety training among smaller and larger construction firms with non-native workers: Evidence of overlapping vulnerabilities. *Safety Science, 103*(September 2017), 62–69. 10.1016/j.ssci.2017.11.011

Dong, X. S., Fujimoto, A., Ringen, K., & Men, Y. (2009). Fatal falls among Hispanic construction workers. *Accident Analysis & Prevention, 41*(5), 1047–1052. 10.1016/j.aap.2009.06.012

Flynn, M. A. (2014). Safety & the diverse workforce: Lessons from NIOSH's work with Latino immigrants. *Professional Safety, 59*(6), 52–57. http://www.ncbi.nlm.nih.gov/pubmed/26566296

Golovina, O., Kazanci, C., Teizer, J., & König, M. (2019). Using serious games in virtual reality for automated close call and contact collision analysis in construction safety. *ISARC. Proceedings of the International Symposium on Automation and Robotics in Construction, 36*, 967–974.

Huang, T., Kong, C. W., Guo, H., Baldwin, A., & Li, H. (2007). A virtual prototyping system for simulating construction processes. *Automation in Construction, 16*(5), 576–585.

Hussain, R., Pedro, A., Lee, D. Y., & Park, C. (2018). Impact of safety training and interventions on training-transfer: Targeting migrant construction workers. *International Journal of Occupational Safety and Ergonomics, 4*(2), 56414.

Jayaram, S., Connacher, H. I., & Lyons, K. W. (1997). Virtual assembly using virtual reality techniques. *Computer-Aided Design, 29*(8), 575–584.

Jin, D., Yan, F., & Ito, Y. (2007). Applications of virtual reality to civil and architectural engineering projects. *International Symposium on Innovation \& Sustainability of Structures in Civil Engineering*, 1–10.

Kaluschke, M., Yin, M. S., Haddawy, P., Suebnukarn, S., & Zachmann, G. (2022). The Impact of 3D Stereopsis and Hand-Tool Alignment on Effectiveness of a VR-based Simulator for Dental Training. *2022 IEEE 10th International Conference on Healthcare Informatics (ICHI)*, 449–455.

Le, Q. T., Pedro, A., & Park, C. S. (2015). A social virtual reality based construction safety education system for experiential learning. *Journal of Intelligent & Robotic Systems*, *79*(3), 487–506.

Lewis, G. N., Woods, C., Rosie, J. A., & Mcpherson, K. M. (2011). Virtual reality games for rehabilitation of people with stroke: Perspectives from the users. *Disability and Rehabilitation: Assistive Technology*, *6*(5), 453–463.

Life, R. F. (2012). ESL: Are we really communicating? *International Journal of Business, Humanities and Technology*, *2*(2).

Lin, K.-Y., Lee, W., Azari, R., & Migliaccio, G. C. (2018). Training of low-literacy and low-English-proficiency Hispanic workers on construction fall fatality. *Journal of Management in Engineering*, *34*(2), 5017009.

Lucas, J. D., & Thabet, W. (2008). *Implementation and evaluation of a VR task-based training tool for conveyor belt safety training.*

Morando, M., & Brullo, L. (2022). Promoting safety climate training for migrant workers through non-technical skills: A step forward to inclusion. *Merits*, *2*(1), 26–45.

Peiró, J. M., Nielsen, K., Latorre, F., Shepherd, R., & Vignoli, M. (2020). Safety training for migrant workers in the construction industry: A systematic review and future research agenda. *Journal of Occupational Health Psychology*, *25*(4), 275.

Setareh, M., Bowman, D. A., & Kalita, A. (2005). Development of a virtual reality structural analysis system. *Journal of Architectural Engineering*, *11*(4), 156–164.

Sulbaran, T., & Shiratuddin, M. F. (2006). *A proposed framework of a virtual reality training tool for design and installation of electrical systems. 6th International Conference on Construction Applications of Virtual Reality.*

Thamrin, Y. (2018). A literature review of migrant workers' health and safety. *Jurnal Kesehatan Masyarakat Maritim*, *1*(2).

Trajkovski, S., & Loosemore, M. (2006). Safety implications of low-English proficiency among migrant construction site operatives. *International Journal of Project Management*, *24*(5), 446–452.

Trax, M., Brunow, S., & Suedekum, J. (2015). Cultural diversity and plant-level productivity. *Regional Science and Urban Economics*, *53*, 85–96.

Yap, J. B. H., & Lee, W. K. (2020). Analysing the underlying factors affecting safety performance in building construction. *Production Planning & Control*, *31*(13), 1061–1076.

Zhao, D., & Lucas, J. (2014). Virtual reality simulation for construction safety promotion. *International Journal of Injury Control and Safety Promotion*, *0*(0), 1–11. 10.1080/17457300.2013. 861853

Zhou, Z., Goh, Y. M., & Li, Q. (2015). Overview and analysis of safety management studies in the construction industry. *Safety Science*, *72*, 337–350.

7 Automated Data Retrieval Solution from BIM Using AI Voice Assistant

*Faris Elghaish[1], Jatin Kumar Chauhan[1],
Sandra Matarneh[2], Farzad Pour Rahimian[3], and
M Reza Hosseini[4]*

[1]School of Natural and Built Environment, Queen's University Belfast,
Belfast, UK; [2]Faculty of Engineering, Al Ahliyya Amman University,
Jordan; [3]School of Computing, Engineering & Digital Technologies, Teesside
University, Middlesbrough, UK; [4]School of Architecture and Building,
Deakin University, Geelong, Victoria, Australia

7.1 Introduction

Building Information Modelling (BIM), in its most basic form, is based on the utilisation of detailed digital representations of buildings. These digital representations are made up of geometric data and building component parameters to enable users of obtaining data and their parametric values anytime and anywhere (Preidel et al., 2017; Wu et al., 2019). BIM's unique selling point is premised upon its inherent capability to create new collaborative work practices (Oraee et al., 2021), where project stakeholders (designers, contractors, clients and end users) have continuous access to up-to-date digital building data via a common data environment (CDE) (Hossain and Ng, 2018). This collaborative platform streamlined data flow throughout the entire life cycle of construction in the Architecture, Engineering, Construction and Owner-operated (AECO) industry, which means that data utilisation in BIM software has started to become crucial (Shin et al., 2020b). As a result, information retrieval (IR) in BIM software has begun to be addressed in order to retrieve more precise data (Wang et al., 2021).

BIM models are highly structured instances of a well-defined data schema. Retrieving the required information from BIM models requires using query languages to process this kind of structured data using neutral schema such as the industry foundation classes (IFC) (Wu et al., 2019). However, performing searches on platforms that are based on IFC database encountered many limitations since IFC-based data do not include all the data available in the building information model (Lange et al., 2018). In addition, this type of search requires experts in dealing with IFC-based data to manipulate and process BIM data (Motawa, 2017). There is a need for more resilient solutions to lay the foundation for changing the BIM environment from expert- to customer-oriented by facilitating the usability and accessibility of BIM in the AECO industry.

BIM is considered as the source of digital information for all construction phases and processes (Alizadehsalehi et al., 2020; Pan and Zhang, 2021), therefore, managing data through BIM environment is an essential for successful project delivery (Abdirad, 2021; Beach et al., 2017). Throughout the last few years, there were successful attempts to integrate emerging digital technologies into BIM, for example, coupling BIM and Internet of Things (IoT) to enhance the site health and safety management (Chen et al., 2021; Liu, 2021), integrating blockchain into BIM to automate sharing risk/reward for

DOI: 10.1201/9781003408949-7

Integrated Project Delivery (IPD) projects (Elghaish et al., 2020), using immersive technologies with BIM to efficiently involve the end users in the design development stage (Khalili, 2021).

Similarly, Artificial Intelligence (AI) has been integrated into BIM for several reasons such as automating decision-making (Plaß et al., 2021), optimising design (He et al., 2021; Liu and Jiang, 2021), detecting defects in infrastructure systems (Sresakoolchai and Kaewunruen, 2022). As such, there is a significant level of attention to employ emerging technologies to leverage BIM processes. Alotto et al. (2020) assert that the future of digital construction requires a necessary collaboration between human intelligence and AI. This integration is recommended to enable using interoperability between different devices, for instance, using mobile to interact with BIM model in a server through designated rooms for interaction, this will enable users to manage information automatically and remotely (Lin et al., 2016).

The growth of Automatic Speech Recognition (ASR) technology for virtual assistants provided a wide range of solutions to retrieve information using users' voices (Croft, 2019; Wu et al., 2020). Search engines have recently developed to detect and understand natural human language in addition to accepting keyword-based requests from the human voice (Shin et al., 2020a). Throughout the last few years, ASR technology started to be employed to improve the job efficiencies in terms of retrieving data (Ivanov, 2017). Regarding developing an automated data retrieving approaches for BIM, Motawa (2017) proposed a spoken dialogue BIM system approach to capture the building operation information such as maintenance data, however, this approach relies on extracting information from IFC protocol and storing it in the knowledge base to be provided upon the verbal request, which limits the capabilities of using this system for different types of information; Alotto et al. (2020) integrated AI voice assistant into BIM for training purposes to enable BIM novice users to develop parametric models, however, the proposed solution is lacked in providing two ways voice interaction solution; Shin et al. (2020a) developed AI voice assistant application to retrieve information from BIM-based SQL query, however, the application is limited to enable remote interaction given it is server-based solution. As such, there is a need for solution that is characterised as (1) enabling remote interaction, (2) supporting retrieving different types of BIM data, (3) allowing novice BIM users to develop parametric designs by uttering specific keywords, (4) providing retrieved information verbally to enable blind users to interact with BIM models.

With all above in the mind, this chapter presents a workable solution for BIM users-based AI voice assistant technology to enable interacting with the information model remotely, as well as, giving an opportunity to disable user—visual disabilities—to manage BIM model.

The chapter is structured as follows. Section 7.2 represents the conceptual background, followed by the research methods in Section 7.3. Section 7.4 presents the solution development and Section 7.5 includes the evaluation of the proposed solution. The discussion, significance and limitation are presented in Section 7.6. Finally, Section 7.7 is the research conclusion.

7.2   Contextual Background

7.2.1   *BIM and Data Retrieval*

A huge amount of multi-scale information is contained in a BIM model (component level, attribute level, etc.). BIM users retrieve specific information based on their work requirements, for example, a **BIM** manager retrieves both components and attributes information about a building to verify conformity with requirements and standards (Zhang and El-Gohary, 2015). On the other hand, a facility manager is more interested in retrieving non-geometric information about the building systems and equipment (Matarneh et al., 2020). However, most of the existing retrieval approaches are based on information extracted from components. For example, Gao et al. (2015) developed a semantic search engine based on BIM-oriented ontology for the contextual meaning of terms and local context analysis technique (LCA) for query expansion, which retrieves online BIM documents more quickly and accurately. Liu et al. (2017) presented an enhanced explicit semantic analysis method for product model retrieval in construction industry that enables an automated generation of semantic information from domain-specific knowledge repositories. Gao et al. (2017) focused on constructing a lightweight IFC ontology for their proposed system to retrieve online BIM product documents. In the same vein, Matarneh et al. (2019) proposed a system that extracts specific non-geometric attributes relevant to facilities systems and equipment from BIM model using IFC schema to manipulate the extracted information and generate a spreadsheet that contains all facilities management systems requirements. Shi et al. (2018) developed a file comparison tool IFCdiff to search online BIM documents quickly and accurately. Wu et al. (2019) developed a semantic search engine for intelligent retrieval in the BIM object database. Using the domain ontology and natural language processing technique, their developed engine can understand the real meaning from the natural query sentences of users for BIM components. To focus on the information collaboration issue of real-world facilities and BIM models, Xie et al. (2019) proposed a solution that matches real-world facilities to BIM data using natural language programming (NLP). However, the proposed system supports only simple sentences. Complex sentences containing verbs and operators are not supported yet. Finally, Preidel et al. (2017) concluded in their study after they examined and assessed two common query languages for processing BIM model information that textual query languages are too complex to be employed by typical end users in the construction industry such as architects and engineers. The research discussed above indicates that the existing IR methods generate irrelevant results in many cases which require time and effort searching useful and professional information that meets the specific needs of the user (Liu et al., 2017; Wu et al., 2019; Xie et al., 2019). It is apparent that the increased use of BIM in practice results in more and more engineers and architects applying data retrieval features. This indeed will make BIM models more developed and more complex. For this reason, the need for methods architects and engineers can easily use to extract relevant information from the models will be higher.

7.2.2 IR from a BIM Model

Very few studies were found which focus on retrieving information from a BIM model. A cloud BIM intelligent IR and representation method was proposed by Lin et al. (2016) using the NLP. However, the IR process was difficult to retrieve multiscale information due to the lack of correlation between different types of documents. A BIM model retrieval method using airborne LiDAR point clouds was proposed by Chen et al. (2017). For efficient model retrieval, point cloud queries and building models are encoded using the roof geometry encoding method. However, this retrieval method uses Airborne LIDAR point cloud as input query, so it is necessary to have matching point cloud files of specific buildings to use this retrieval tool. Another study investigated the use of two visual programming languages (QL4BIM and VCCL) to enhance the effectiveness of IR from BIM model (Preidel et al., 2017). Authors concluded that the use of visual programming languages requires experts to retrieve the required information efficiently. One more study attempted to enhance the IR efficiency is conducted by Gui et al. (2019) who proposed a partial data model retrieval method using IFC to extract the required attributes and then using the local model to complete the IR. The research discussed above indicates that BIM IR process has made some progress, but it is still challenging to retrieve information in a BIM model itself.

7.2.3   BIM Data Retrieval in a BIM Software

There have been a great number of technological advances within the field of AI in the past decade. Among many different AI applications, voice recognition technology has been commercialised and applied in many fields to increase work productivity. Currently, AI has also emerged in the field of AECO, yet, little emphasis has been placed on applying voice recognition, and few studies have attempted to use it in BIM software (Motawa, 2017; Shin et al., 2021). However, many of these attempts faced pitfalls in retrieving BIM data in BIM software even when speech recognition systems were used. One of the main pitfalls is the use of keyword-based search engines such as Voice 360 plugin. Autodesk developed Voice 360 plugin to execute Forge Viewer commands. Using the voice recognition, user can give commands instead of taking inputs from mouse and keyboard. However, Voice 360 plugin requires using command-based keywords and is limited in the keywords it accepts (Kim et al., 2018). The limited number of keywords is a major limitation in keyword-based search engines, it is difficult to find accurate results when the precision and recall values used for search are low (Malve and Chawan, 2015). Another limitation with this method is the words with multiple meanings or multiple words with the same meaning which can lead to irrelevant searching results (Malve and Chawan, 2015). Another retrieval system for searching BIM data was developed by Motawa (2017) used the IFC database and developed an NL-based retrieval system for searching BIM data. This method is limited as well since the IFC schema focuses on the relationship between objects, which means that all the detailed data of an element are not converted to IFC format (Lange et al. 2018). To overcome the limitations of these studies (Shin et al., 2021) developed the BIM automatic speech recognition (BIMASR) framework to enable users asking questions based on NL using their voice and to perform searches and manipulations directly in BIM software. However, in this study, the case studies did not target all BIM model data, but rather the walls within the architecture. Table 7.1 shows relevant studies regarding their focus of study, employed method, data input types and limitations.

7.2.4   BIM and Visual Programming Languages

Recently, parametric design approach has gained prevalent attention within AECO robotic research for the development of custom and experimental workflows (Raspall, 2015; Schumacher, 2009). VPLs have proven its value within general robotics in developing various robotic applications (Steinmetz and Weitschat, 2016; Thomas et al., 2013). One of the key value advantages is simplicity of programming, uncomplicated knowledge encapsulation for non-programmers, efficacy in developing geometry-intensive workflows

Table 7.1 Previous studies for ASR in BIM

Author	Focus of Study	Method	Input Type	Limitations
Shin et al. (2020b)	Develop a framework for ASR-based building IR from BIM software	Introducing BIM to RDBMS module Semantic-based BIM	SQL Query	Relies on Oracle database, not on cloud-based database
Alotto et al. (2020)	Building modelling with AI and Speech recognition for learning purpose	Developed a prototype for automated modelling in BIM using a voice assistant	Natural Language understanding	Focused on parametric modelling using voice assistant
Motawa (2017)	Develop a Spoken Dialogue BIM system to capture building operation knowledge	Integrating cloud-based spoken dialogue system and case-based reasoning BIM system.	Natural Language queries	Only IFC protocol-based knowledge IR from the BIM model
Kim et al. (2018)	IR from BIM and modification for BIM data in Revit	Using the conventional AI technology and the Algorithm-based BIM	Keyword-based commands	User requires knowledge of customised commands

and modular architectures (Peters and Peters, 2013). Even though these advantages are well acknowledged, the application of VPL-based parametric modelling is limited to project-specific scripts, graphs, functions or prototypical workflows within AECO industry (Neythalath et al., 2021). BIM applications have evolved over the last decade to develop several applications supporting modelling, generation of documentation and visualisation (Jay et al., 2009). Recently, other BIM applications, such as parametric design, generation of analysis graphs, and emission estimation, have been developed using specific line code programming applications (Farooq et al., 2014), which were incorporated into BIM software. These programming applications create algorithms that identify several process structures by identifying command lines of instructions (Zamora-Polo et al., 2019). Yet, the high programming skills involved in these applications are challenging for AECO industry practitioners (Collao et al., 2021). To overcome this challenging issue, BIM software development companies have created VP tools.

The VP tools create algorithms based on visual expressions to create process sequence scripts (Seghier et al., 2017). Instead of using the code lines of text programming, these algorithms are represented as process charts. One of the most common VP tools in the AECO industry Dynamo which is associated with Autodesk Revit software (Kensek et al., 2016). Several researchers have adopted Dynamo as a VP tool in their studies. For example, an dynamic workflow was established in Dynamo to combine the linked sensors' maintenance-related information to a cloud-based tool to facilitate efficient communications between the facility management team and IoT companies for intelligent sensor management (Valinejadshoubi et al., 2021). To present the efficient use of BIM information, bespoke queries have been generated using Dynamo to allow searching, isolating and visualising of the resulting elements (Santoni et al., 2021). A sensor integration method was developed by (O'Shea and Murphy, 2020) using Revit and Dynamo to monitor the structural health of existing structures. A procedural model for the design of a road infrastructure was presented by (Biancardo et al., 2020) using Dynamo to adjust the values of the input parameters.

7.2.5   Current State of IR in BIM Model

Currently, BIM is widely used in the AECO industry. IR is the essence of the BIM utilisation. Most of the current research on BIM IR focuses on using IFC schema for data manipulation and IR from BIM model or using keywords for searching. Both methods required experience in BIM commands and coding (Gao et al., 2017; Gui et al., 2019; Matarneh et al., 2019; Motawa, 2017; Shi et al., 2018). Very few research focused on IR in BIM model (Kim, 2018; Motawa, 2017; Shin et al., 2021). In addition, most of existing studies used the natural language-based approach (NL) for IR. To enhance BIM IR process, there is a need for developing a workable IR system capable of retrieving multi-scale information in BIM model by different users with different experience levels in BIM application tools. Furthermore, the advent ASR technology gained more attention in facilitating IR in different fields. Thus, this study aims to utilise the visual language (VL) processing technology (Dynamo) and the automatic voice recognition technology (Alexa) to facilitate information manipulation and retrieval in BIM model for inexperienced BIM users.

7.3   Methods

The objective of this research is to develop workable solution based on integrating AI voice assistant technology into BIM in order to enable users to interact with BIM remotely. Snyder (2019) states that literature review methodology plays an important role to develop a conceptual background, create guidelines and enabling evaluating of existing practices. Therefore, the literature review was employed in this research to evaluate existing solutions-based AI voice assistants to interact with BIM model, then, highlighting deficiencies and gaps in these solutions. In order to understand the interoperability among different tools in the software architecture, a high-level architecture model should be developed (Jaiswal, 2019; Shen et al., 2019). That's why, a high-level architecture of the proposed an automated AI-voice assistant-Based BIM interaction model was developed to check the interoperability and compatibility of AI voice assistant skill and BIM environment. Knepell and Arangno (1993) asserted that the computerised simulation of an existing real-life problem is the best approach to check the validity of a proposed conceptual model. In order to test the created computerised solution, valid and reliable data using a real-life case study should be used to test the applicability, validity and practicability of created platforms (Dasso, 2006).

Figure 7.1 shows the processes of developing AI-voice assistant-based BIM solution including data collection tasks and solution development processes to enable users (1) retrieve information automatically and remotely from the BIM model using any type of device (mobile or tab), (2) requesting BIM platform to develop information/design elements automatically. Amazon Alexa is selected to develop the AI-voice assistant skill given it is a user-friendly platform, as well as, it has a high interoperability level with a wide range of platforms (Kita et al., 2019). Moreover, Autodesk Dynamo is employed to convert CSV files and ask 3D BIM platform to perform the verbally requested task (Cavalliere et al., 2017; Shin et al., 2020b).

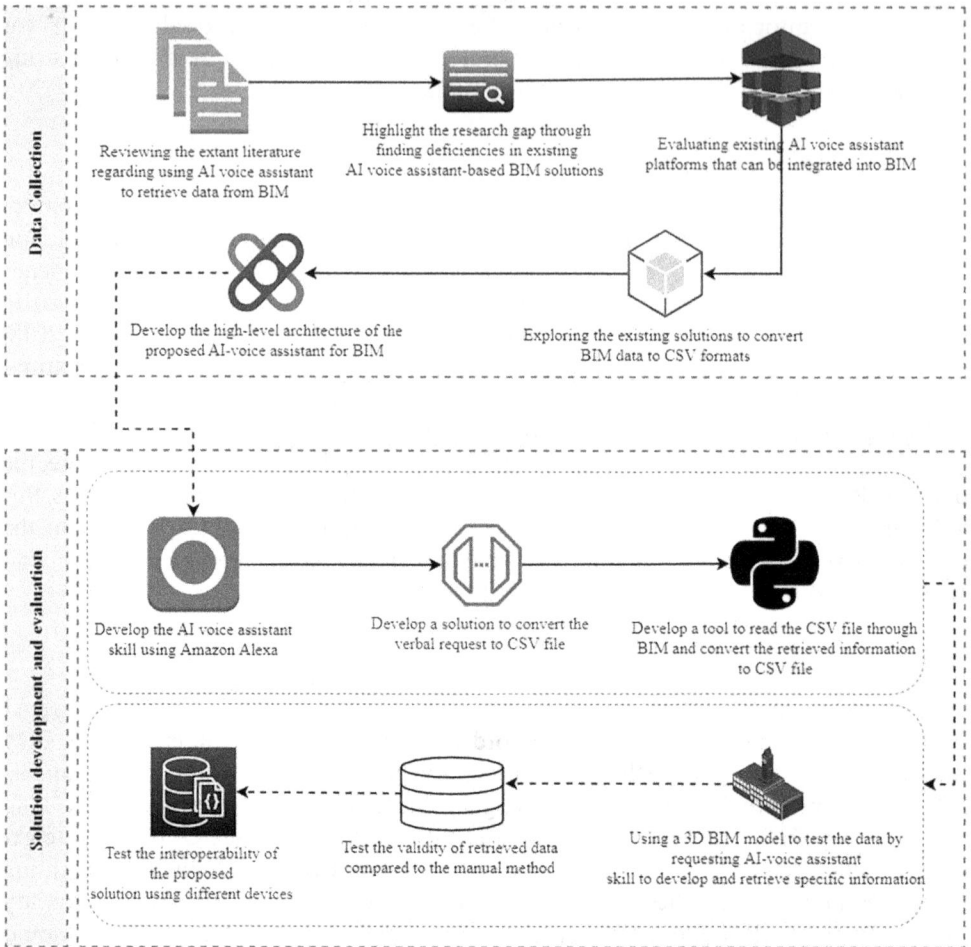

Figure 7.1 Research methods and design.

7.4   Solution Development

The transition to construction 4.0 requires integrating a wide range of technologies such as AI and IoT in order to achieve the aim of minimising human interference to reduce errors, as well as, optimising consuming resources. Adopting BIM in construction industry has been transmitted from the zone of complying standards and using CAD platforms to the zone of automating tasks and full integration of its dimensions through integrating emerging digital technologies into BIM process. In the present research, AI voice assistant is employed to enable BIM users to interact with the information model remotely as well as retrieving data without a need to follow compound steps. In order to develop BIM-based AI voice assistant, three environments should be developed, namely, building AI-voice assistant skill, the mediation environment to link between AI-voice assistant skill and BIM model, tool to read BIM information. Figure 7.2 shows the process of developing mentioned three environments, all components and interrelationships of each environment are presented.

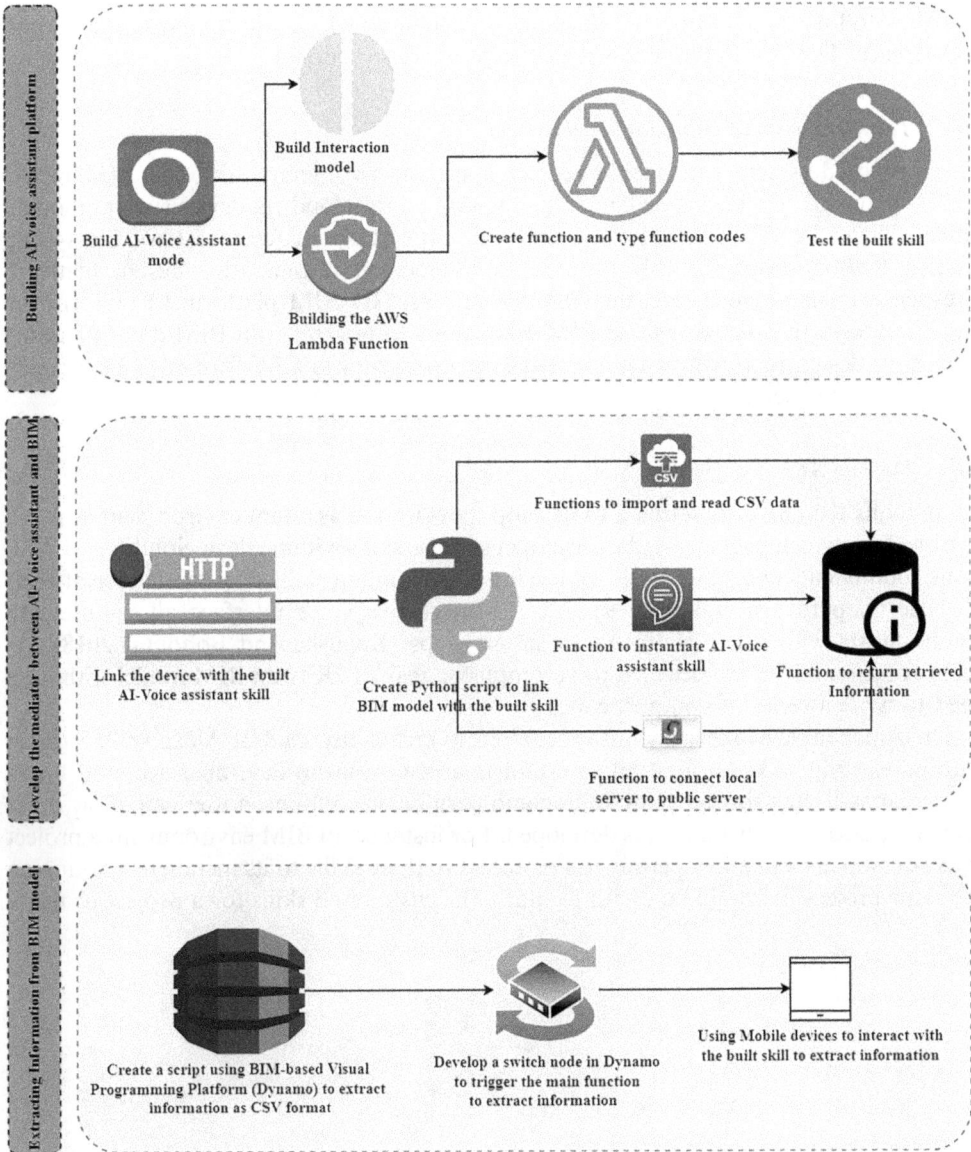

Figure 7.2 Components of AI-voice assistant-based BIM environment.

In order to build the AI-voice assistant environment, the developer should build an interaction model that includes the logic for the skill, as well as, defining the user interface. And developing the functions that AI-voice assistant skill was developed to perform (see Figure 7.2). The built AI-voice assistant skill requires a mediator environment to (1) link between the cloud skill and server, (2) send and read information from the BIM model. Visual programming platform is then needed in order to interact with 3D BIM model (i.e., Revit), this visual programming platform (i.e., Dynamo) will be used to (1) read the CSV file that issued by Ai-voice assistant skill, (2) convert this to an order to 3D BIM platform to perform the task (see Figure 7.1). All mentioned processes will be performed automatically, and they follow the concept of 'Machine to Machine', which will minimise the human interference. This is a step towards the concept of construction 4.

7.4.1   *The Mechanism of Using Proposed System*

This proposed system can enable project managers to interact with the BIM model remotely, the process begins by launching the skill on the cloud via device such as mobile/tablet. Then, the AI-voice assistant skill will convert the request as CSV file—using created Python script for this purpose. Subsequently, a generative design platform (Autodesk Dynamo) will read the CSV file and ask 3D BIM platform to perform the requested task. In order to enable BIM blind users to interact with BIM data, the same generative design platform will convert the retrieved data to CSV format to be verbally returned to the user device (see Figure 7.3).

7.4.2   *Building AI-Voice Assistant Skill*

Alexa Skills Kit (ASK) is selected to develop the AI-voice assistant environment as it is a flexible App, which provides a platform to customise skills within Alexa cloud-based API, skills components and other tools to develop and maintain skills over the course of their lifecycle (Lopatovska et al., 2019). Alexa's interactive voice interface allows users to communicate with other platforms in simple steps (Kepuska and Bohouta, 2018). As such, Amazon Alexa was selected to accomplish a task or IR from BIM model without a need to write a series of complex codes.

The Amazon developer account is required to create any kind of Alexa skill. Sign up with an existing Amazon account or establish a new Amazon developer account. For a particular skill, by using account linking and permissions options, a user can allow other users to access their custom skills developed. For instance, in BIM environment, a project lead or company can allow permission or access to these skills to its members working on the same project. It would be useful to utilise the customised skills for a project or team.

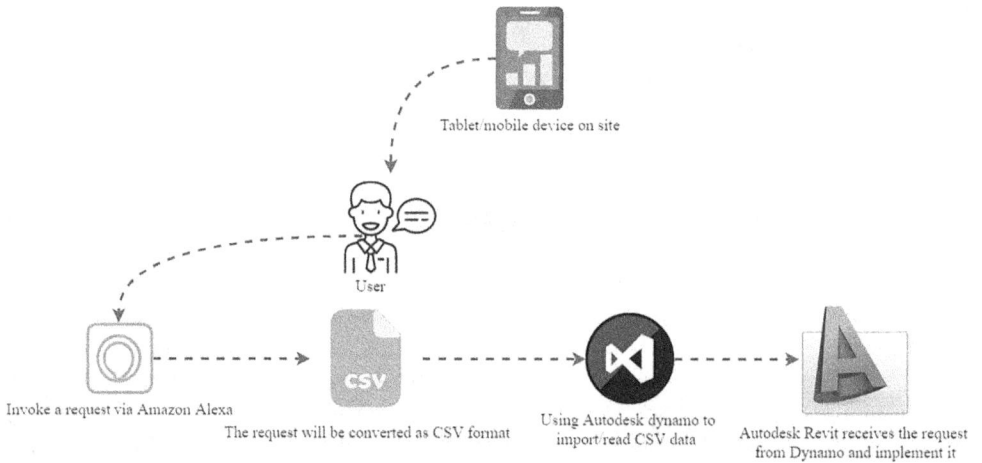

Tablet/mobile device on site

User

Invoke a request via Amazon Alexa

The request will be converted as CSV format

Using Autodesk dynamo to import/read CSV data

Autodesk Revit receives the request from Dynamo and implement it

Figure 7.3 The mechanism of adopting AI-voice assistant for BIM data management.

7.4.2.1 Developing a Custom Skills-Based Amazon Alexa

To build a custom skill, an interaction model is first defined by the developer according to the skill set he desires to build. The user's spoken input translates to the requests, or intents, that your skill can handle, according to your particular interaction model. When a user talks to your skill, Alexa determines the underlying intent using your interaction model and passes it to your skill application logic.

Amazon provides many ways to start creating a skill. These are custom models or existing built models. Pre-built models are the developed packages of intents and utterances that you can be selected to create a new skill in the Amazon's developer console. The custom model is the most appropriate model to be fabricated according to special cases.

The first step is to develop the interaction model, which refers to a set of intents, sample utterances, slots and dialogue models. User requests by which the skill can be invoked are represented by intents. The user should define the intents for skill's specific features and can also use built-in intents for basic actions like stopping, cancelling and asking for assistance.

Slots as a part of Skill Alexa development are optional parameters or variables that can be added to intents. Each slot in the interaction model gets a name and a slot type. In order to improve recognition accuracy, the slot type comprises a list of representative values for the slot. For widely used arguments like dates and numbers, The user can use a developed slot type or create a bespoke slot for the case.

Figure 7.4 depicts a voice command with relevant items of interaction model. Explained with an example, if the user wants to create a room schedule using Alexa, the user can utter command like 'Alexa, ask project 101 to create a room schedule'. Here, the word 'Alexa' is the wake word to enable the Alexa service, 'ask' as the launch or action word, 'project 123' is the invocation name and 'to create a room schedule' is a complete utterance consisting of a slot 'room'. Here slot can be a 'door', or 'window', which can be known as variable for a request for the schedule.

For example, it can be 'name of floor finish on 3rd level', in case user wants to enquire about floor materials. Within these utterances you can provide slots which are variables to define a specific request like in the utterances 'name of floor finish on 3rd level', third floor can be put as slots with the option provided in the interface. It can be second or fourth floor. There is no need for the developer to create separate intents for these variables.

Figure 7.5 shows the process of developing the AI-Voice assistant platform to interact with BIM using Amazon Alexa platform. The first figure shows the development steps that have been adopted to develop the skill including customising the skill's intent and slot, building the interaction model and subsequently, the endpoint for skill was selected as 'HTTPS' to deploy your code on the public IP address. When the web server port (public IP address) is provided by port connector, make sure to insert it in this section.

Figure 7.6 shows the created 'intents' to test the proposed solution, it can be seen that the intent is named 'CreateSchedule' and three utterances were added to ask Alexa to

Figure 7.4 Sample voice command.

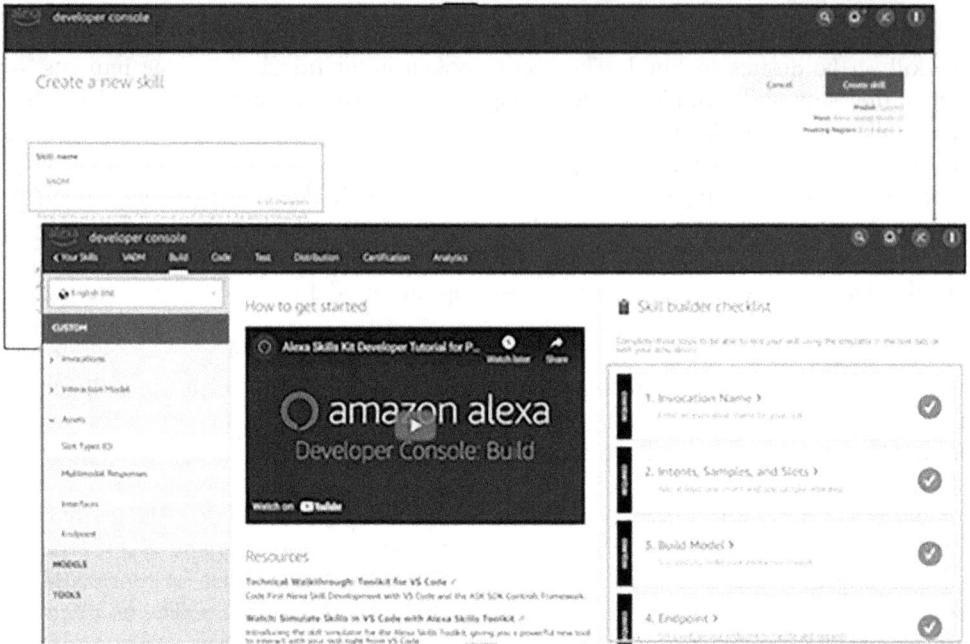

Figure 7.5 The developed Alexa skill interface.

run the request, namely, 'to create a room schedule', 'to make a room schedule', 'please create room schedule'. All these utterances are coded in a JSON file as presented in Figure 7.5. Subsequently, codes should be deployed, and the BIM user should request the AI voice assistant skill to perform the task by calling the coded utterances.

In order to connect the local server that contains the BIM model to a public IP address which would connect to Alexa Cloud Services, a port forwarding application Ngrok is employed. Figure 7.7 shows how the 'https' links were copied from Ngrok to be used in the endpoint. Ngrok received JSON data, processed it, and responded with JSON data from the Python script, that was used for this demonstration. After inserting the endpoints as provided by Ngrok application, it saves the endpoint in Alexa skill builder.

7.4.2.2 *Building Mediator Environment*

The python script was used to link four primary data platforms: the port connector, the Flask-ask extension, the Alexa input from AWS, and the CSV file. To deal with requests, responses, and modifications to a CSV file, the Python script was developed. The script allowed data to be redirected to an HTTPS server, which was read by Alexa voice.

Figure 7.8 shows the developed Python code, which is structured to (1) import CSV, as well as flask-ask initialisation, (2) switch function to read a CSV file, (3) a function to Create Schedule via manipulating CSV and generates a response message using Alexa's intent, (4) Port for Ngrok: To connect local server to public server.

A 'Switchboard_Test' CSV file was created (see Figure 7.8) to reflect changes made by Python script if the affirmative signal is received from the intent created in the Alexa skill

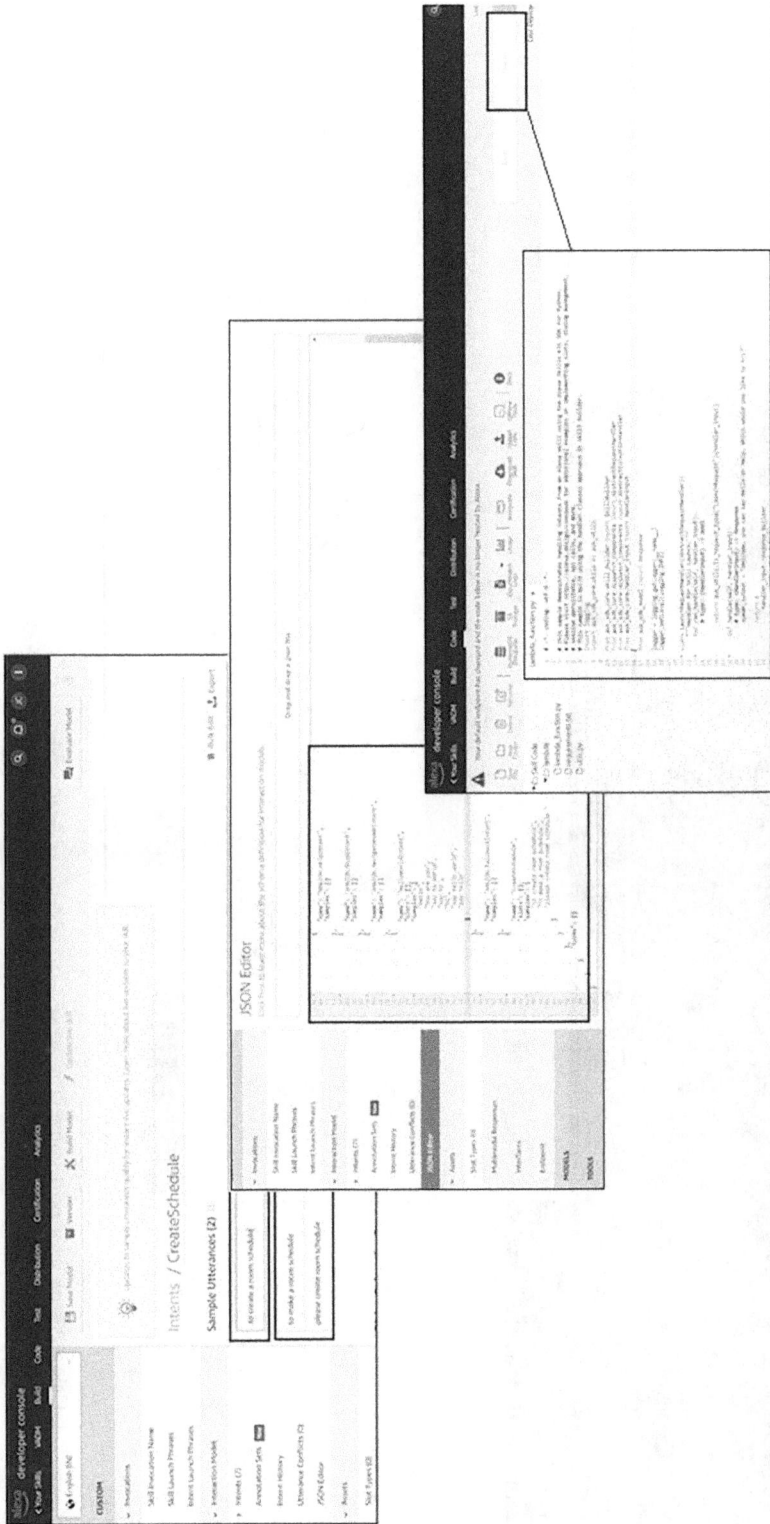

Figure 7.6 JSON structure for the intent and deploying the created skill.

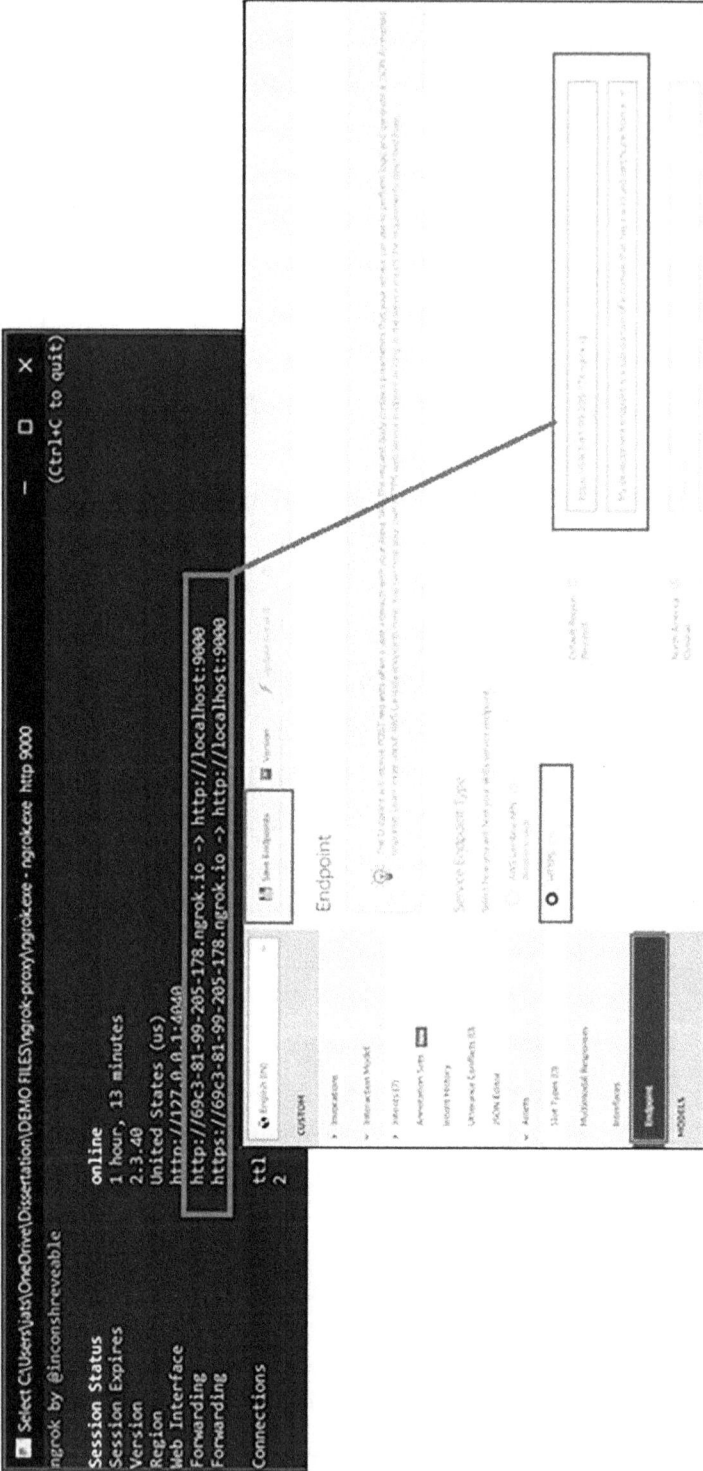

Figure 7.7 Using Ngrok as a port connector.

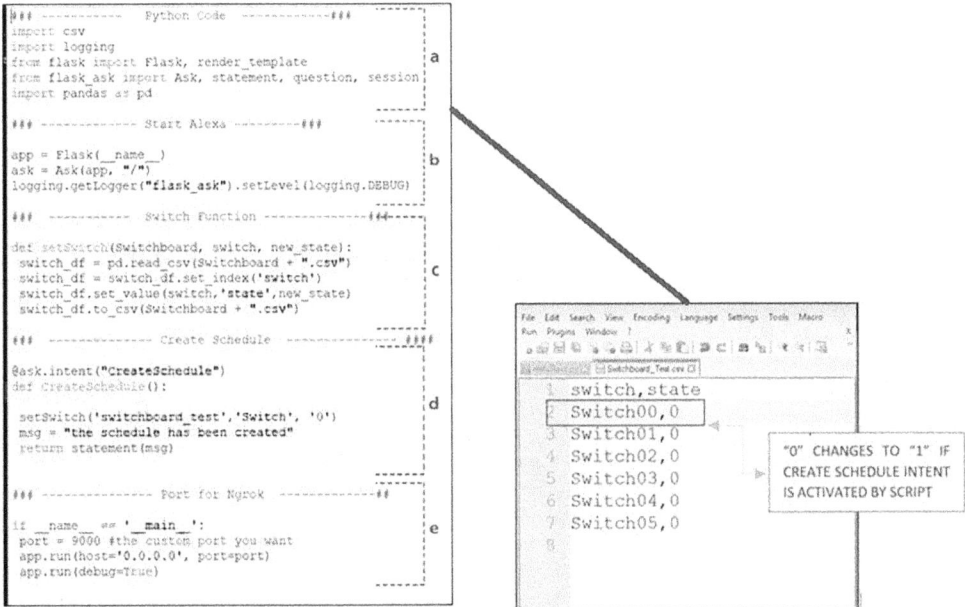

Figure 7.8 Python codes to interact between AI-voice assistant skill and BIM model.

builder. This change is reflected by the change of State 0 to 1 in Switch00. '1' here is read by Dynamo script as True which turns on the Switch Node in the Dynamo (later explained in this demonstration).

7.4.2.3 AI-Voice Assistant to BIM

The data can be retrieved from the BIM model as a CSV file, however, there is a need for a mediator to link between AI-voice assistant-created skill (Amazon Alexa) and the 3D BIM model (Revit). The visual programming software like Dynamo can be used in Revit environment to manage data in BIM model through CSV files and it can also manipulate data in model files after the changes are reflected in CSV file. The Visual programming language script can be designed for both ways depending on the task.

A Dynamo script consists of nodes which are designed to read data from CSV file and ask 3D BIM model platform to perform the task. In this 'Proof of Concept' demonstration, a room schedule in Revit model is automatically created by using a Dynamo script linked to a CSV file. Figure 7.9 shows the structure of the created Dynamo script to read the CSV file from Amazon Alexa and ask 3D BIM platform (Revit) to perform the task, subsequently, return the results to the Alexa skill to enable user to get results remotely. The created scrip will be performed automatically, Dynamo script is structured as follows:

- Import CSV file – The CSV file that was created as 'Switchboard_Test' is linked with this script.
- Switch – This custom node acted as a switch function to trigger 'create schedule' node.

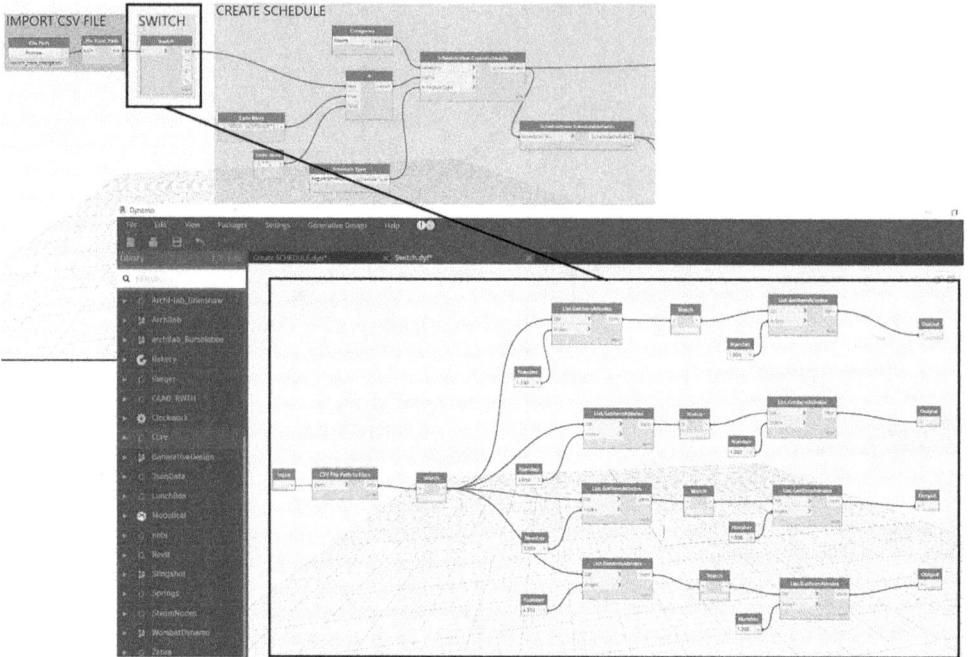

Figure 7.9 Extracting BIM information using Dynamo visual programming.

- Create Schedule – Script designed for the automated creation for room schedule in the Revit model if the switch is activated by Python script.

7.5 'Proof of Concept' Demonstration and Evaluation

Given that the process requires an integration between different platforms, therefore, there is a need for an interoperability and workflow flowchart as presented in Figure 7.10 to enable users to understand the developed compound application.

Figure 7.11 shows the steps that have been conducted in order to develop a 'schedule for a room'. The process begins by uttering Amazon Alexa to 'Create a room schedule', then, the developed Python script will convert the request to a CSV file, which is readable by Autodesk Dynamo. The created Dynamo script includes a node that works as a switch function to trigger creating a 'room schedule'. A 3D BIM model as presented in Figure 7.11 is used to check the 'proof of concept' validity, scalability and workability, the results show that a 'room schedule' is developed properly including all parameters that mentioned in the Dynamo script – around 43 parameters were included according to the nature of the extracted information such as area, volume, number, level, etc. As such, the created script can deal with all BIM model sizes given there are 43 parameters listed in the Dynamo script.

Given, the proposed solution is mainly designed to facilitate extracting information remotely, as well as, enabling those who has disability to read to be able to interact with BIM model through sending and receiving information verbally. Therefore, the created system-based AI voice assistance enables to convert the produced information as CSV file, then users can hear information 'row by row', regardless of the location.

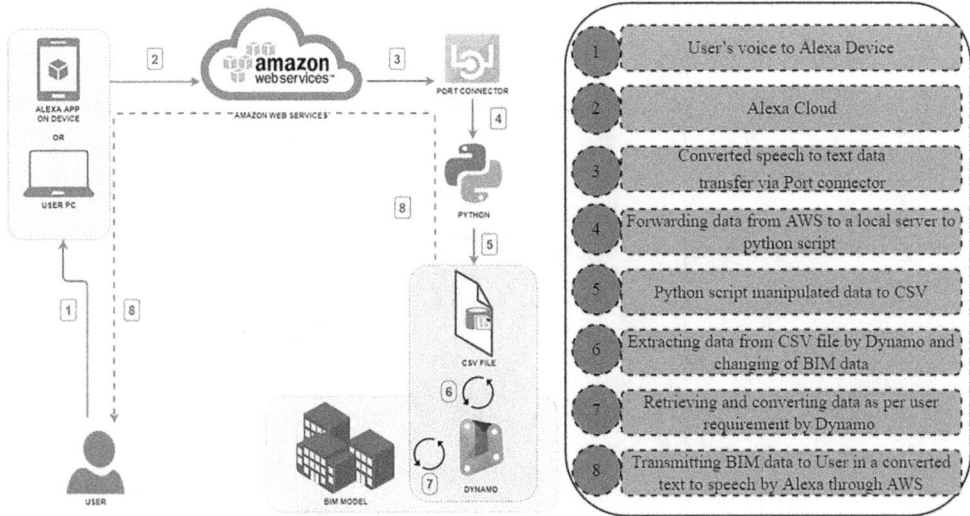

Figure 7.10 The model interoperability and workflow.

Different users can interact with the created system, as well as information can be then sent to all stakeholders. As such, this tool can facilitate decision-making on-site in case that users are not able to access the 3D BIM platform, or the user is not familiar with extracting data from BIM models. All project team will be able to get benefits from BIM data, regardless of their personal capabilities and knowledge in using BIM platform, particularly, the management team since project management tasks requires retrieving data from design repetitively.

7.6   Discussion, Significance and Limitation

This study came to respond to an intensive demand by researchers to enable the real-time retrieval of data from BIM models remotely such as (El Ammari and Hammad, 2019; Singh et al., 2011; Svalestuen et al., 2017). The proposed AI voice assistant-based BIM system is developed to enable users to interact with BIM models (3D, 4D and 5D) remotely and verbally. This can save time while working on site, as well as, enabling non-BIM users to retrieve information from the design without a need for BIM knowledge and competency. Moreover, users with reading disability will be able to interact with BIM model without a need to follow complex and compound steps and information will be also provided verbally.

The proposed system provides data in different formats to enable re-using it for multiple purposes, for example, the data is initially retrieved as CSV file, which is a flexible format that can be converted to Excel format, therefore, this can facilitate integrating the retrieved data into other platforms for further process (i.e., cost estimation).

Even though there are a few attempts to use AI voice assistant technology to interact with BIM data such as systems that were developed by Motawa (2017), however, the BIM model was not linked with AI voice assistant platform to enable users to interact verbally. Another research was conducted by Shin et al. (2020a; 2020b) to develop a framework based on integrating use of SQL and Dynamo to retrieve BIM data verbally, however, the

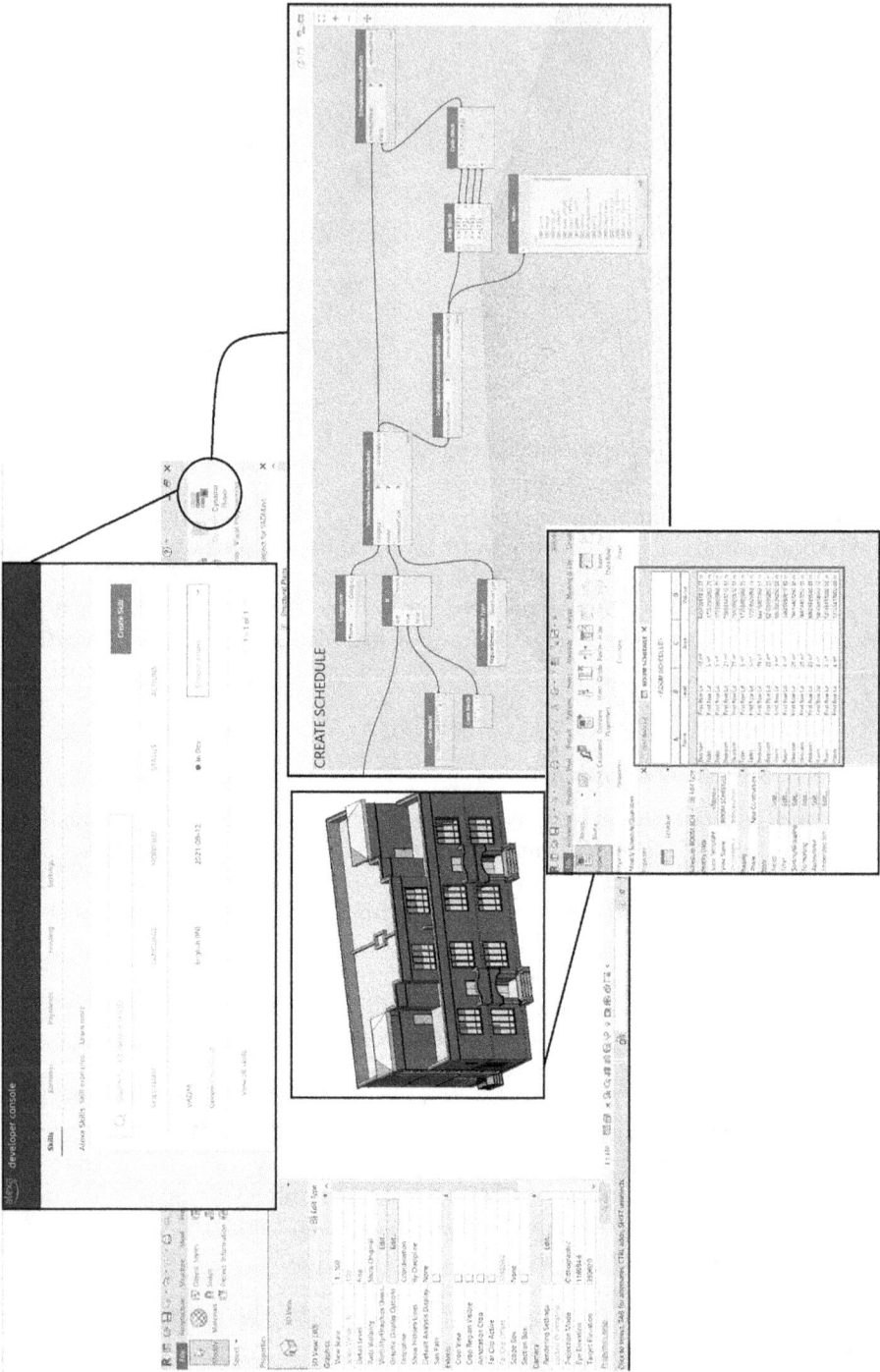

Figure 7.11 Demonstration of creating 'room schedule-based AI-voice assistant'.

system is lacked in conducting the process in an integrated environment, as well as, there is no proposed interface AI voice assistant skill to enable users to send queries easily. As such, the proposed platform dealt with raised limitations in relevant studies by (1) using Amazon Alexa to develop a user-friendly interface to interact with BIM model, (2) developing an automated mediator environment to read CSV data from AI voice assistant skill and request 3D BIM model to perform the task, (3) the proposed solution enables to provide the data verbally, which enables users with disability to deal with retrieved data.

The proposed methodology of performing tasks remotely can be considered as a point of departure for an automated generative design-based AI voice assistant to develop a complete design using standard utterances. This can be a comprehensive solution for blindness and vision loss users to develop design and management documents using BIM platforms given users will be able to send request to develop tasks, then either receiving a verbal message that the task is performed successfully or retrieving the information verbally.

Recently, there are several attempts to reducing the consumed time to perform repetitive tasks while developing BIM models (Daria and Philipp, 2019; Sheikhkhoshkar et al., 2019), however, most attempts rely on using Autodesk Dynamo to develop scripts to interact with Autodesk Revit to develop complex/repetitive tasks automatically for specific design elements and the user should manually perform the task. The proposed system using AI voice assistant which enables the full automation of the process, which will significantly foster the transformation to construction 4.0.

The simplicity of performing automation tasks is an essential to ensure the scalability of the solution, therefore, the design of such solutions considers this properly by developing a user-friendly AI-voice assistant skill using Amazon Alexa platform, as well as, automating all further tasks including switch node from CSV file to visual script in Autodesk Dynamo, as well as extracting data and convert it as CSV file to resend it to the amazon Alexa platform automatically.

Given, the proposed system interacts with BIM model through sensing queries as CSV file to the user device that holds the BIM model, however, this system can be extended to interact with the BIM data in a cloud such BIM 360 to enable multiple users to retrieve information from BIM models. Moreover, the proposed solution is tested to develop a project schedule and retrieve the created schedule verbally, however, in order to globalise the benefits of using this solution, multiple dynamo scripts should be developed to enable performing and retrieving a wide range of tasks/data according to each project needs.

7.7   Conclusion

This chapter presented a novel solution to retrieve information from BIM models remotely, as well as, converting retrieved texts and numbers to verbal data to enable BIM users with a blindness disability to interact with BIM environment. The proposed solution works automatically, in other words, the BIM user can request from the AI voice assistant created skill – Amazon Alexa – to perform a specific task, then, the verbal request will be converted as a CSV file, which, will be readable by Autodesk Dynamo platform. Finally, the data will be retrieved from 3D BIM model and then converted as CSV file to be presented verbally to the BIM user via the Amazon Alexa Skill (AI-Voice assistant skill).

The proposed 'Proof of Concept' is the first of its kind to use AI-voice assistant platform such as Amazon Alexa in conjunction with generative design tool – Autodesk dynamo – to interact with 3D BIM model to either retrieving information of an existing model or performing tasks to develop a design. In case that the user utilises the system to retrieve information, the AI voice assistant skill can read the CSV data from the Autodesk Dynamo platform.

This solution is a step towards construction 4.0 given AI and IoT tools and concepts are integrated into BIM to automate the entire process, as well as, minimising the human interference to reduce errors and wastes.

The created 'Proof of Concept' is validated using an illustrative case study to retrieve information from a 3D BIM model (developing a room schedule) in order to check its validity, workability and scalability, the results proved that the entire process is robust including the AI-voice assistant interface-based Amazon Alexa, the mediator environment that converts the verbal request to a CSV file to be readable by the generative design platform (Autodesk Dynamo) and the retrieved information is correct and reliable.

Even though the proposed solution works automatically without a need for a human to link between its processes, however, the Dynamo script to perform specific tasks, as well as, retrieving specific types of data requires should be developed for each case, which requires to develop a wide range of scripts to enable the scalability of the proposed solution. In order to enable multiple users to interact with BIM model without a need to connect the device to the AI-voice assistant skill, the created solution could be extended to link between the Amazon Alexa skill and the BIM model in a cloud such as BIM 360.

References

Abdirad, H. (2021). Managing digital integration routines in engineering firms: cases of disruptive BIM cloud collaboration protocols. *Journal of Management in Engineering, 38*(1), 05021012.

Alizadehsalehi, S., Hadavi, A., & Huang, J. C. (2020). From BIM to extended reality in AEC industry. *Automation in Construction, 116*, 103254.

Alotto, F., Scidà, I., & Osello, A. (2020). Building modeling with artificial intelligence and speech recognition for learning purpose. Proceedings of 7th EDULEARN20 Conference.

Beach, T., Petri, I., Rezgui, Y., & Rana, O. (2017). Management of collaborative BIM data by federating distributed BIM models. *Journal of Computing in Civil Engineering, 31*(4), 04017009.

Biancardo, S. A., Capano, A., de Oliveira, S. G., & Tibaut, A. (2020). Integration of BIM and procedural modeling tools for road design. *Infrastructures, 5*(4), 37.

Cavalliere, C., Dell'osso, G. R., & Leogrande, M. A. (2017). Automatic workflow for 4D-BIM based modelling. Progress (es), Theories and Practices: Proceedings of the 3rd International Multidisciplinary Congress on Proportion Harmonies Identities (PHI 2017), October 4–7, 2017, Bari, Italy. CRC Press, 405.

Chen, H., Hou, L., Zhang, G. K., & Moon, S. (2021). Development of BIM, IoT and AR/VR technologies for fire safety and upskilling. *Automation in Construction, 125*, 103631.

Chen, Y. C., Lin, B. Y., & Lin, C. H. (2017). Consistent roof geometry encoding for 3D building model retrieval using airborne LiDAR point clouds. *ISPRS International Journal of Geo-Information, 6*(9). 10.3390/ijgi6090269

Collao, J., Lozano-Galant, F., Lozano-Galant, J. A., & Turmo, J. (2021). BIM visual programming tools applications in infrastructure projects: A state-of-the-art review. *Applied Sciences (Switzerland), 11*(18). 10.3390/app11188343

Croft, W. B. (2019). The importance of interaction for information retrieval. *SIGIR*, 1–2.

Daria, S., & Philipp, S. (2019). Revit Dynamo: Designing objects of complex forms. Toolkit and process automation features. *Architecture and Engineering, 4*(3), 30–38.

Dasso, A. (2006). *Verification, validation and testing in software engineering.* IGI Global. ISBN: 1591408539

El Ammari, K., & Hammad, A. (2019). Remote interactive collaboration in facilities management using BIM-based mixed reality. *Automation in Construction, 107,* 102940.

Elghaish, F., Abrishami, S., & Hosseini, M. R. (2020). Integrated project delivery with blockchain: An automated financial system. *Automation in Construction, 114,* 103182.

Farooq, M. S., Khan, S. A., Ahmad, F., Islam, S., & Abid, A. (2014). An evaluation framework and comparative analysis of the widely used first programming languages. *PLoS One, 9*(2), e88941. 10.1371/journal.pone.0088941

Gao, G., Liu, Y.-S., Wang, M., Gu, M., & Yong, J.-H. (2015). A query expansion method for retrieving online BIM resources based on Industry Foundation Classes. *Automation in Construction, 56,* 14–25. 10.1016/j.autcon.2015.04.006

Gao, G., Liu, Y. S., Lin, P., Wang, M., Gu, M., & Yong, J. H. (2017). BIMTag: Concept-based automatic semantic annotation of online BIM product resources. *31,* 48–61. 10.1016/j.aei.2015.10.003

Gui, N., Wang, C., Qiu, Z., Gui, W., & Deconinck, G. (2019). IFC-based partial data model retrieval for distributed collaborative design. *33*(3). 10.1061/(ASCE)CP.1943-5487.0000829

He, R., Li, M., Gan, V. J., & Ma, J. (2021). BIM-enabled computerized design and digital fabrication of industrialized buildings: A case study. *Journal of Cleaner Production, 278,* 123505.

Hossain, M. U., & Ng, S. T. (2018). Critical consideration of buildings' environmental impact assessment towards adoption of circular economy: An analytical review. *205,* 763–780. 10.1016/j.jclepro.2018.09.120

Ivanov, R. (2017). An approach for developing indoor navigation systems for visually impaired people using Building Information Modeling. *Journal of Ambient Intelligence and Smart Environments, 9*(4), 449–467.

Jaiswal, M. (2019). Software architecture and software design. *International Research Journal of Engineering and Technology (IRJET) e-ISSN,* 2395-0056.

Jay, G., Hale, J., Smith, R., Hale, D., Kraft, N., & Ward, C. (2009). Cyclomatic complexity and lines of code: Empirical evidence of a stable linear relationship. *JSEA, 2,* 137–143. 10.4236/jsea.2009.23020

Kensek, K., Ding, Y., & Longcore, T. (2016). Green building and biodiversity: Facilitating bird friendly design with building information models. *Journal of Green Building, 11*(2), 116–130. 10.3992/jgb.11.2.116.1

Kepuska, V., & Bohouta, G. (2018). Next-generation of virtual personal assistants (Microsoft Cortana, Apple Siri, Amazon Alexa and Google Home). 2018 IEEE 8th annual computing and communication workshop and conference (CCWC), IEEE, 99–103.

Khalili, A. (2021). An XML-based approach for geo-semantic data exchange from BIM to VR applications. *Automation in Construction, 121,* 103425.

Kim, H. (2018). A research on utilization of speech recognition artificial intelligence for efficient BIM works. *Designs in Convergence Study, 17*(1), 1–16. 10.31678/SDC.68.1.

Kim, K., Cho, Y., & Kim, K. (2018). BIM-driven automated decision support system for safety planning of temporary structures. *Journal of Construction Engineering and Management, 144*(8), 04018072.

Kita, T., Nagaoka, C., Hiraoka, N., & Dougiamas, M. (2019). Implementation of voice user interfaces to enhance users' activities on Moodle. 2019 4th International Conference on Information Technology (InCIT). IEEE, 104–107.

Knepell, P. L., & Arangno, D. C. (1993). Simulation validation: A confidence assessment methodology. IEEE Computer Society Press.

Lange, H., Johansen, A., & Kjærgaard, M. B. (2018). *Evaluation of the opportunities and limitations of using IFC models as source of building metadata.* Paper presented at the Proceedings of the 5th Conference on Systems for Built Environments, Shenzen, China. 10.1145/3276774.3276790

Lin, J. R., Hu, Z. Z., Zhang, J. P., & Yu, F. Q. (2016). A natural-language-based approach to intelligent data retrieval and representation for cloud BIM. *Computer-Aided Civil and Infrastructure Engineering, 31*(1), 18–33. 10.1111/mice.12151

Liu, H., Liu, Y. S., Pauwels, P., Guo, H., & Gu, M. (2017). Enhanced explicit semantic analysis for product model retrieval in construction industry. *IEEE Transactions on Industrial Informatics, 13*(6), 3361–3369. 10.1109/TII.2017.2708727

Liu, P. (2021). Mountain rainfall estimation and BIM technology site safety management based on Internet of things. *Mobile Information Systems, 2021.*

Liu, Z., & Jiang, G. (2021). Optimization of intelligent heating ventilation air conditioning system in urban building based on BIM and artificial intelligence technology. *Computer Science and Information Systems* (00), 27–27.

Lopatovska, I., Rink, K., Knight, I., Raines, K., Cosenza, K., Williams, H., Sorsche, P., Hirsch, D., Li, Q., & Martinez, A. (2019). Talk to me: Exploring user interactions with the Amazon Alexa. *Journal of Librarianship and Information Science, 51*(4), 984–997.

Malve, A., & Chawan, P. (2015). A comparative study of keyword and semantic based search engine. *International Journal of Innovative Research in Science, Engineering and Technology, 4*(11), 11156–11161.

Matarneh, S., Danso-Amoako, M., Al-Bizri, S., Gaterell, M., & Matarneh, R. (2019). BIM-based facilities information: Streamlining the information exchange process. *Facilities, 17*(6), 1304–1322. 10.1108/JEDT-02-2019-0048

Matarneh, S. T., Danso-Amoako, M., Al-Bizri, S., Gaterell, M., & Matarneh, R. T. (2020). BIM for FM: Developing information requirements to support facilities management systems. *Facilities, 38*(5–6), 378–394. 10.1108/F-07-2018-0084

Motawa, I. (2017). Spoken dialogue BIM systems – An application of big data in construction. *Facilities, 35*(13–14), 787–800. 10.1108/F-01-2016-0001

Neythalath, N., Søndergaard, A., Kumaravel, B., Bærentzen, J. A., & Naboni, R. (2021). Applying software design patterns to graph-modelled robotic workflows. *Automation in Construction, 132.* 10.1016/j.autcon.2021.103965

Oraee, M., Hosseini, M. R., Edwards, D., & Papadonikolaki, E. (2021). Collaboration in BIM-based construction networks: A qualitative model of influential factors. *Engineering, Construction and Architectural Management.* 10.1108/ECAM-10-2020-0865

O'Shea, M., & Murphy, J. (2020). Design of a BIM integrated structural health monitoring system for a historic offshore lighthouse. *Buildings, 10*(7), 131.

Pan, Y., & Zhang, L. (2021). A BIM-data mining integrated digital twin framework for advanced project management. *Automation in Construction, 124*, 103564.

Peters, B., & Peters, T. (2013). Inside SmartGeometry: Expanding the architectural possibilities of computational design. John Wiley & Sons.

Plaß, B., Prudhomme, C., & Ponciano, J. (2021). BIM on artificial intelligence for decision support in E-Health. *The International Archives of Photogrammetry, Remote Sensing and Spatial Information Sciences, 43*, 207–214.

Preidel, C., Daum, S., & Borrmann, A. (2017). Data retrieval from building information models based on visual programming. *Visualization in Engineering, 5*(1). 10.1186/s40327-017-0055-0

Raspall, F. (2015). A procedural framework for design to fabrication. *Automation in Construction, 51*, 132–139. 10.1016/j.autcon.2014.12.003

Santoni, A., Martín-Talaverano, R., Quattrini, R., & Murillo-Fragero, J. I. (2021). HBIM approach to implement the historical and constructive knowledge. *The Case of the Real Colegiata of San Isidoro (León, Spain). 2021, 12*(24), 17. 10.4995/var.2021.13661

Schumacher, P. (2009). Parametricism: A new global style for architecture and urban design. *Architectural Design, 79*(4), 14–23. 10.1002/ad.912

Seghier, T. E., Lim, Y. W., Ahmad, M. H., & Samuel, W. O. (2017). Building envelope thermal performance assessment using visual programming and BIM, based on ETTV requirement of green mark and GreenRE. *International Journal of Built Environment and Sustainability*, 4(3). 10.11113/ijbes.v4.n3.216

Sheikhkhoshkar, M., Rahimian, F. P., Kaveh, M. H., Hosseini, M. R., & Edwards, D. J. (2019). Automated planning of concrete joint layouts with 4D-BIM. *Automation in Construction*, *107*, 102943.

Shen, G., Arnold, N., Benes, S., Jarosz, D., Johnson, A., Stasic, D., Usmani, I., & Veseli, S. (2019). High-level application architecture design for the APS upgrade. 17th International Conference on ACC and Large Exp Physics Control Systems, 1436–1440.

Shi, X., Liu, Y.-S., Gao, G., Gu, M., & Li, H. (2018). IFCdiff: A content-based automatic comparison approach for IFC files. *Automation in Construction*, *86*, 53–68. 10.1016/j.autcon.2017.10.013

Shin, S., Lee, C., & Issa, R. R. (2020a). Advanced BIM platform based on the spoken dialogue for end-user. International Conference on Computing in Civil and Building Engineering. Springer, 123–132.

Shin, S., Lee, C., & Issa, R. R. (2020b). Framework for automatic speech recognition-based building information retrieval from BIM software. *Construction Research Congress 2020: Computer Applications*. American Society of Civil Engineers Reston, VA, 992–1000.

Shin, S., Lee, C., & Issa, R. R. A. (2021). Advanced BIM platform based on the spoken dialogue for end-user. *98*, 123–132. 10.1007/978-3-030-51295-8_11

Singh, V., Gu, N., & Wang, X. (2011). A theoretical framework of a BIM-based multi-disciplinary collaboration platform. *Automation in Construction*, *20*(2), 134–144.

Snyder, H. (2019). Literature review as a research methodology: An overview and guidelines. *Journal of Business Research*, *104*, 333–339.

Sresakoolchai, J., & Kaewunruen, S. (2022). Integration of building information modeling (BIM) and artificial intelligence (AI) to detect combined defects of infrastructure in the railway system. In *Resilient Infrastructure*. Springer, 377–386.

Steinmetz, F., & Weitschat, R. (2016). Skill parametrization approaches and skill architecture for human-robot interaction. 2016 IEEE International Conference on Automation Science and Engineering (CASE), 21–25 Aug. 2016, 280–285. 10.1109/COASE.2016.7743419

Svalestuen, F., Knotten, V., Lædre, O., Drevland, F., & Lohne, J. (2017). Using building information model (BIM) devices to improve information flow and collaboration on construction sites.

Thomas, U., Hirzinger, G., Rumpe, B., Schulze, C., & Wortmann, A. (2013). A new skill based robot programming language using UML/P Statecharts. 2013 IEEE International Conference on Robotics and Automation, 6–10 May 2013, 461–466. 10.1109/ICRA.2013.6630615

Valinejadshoubi, M., Moselhi, O., & Bagchi, A. (2021). Integrating BIM into sensor-based facilities management operations. *Journal of Facilities Management, ahead-of-print*(ahead-of-print). 10.1108/JFM-08-2020-0055

Wang, J., Gao, X., Zhou, X., & Xie, Q. (2021). Multi-scale information retrieval for BIM using hierarchical structure modelling and natural language processing. *Journal of Information Technology in Construction: ITcon*, *26*, 409–426. 10.36680/j.itcon.2021.022

Wu, S., Shen, Q., Deng, Y., & Cheng, J. (2019). Natural-language-based intelligent retrieval engine for BIM object database. *108*, 73–88. 10.1016/j.compind.2019.02.016

Wu, S., Hou, L., & Zhang, G. K. (2020). Integrated application of BIM and eXtended reality technology: A review, classification and outlook. International Conference on Computing in Civil and Building Engineering. Springer, 1227–1236.

Xie, Q., Zhou, X., Wang, J., Gao, X., Chen, X., & Chun, L. (2019). Matching real-world facilities to building information modeling data using natural language processing. *IEEE Access*, *7*, 119465–119475. 10.1109/ACCESS.2019.2937219

Zamora-Polo, F., Martínez Sánchez-Cortés, M., Reyes-Rodríguez, A. M., & García Sanz-Calcedo, J. (2019). Developing project managers' transversal competences using building information modeling. *Applied Sciences*, *9*(19), 4006.

Zhang, J., & El-Gohary, N. M. (2015). Automated information transformation for automated regulatory compliance checking in construction. *Journal of Computing in Civil Engineering*, *29*(4), B4015001.

8 BIM-VR Integration for Infrastructure Asset Management

A Systematic Review

Rhijul Sood and Boeing Laishram

Department of Civil Engineering, IIT Guwahati, Assam, India

8.1 Introduction

Infrastructure includes all physical assets, equipment, and facilities of interconnected systems and critical service providers who deliver linked commodities and services to the public to enable, maintain, or improve social living conditions (Weber, Staub-Bisang and Alfen, 2016). The majority of countries' economic progress is dependent on infrastructure development, and technology advances have accelerated construction in many countries. An infrastructure project's lifecycle includes several stages, including conception, design, construction, operation and maintenance (O&M), and demolition. The financial management of the O&M phase is the true challenge when a project is over, as it accounts for 50–70% of total annual facility running costs and 85% of total lifespan costs, respectively (Lewis, Riley and Elmualim, 2010; Love et al., 2015). Repairing defective structural components found in O&M might add up to 5% to the total cost of building a project (Boukamp and Akinci, 2004). The greater the complexity of the buildings, the more frequently O&M difficulties arise, which in turn drives up operational costs. Multiple stakeholders are involved in the construction industry's data-intensive operations, where generated data must be coordinated and kept up-to-date throughout the project's life cycle (Salem, Samuel and He, 2020). From the point of view of building owners and investors, it is crucial to address concerns linked to O&M if the project is to advance over the long term. Moreover, it has become evident that the current infrastructures need to be maintained and rehabilitated in order to keep operating at the level of efficiency for which they were originally intended. For this reason, government agencies must devote substantial resources to infrastructure maintenance, renewal, and repair (Schraven, Hartmann and Dewulf, 2011). As per (Flintsch and Medina, 2001), systematic infrastructure asset management (IAM) is the best way to balance growing needs, maintaining existing infrastructures, and limited resources.

IAM aims at effectively managing physical assets throughout their lifecycle through a deliberate and methodical approach (Akofio-Sowah et al., 2014). IAM systems are commonly used to aid in the evaluation of an asset's existing state, forecasting of its future degradation, selection of maintenance and repair procedures, enhancement of its state following repair, asset priority, and allocation of funds. (Elhakeem and Hegazy, 2012; Al-Kasasbeh, Abudayyeh and Liu, 2021). Most asset managers rely on data that they have compiled themselves, which is usually duplicate, unreliable, and lacking in significance (Aldowayan, Dweiri and Venkatachalam, 2020). The management of infrastructures has been transformed in recent years as a result of the broad

DOI: 10.1201/9781003408949-8

use of digital and novel technologies with the aim of providing a more energy-efficient, pleasant, sustainable, and lucrative built environment (Bolpagni, Gavina and Ribeiro, 2022). This transition has allowed infrastructures to share data and engage with its users (Casini, 2022) resulting in benefiting asset management from the application of novel artificial intelligence and information technologies (Molinari, Paganin and Talamo, 2002).

The Architecture, Engineering and Construction (AEC) and O&M sectors are digitalizing to improve IAM, assist trade personnel, and create a more efficient working environment. One such technology is Building Information modeling (BIM) which construction industry was driven to adopt to overcome its issues related to low productivity and less sustainable practices (Froese, 2010). It spreads over various maturity levels (Level 0 to 3) and dimensions (3D, 4D, 5D, 6D, and 7D) with each dimension having a particular role in infrastructure's lifecycle (Sood and Laishram, 2021) such that intelligent three-dimensional (3D) models are becoming frequently available to support design, construction, project delivery, and facility management (Hardin and McCool, 2015). While BIM excels at offering product-oriented solutions, it falls short of meeting the need for a shared platform for all stakeholders to communicate and exchange information in order to achieve collaborative decision-making (Alreshidi, Mourshed and Rezgui, 2017). In most cases, switching back and forth between 2D drawings and 3D models frequently is necessary for gaining a thorough comprehension of an issue, determining its impact, and providing feedback. With the increasing use of BIM, viewing virtual BIM content for identifying maintainability concerns can be challenging, and maintainability issues can easily be hidden within the 3D model (Liu and Issa, 2013). Because of this, BIM has recently undergone a number of enhancements that make it better at providing process-based solutions, especially during the O&M stage of an infrastructure. These include BIM with Geographic Information System (GIS), Cloud BIM, BIM with the Internet of Things (IoT), and BIM with Virtual Reality (VR) and Augmented Reality (AR) (Salem, Samuel and He, 2020).

Among the technologies mentioned above, the applicability of VR to facilitate maintainability-focused design reviews with O&M professionals has not been thoroughly investigated. The asset managers have even less time (often only a day or two) to thoroughly examine all of the project paperwork during the O&M phase (Kim et al., 2018). As a result, they are unable to evaluate the project documentation in detail due to time constraints. These limitations of maintainability-focused design evaluations encourage the investigation of employing an immersive platform like VR, with its established advantages of enhancing spatial comprehension and shortening the time required to make decisions. There is a growing body of literature on VR technology and its use in O&M phase in the industry, although it is still confined in scope and short on details. Integrating BIM with VR technology is a relevant area of study that needs more research, practical application, and software adaptation (Sampaio, 2018).

Hence, this study aims to address the research gap by assessing the usability of BIM-VR for O&M practitioners to provide maintainability-focused design inputs. Specifically, this chapter addresses the following research questions:

1 What are the existing BIM and VR technologies in construction industry?
2 How BIM and VR integration can help the asset managers during O&M stage?

8.2   Research Methodology

In the present study, we drew on a systematic literature review (SLR) as our research technique because it offered the best overview of the current state of knowledge on the topic (Sartor et al., 2014). This is because systematic reviews have been shown to be a useful scientific study approach for evaluating, synthesizing, and communicating the findings and implications of a large body of research articles on a certain topic (Booth, 2008). While traditional literature reviews tend to be broad and basic, SLRs employ rigorous criteria, conduct in-depth critical analyses, and synthesize all relevant literature in a given field to provide a coherent whole that sheds light on the topic under discussion (Cronin, Ryan and Coughlan, 2008). The approach for the review was divided into three steps, which are explained next.

8.2.1   Planning the Review

During the planning phase, a review protocol is created that details the research questions that will be answered, the population (or sample) that will be studied, the search strategy that will be used to locate relevant studies, and the criteria that will be used to select which studies will be included and which will be excluded (Davies and Crombie, 1998). After identifying the research problem and its formulation, determine the proper data sources and keywords to develop the search strategy (Tjahjono et al., 2010). Hence, a literature search in Scopus, Web of Science, and EBSCO host was then used to select the articles to be included in the review. These databases were selected because of the range of their coverage as they contain the great majority of peer-reviewed journals devoted to construction automation, megaprojects, and project management. After finalizing the keywords, the search was undertaken using building blocks, a strategy that is popular among literature reviewers (Booth, 2008). Because of this, the study topic/problem is broken down into its constituent parts, each of which is then connected to the others via Boolean operators like "AND" and "OR." Snowballing and trial-and-error searches yielded the following keywords for the current study: "*BIM, Building Information Modeling, Virtual Reality, VR, Infrastructure asset management, IAM, * Asset Management.*"

8.2.2   Conducting the Review

The initial step that was performed was a keyword search in the aforementioned three databases. All three databases were searched, and a total of 433 articles were found that were relevant to the topic from 2006 till February 2023. After eliminating duplicates and restricting our focus to publications written in English, we were left with 348 articles. As a second stage, we used the database's built-in filtering and selection tools to critically evaluate the titles and abstracts of relevant publications. Articles that focused on only one area of study were disqualified, as were those that did not relate directly to the building industry; and finally, papers that addressed the integration of BIM and Virtual Reality (VR) were accepted. After applying this backward filter, a total of 143 papers were selected. Finally, a list of 75 papers was generated for the systematic review study after exporting the filtered publications and doing a full-text manual analysis (Figure 8.1). The list of inclusion and exclusion criteria is shown in Table 8.1. The vast majority of this research can be found in peer-reviewed publications focusing on topics like digital and automated construction, construction safety, IT in construction, cultural heritage, and

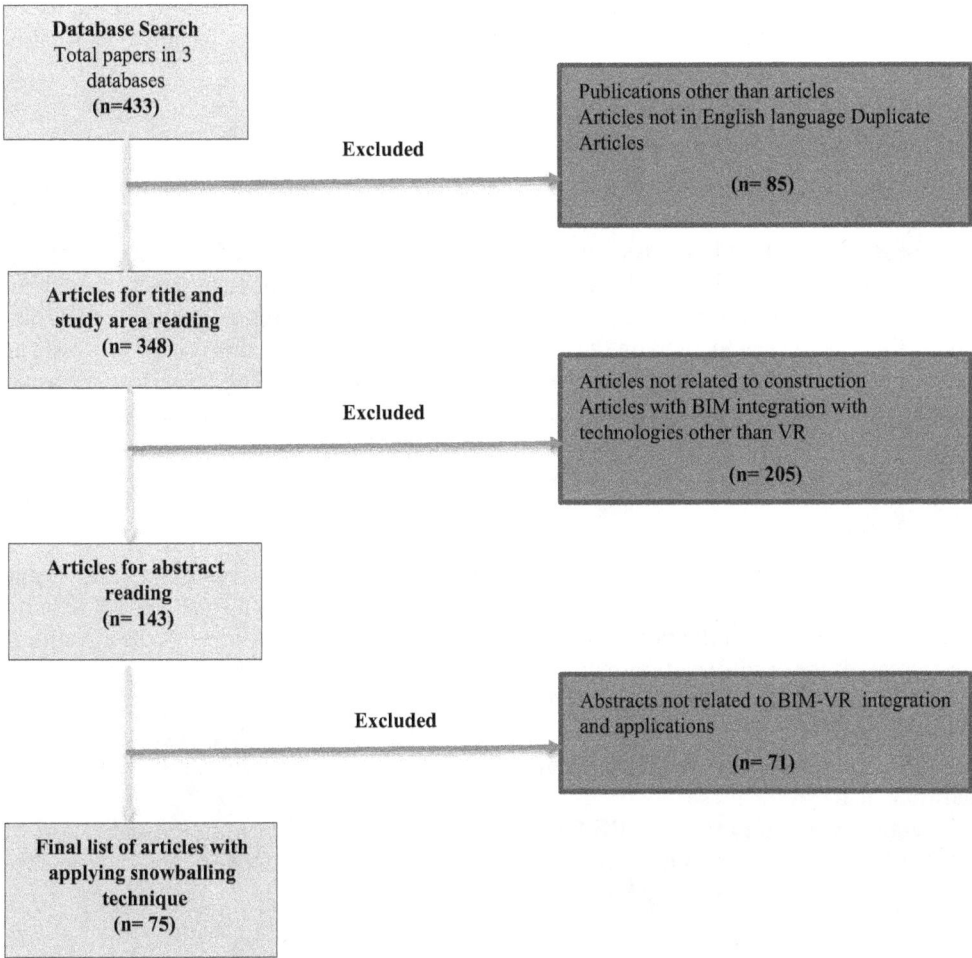

Figure 8.1 Flowchart of systematic literature review process.

Table 8.1 Inclusion and exclusion criteria for SLR

Inclusion Criteria	Exclusion Criteria
Articles published between 2006 and 2022	Any publication before the year 2006
Articles published in well-known databases: Scopus, Web of Science and EBSCO host	Articles published in nonacademic databases
Fully accessed articles	Non-fully accessed articles
Articles published in academic journals	Books and articles published in online sites and gray literature (conferences, master's theses, doctoral dissertations, textbooks, newsletters, working papers, project deliverables, PowerPoint presentations, etc.)
Only international peer-reviewed articles	Articles that were not peer-reviewed
Articles in area of construction industry	Articles focusing on unrelated domain appeared (e.g., nursing, medicine, applied economics, etc.)
Articles highlighting BIM and VR for infrastructure management	Articles highlighting BIM for only design stage
Articles written in the English language	Articles written in any language other than English

sustainable development. It was also found that there has been a rise in the number of studies examining the integration of BIM and VR with studies on Augmented Reality (AR) and Mixed Reality (MR).

8.2.3   *Analyzing and Reporting the Review*

In this step, we did a descriptive analysis of the papers to offer publishing statistics by year, research area, and country. In the following section, we use visual aids like charts and tables to present the results of the descriptive analysis. A uniform unit of analysis is used in inductive content analysis, and classes are derived from the contents themselves through an iterative coding process, category construction, and re-examination (Seuring and Gold, 2012). Since the review focused on the text of the entire manuscript, this is what was used as the basic unit of analysis. Multiple rounds of back-and-forth reading of each publication were necessary to extract, cluster, and ultimately classify the codes contained inside.

8.3   Analysis and Results

8.3.1   *Descriptive Analysis*

From 2006 on, as can be seen in Figure 8.2, the descriptive analysis includes a look at how publications have been spread out by year. The assessment covers a time span of around 16 years and focuses on the proliferation of BIM and VR around the world as a result of the increased use of information technology (IT) in the construction industry. The graph shows that the number of publications in the field of IAM was low up until 2017, but has since increased in tandem with the rise of technological innovation, the incorporation of more advanced tools and techniques into the construction industry, and the increased maturity of BIM.

The most common areas for BIM-VR integration include design and visualization of structures throughout the lifecycle (27 papers), to visualize the safety of construction sites

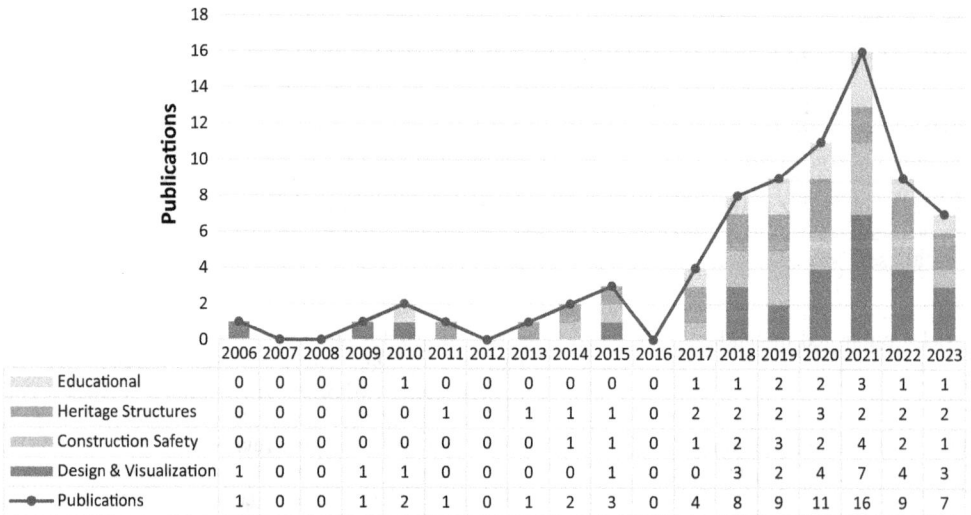

	2006	2007	2008	2009	2010	2011	2012	2013	2014	2015	2016	2017	2018	2019	2020	2021	2022	2023
Educational	0	0	0	0	1	0	0	0	0	0	0	1	1	2	2	3	1	1
Heritage Structures	0	0	0	0	0	1	0	1	1	1	0	2	2	2	3	2	2	2
Construction Safety	0	0	0	0	0	0	0	0	1	1	0	1	2	3	2	4	2	1
Design & Visualization	1	0	0	1	1	0	0	0	0	1	0	0	3	2	4	7	4	3
Publications	1	0	0	1	2	1	0	1	2	3	0	4	8	9	11	16	9	7

Figure 8.2 Year-wise publications.

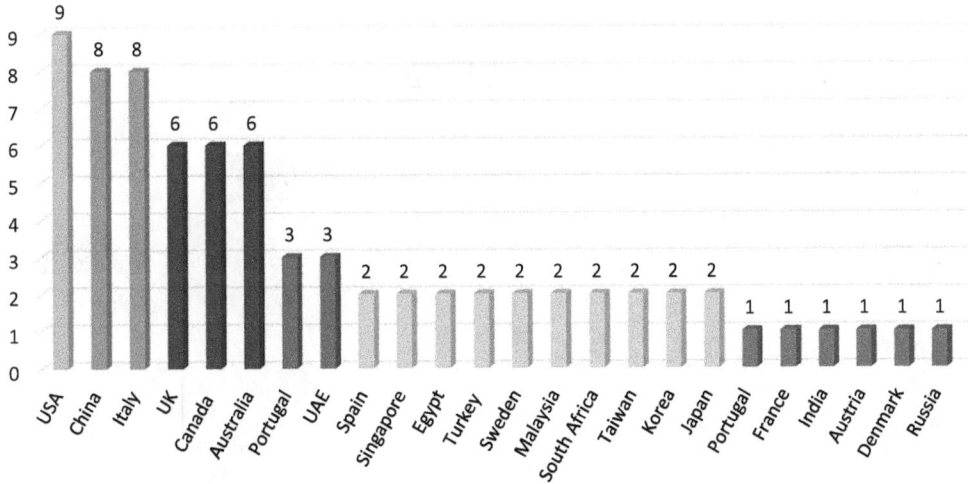

Figure 8.3 Country-wise publication.

and structures (17 papers), analysis of heritage structures (19 papers), and educational purposes (12 papers). Similarly, Figure 8.3. shows the country-wise distribution of research articles. The maximum number of articles were published in the USA (nine papers) followed by China (eight papers), and Italy (eight papers). UK, Canada, and Australia had six papers each followed by Portugal (three papers), UAE (three papers), and others.

8.3.2 Content Analysis

This section presents the findings of the content analysis, which includes a thorough evaluation of all relevant research papers. BIM technology, VR, and BIM-VR integrated applications and potential challenges are all discussed in the following Sections 8.3.2.1, 8.3.2.2, and 8.3.2.3 respectively.

8.3.2.1 Building Information Modeling (BIM)

BIM is basically a digital representation of a structure that helps the AEC stakeholders in effective designing, construction, O&M, and demolition. BIM is the process of creating building information in a centralized virtual model that can be controlled to allow interoperable and reusable data exchange throughout project's lifecycle (Azhar, 2011). Hence, BIM can be called as a technology as well as a process (Mahalingam, Yadav and Varaprasad, 2015). BIM aids in the integration of the scattered AEC industry by reducing waste, facilitating better communication and cooperation, and increasing output. The model may be analyzed for different kinds of input, and used to provide additional insight. The model's abundance of information proves useful in making decisions and enhancing the design's creation and distribution processes. BIM has various dimensions and maturity levels that define the level of detail and development of a model. The maturity levels vary from level 0 to 3 whereas the dimensions vary from 2D to 7D BIM (Figure 8.4). For the creation of n-D/BIM applications, the usage of BIM data in the development of numerous tasks is taken into consideration (Sampaio, 2018).

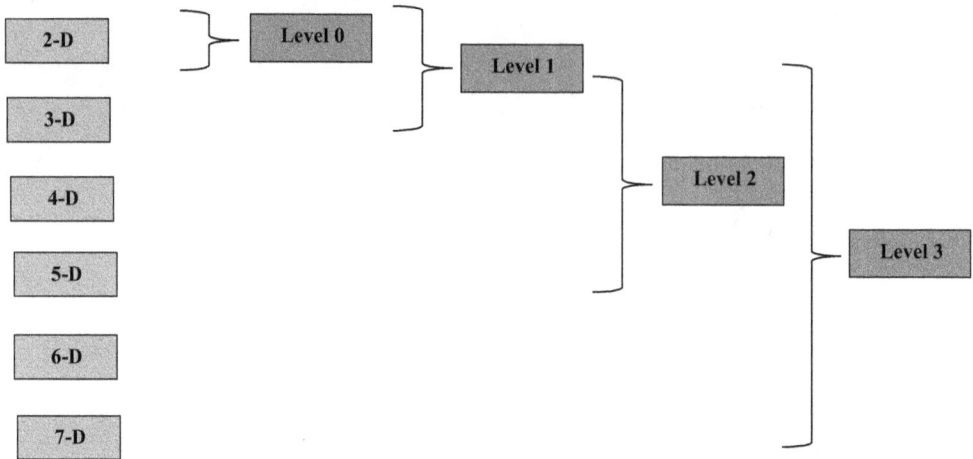

Figure 8.4 BIM dimensions and maturity levels.

The 3D BIM model refers to the modeling process (x, y, and z coordinates) of all disciplines (architecture, civil, and MEP – mechanical, electrical, and plumbing), and because it comprises parametric items, it helps to explain the inter-connections and model attributes. A 4D BIM model is one that allows for a visual simulation of the building construction process and its management. The 4D BIM associates groups of 3D objects with their corresponding schedules, so that users may evaluate how well their actual workflow matches their ideal one (accommodates alterations throughout time and permits the lasting implementation of plans) (Sampaio, 2018). When it comes to 5D BIM, the model is connected to the estimation of costs (management of the budget according to the amount of resources used at each stage of construction) (Bruno et al., 2022). The 6D BIM model helps in determining the performance of infrastructure's energy and sustainability assessment through Life cycle Analysis (LCA) from design to demolition stage (the estimation of energy use throughout the project and then monitoring, measuring, and verifying of energy use by the occupants) (Kivits and Furneaux, 2013) and the 7D/BIM model is being created to facilitate in building maintenance and to the implement the provision of distinct facilities required by the structure while it is occupied (Sampaio and Simões, 2015).

8.3.2.2 *Virtual Reality (VR)*

The word "virtual reality" (VR) refers to a computer-generated, 3D, immersive world that a user may move around in and interact with in the real-time (feedback from actions is given without noticeable pause) (Whyte, 2002) meaning to explore digitally modeled architecture by providing a spatial-visual depiction of the product's design. Similarly, VR is defined as computer-generated simulations of 3D objects that appear genuine and require physical human engagement (Dionisio and Burns III, 2013). The user is immersed in a setting through the use of VR technology, providing the user with a sense of presence in a location at which they are not actually present. It is possible to have a non-immersive experience with VR content by using only a desktop or tablet PC. On the other hand, an immersive experience can be achieved by utilizing various VR technologies including head-mounted display (HMD) such as Gear VR or Oculus Rift, hologram projections; Smart Reality and Visidraft apps; PrioVR full or upper body suit; and CAVE platform (Kreutzberg, 2015). VR-HMD screens may facilitate better pre-construction communication between architects and engineers. The HMD immerses the user in the project model, allowing them to better grasp new design concepts and evaluate alternatives developed with BIM modeler tools and techniques. Similar to the thrill one receives from using an arcade game or other gaming device, donning a set of HMDs and disconnecting from the real world may be exciting and thrilling to certain people.

Since the 1980s, multiple efforts were made in order to develop and bring VR technology to the masses (Zaker and Coloma, 2018). VR attempts to improve interactivity by stereoscopic display, tracking head, body, and hand gestures, and allowing manipulation of scene objects (Paes and Irizarry, 2018). As a digital duplicate of the final product, virtual reality has the potential to enable AEC professionals to experience project concepts prior to their construction (Zaker and Coloma, 2018). Another advantage of employing VR at various stages of project development and construction is its ability to engage people and increase collaboration which can result in faster decision-making. The participants in a cooperation or project presentation session may find that many of the sessions are tiresome and uninteresting, thus decreasing decision-making time significantly (Majumdar, Fischer and Schwegler, 2006). In this regard, the VR technology helps to make the presentation more interesting and appealing to various stakeholders. As a result, VR for design review sessions has been found to improve user concentration on the review task and knowledge of issues under investigation (Bhonde, Zadeh and Staub-French, 2022). VR technologies have been shown to be beneficial in a number of areas related to construction, including design, project schedule control, site layout optimization, O&M of built facilities (including historic structures), accelerated site training and protection, coordination with communication with involved parties from owner down to workers, construction safety training for a wide range of on-site participants and workers and lower project costs (Behzadi, 2016; Fu and Liu, 2018; Alizadehsalehi, Hadavi and Huang, 2020). The acceptance of VR use in construction field has been growing and therefore, incorporating this technology has the potential to vastly enhance the entirety of the various construction phases. VR technology can improve BIM methodology, as it allows interaction with 3D/BIM models in two essential ways:

- One of the most common types of 3D visualization is called a **"walkthrough,"** and it allows the user to explore the building from various perspectives whether they are inside or outside;
- **Consulting data** which is basically ability to retrieve information that has been centralized in a BIM model, specifically data associated with parameters that comprise the parametric objects utilized in the modeling process.

VR-BIM is a multimedia technique that allows users to interact with digital objects, giving them a sense of being physically present in an improved digital environment (Biocca, Kim and Levy, 1995). AEC management professionals have noticed that VR technologies make it easier for clients to see ideas earlier, hence reducing material prices and the number of personnel required for projects. There are many other benefits of integrating VR with BIM such as enhanced collaboration among stakeholders (Majumdar, Fischer and Schwegler, 2006; Fernando, Wu and Bassanino, 2013), better spatial understanding of virtual prototypes in comparison to 2D drawings and 3D

models (Paes, Kim and Irizarry, 2017), the anticipation of design decisions and identification of issues otherwise not possible (Bassanino et al., 2010; Dunston et al., 2011),
and prediction of human-building interactions that can provide designers with reliable
user behavior data (Bassanino et al., 2010; Adi and Roberts, 2014; Heydarian et al.,
2015; Kuliga et al., 2015). The benefits of BIM-VR integration are themed under
various categories (Table 8.2) such as design and visualization, educational, Operation
& Management (O&M), heritage structures, and construction health and safety. There
are more than 50 VR software and hardware tools available right now that can redefine
the way BIM is implemented and interpreted. To better integrate VR tools with BIM
software few software tools and plugins are available such as Autodesk 360; viewer
Revit plugin; Enscape Revit plugin; Augment; Stingray; Covise; web3D; EON Icube
(Kreutzberg, 2015).

8.3.2.3   *BIM and VR Integration for IAM*

Infrastructure management consists of O&M phase which involves a wide range of services, skills, methodologies, and instruments to guarantee that they work as intended over
their lifetime and reduce energy consumption and upkeep costs at the urban and building
scale (Rondeau, Brown and Lapides, 2006). Common O&M tasks include inspections,
repairs, and replacements; energy management of the constructed asset; emergency
management; service and use adjustments and relocations; safety and security measures;
waste management (hazardous and non-hazardous); and information and communications technology (ICT) (Rondeau, Brown and Lapides, 2006). As we have discussed in the
previous section, there were few studies related to BIM-VR integration in the O&M stage.
Several software companies have requested that BIM be integrated with VR plugins in
order to promote the development of n-D BIM activities. BIM linked to VR brings a
virtual digital duplication of a final product. With this, AEC professionals experience the
project design before it is built and simulate the on-site tasks (Zaker and Coloma, 2018).
As a first step, the primary foundation of BIM-VR integration is cooperation, but the
ability to examine data while walking through the building dramatically improves the
usage of BIM in design, construction, maintenance, and management (Bhonde, Zadeh
and Staub-French, 2022). One of the fundamental concepts of BIM is to involve project
stakeholders in the early stages of design, and VR can be an effective medium for this. A
VR tour can be used in a BIM model to examine, from an IAM viewpoint, the maintenance schedule (7D BIM model) or from a project team perspective, the design and
construction review (4D/BIM model), thereby facilitating decision-making. This particular section discusses how this integration can be helpful in managing the infrastructure
assets through 7D BIM and VR technology. From the perspective of facility manage-
ment/asset management or maintenance, a VR tour of a 7D BIM program can help with
decision-making. By fusing a VR device's use over a BIM model with a physical
inspection of the site itself, facility managers can gain a fresh perspective on their work
(He and Liu, 2022). VR has the potential to enhance the ease with which project managers and construction employees may obtain virtual information on the existing project.
At any point before, during, or after a physical site inspection, inspectors can access and
examine the BIM model and its multitude of virtual data. Addressing construction difficulties in real time and visualizing internal assets to assist in installation and planning
reduces costs and improves efficiency. Since the data can be viewed, discussed, and even
altered while the team and client are inside a seemingly constructed build, integrating

Table 8.2 Various BIM-VR application areas

BIM-VR Application Area	Description	Authors
Design and Visualization	BIM and VR design evaluation for construction mockups.	(Bassanino et al., 2010; Dunston et al., 2011; Paes and Irizarry, 2018)
	4D BIM planning and scheduling method in the virtual project planning system & constructability analysis meeting.	(Tallgren, Roupé and Johansson, 2021; Boton, 2018)
	Design simulation and communication system based on BIM and VR for transportation infrastructure (DSC-BV-TI system).	(Hao, Zhang and Zhao, 2021)
	VR integrated collaboration workflow to perform a clash-detection in MEP systems via simulation.	(Zaker and Coloma, 2018)
	A case study to increase ventilation in a cluster of modular homes in Singapore's urban region and to aid in making better decisions during the planning stages of the construction project.	(Gan, Liu and Li, 2022)
	Using a VR-based strategy on a BIM-based digital twin to gain superior knowledge and firsthand experience of the circular economy (CE) in building construction.	(O'Grady et al., 2021)
	Using BIM-enabled VR technology, design possibilities are assessed in terms of cost and sustainability.	(Kamari, Paari and Torvund, 2021)
	Integration of laser scanning and VR into a BIM workflow for enhanced project collaboration.	(Dinis et al., 2020)
	Examining how IoT, RFID, AR/VR, drones, Laser Scanning, BIM, and others reduce infrastructure project variances.	(Noruwa, Arewa and Merschbrock, 2020; Froese, 2010; Srivastava et al., 2022; Safikhani et al., 2022)
	Building a database-supported VR/BIM-based communication and simulation (DVBCS) system that combines VR, game engines, and BIM for the design of healthcare facilities.	(Lin et al., 2018)
	Designing a BIM-based lighting design feedback (BLDF) prototype to improve lighting design and energy savings for future occupants.	(Natephra et al., 2017)
	This study used IFC to analyze the envelope shape of amorphous structures to develop alternatives and evaluate energy performance during design for sustainability.	(Choi and Lee, 2023)
	This study created a system that integrates BIM with VR so that clients may adjust and immediately inspect house styles in immersive VR during design phase and decide if the modifications match their needs, which can accurately and quickly estimate cost.	(Wang and Tung, 2023)

(Continued)

Table 8.2 (Continued)

BIM-VR Application Area	Description	Authors
	The research proposes a framework of an automated construction progress monitoring system that incorporates BIM, Digital Twin, and extended reality (VR, AR, and MR) technologies to build, acquire, generate, analyze, manage, and visualize construction progress data, information, and reports.	(Alizadehsalehi and Yitmen, 2023)
Educational	This study looked at the use of cutting-edge technologies including AR, VR, and BIM in architectural education, and developed an integrated model to be utilized in a third-year "Building Construction Project" course.	(Seyman Guray and Kismet, 2023)
	This study established an immersive pedagogical framework that can improve students' AEC knowledge and skills while also fostering a more engaging teaching and learning environment.	(Hajirasouli et al., 2023)
	VR-BIM implementation throughout the curriculum of the construction management undergraduate Program.	(Ghanem, 2022)
	To use big data and artificial intelligence (AI) to derive useful insights from construction industry occurrences.	(Seyman Guray and Kismet, 2022)
	To evaluate BIM-into-VR applications in literature and real life by measuring students' learning performance using the integrated definition function (IDEF) paradigm.	(Alizadehsalehi, Hadavi and Huang, 2021)
	Future construction professionals depend on the quality and depth of undergraduate construction education, which prioritizes student-centered learning, BIM, sustainability, VR, project management, and lean construction (LC).	(Aliu and Aigbavboa, 2021)
	VR in BIM offers a unique, supportive environment that fosters active and collaborative learning in simulated real-world scenarios.	(Wong et al., 2020)
	Desktop-based VR, immersive VR, 3D game-based VR, and BIM-enabled VR have been embraced for CEET (construction engineering education and training) over time.	(Wang et al., 2018)
Operation & Management (O&M)	O&M practitioners reported that VR eased use and boosted trust in their inputs since it can allow maintainability-focused design.	(Bhonde, Zadeh and Staub-French, 2022)
	VDC and digital twin using BIM coupled with AR and VR can give real-time monitoring and prediction of any physical structure at any stage of work.	(Casini, 2022)
	The synergies between BIM, the Internet of Things (IoT) and VR for developing an immersive application for real-time monitoring of thermal comfort conditions.	(Shahinmoghadam, Natephra and Motamedi, 2021)

Category	Description	Citation
	The digital twins' technique is applied to all stages of building construction using physical construction site entity (PCSE) modeling, digital twins' virtual body (DTVB) modeling, and virtual and real interaction (VRI) modeling.	(Wang et al., 2022)
	Integrating BIM and VR to analyze and discuss project management for assembled buildings.	(He and Liu, 2022)
Heritage Structures	This study summarizes the inspection of three research case studies at archeological, architectural, and infrastructure scales to raise awareness of Roman-built heritage using Unmanned Aerial Vehicle (UAV) and representing them through 3D digital techniques like HBIM (historic building information modeling), scan-to-BIM approach, and interactive forms like virtual and augmented reality (VR-AR).	(Banfi et al., 2023)
	Study of VERBuM (Virtual Enhanced Reality for Building Modeling) to examine a central Virtual Technical Tour (VTT) for a continuous stream of information when VR and BIM are merged in refurbishment/restoration of cultural and architectural assets.	(Bruno et al., 2022)
	BIM, VR, 3D digital survey, and computer graphics are used to apply new Scan-to-HBIM-to-VR requirements that translate point clouds into mathematical models and digital information.	(Banfi, 2020)
	Facilitating a multi-year repair of Parliament Hill National Historic Site of Canada using historical records and geo-referenced point cloud data, BIM, and VR for public involvement.	(Graham, Chow and Fai, 2019)
	BIM, VR and video analyzing technology are used to develop a maintenance and virtual training system for ancient architecture in China.	(Ji, Fang and Shim, 2018)
	To generate historic building information models (HBIMs) for an Italian castle using BIM and innovative visualization techniques such as AR and VR.	(Osello, Lucibello and Morgagni, 2018)
Construction Health and Safety	A framework was developed to aid in the creation of BIM and VR-based safety training scenarios for tower crane lifting processes, and their applicability was assessed.	(Shringi et al., 2023)
	The authors suggest a site-specific object library for use in BIM and VR site scenarios for the purpose of instructing construction personnel in safe work practices.	(Getuli et al., 2022)
	Implementing BIM's safety applications (such as risk prevention and site operations) utilizing AR or VR data flow architectures.	(Schiavi et al., 2022)

(Continued)

Table 8.2 (Continued)

BIM-VR *Application Area*	*Description*	*Authors*
	This research intended to reduce accident risks in high-rise building projects by integrating BIM with developing digital technologies as photogrammetry, GPS, RFID, AR, VR, and drones.	(Manzoor et al., 2021)
	BIM, VR, AR, GIS, and gaming technologies are changing document-based safety procedures into digital ones. Replicating the precise scenario where the task will be done, incorporating risks and anticipated incidents, systematically constructing cause-effect links, and including a narrative that gives customers emotional significance.	(Afzal, Shafiq and Al Jassmi, 2021), (Mora-Serrano, Muñoz-La Rivera and Valero, 2021)
	Firefighters may use the information provided by the IoT, BIM, VR, and AR system to pinpoint the location of flames inside buildings more quickly and efficiently.	(Chen et al., 2021)
	Creating a universally accepted framework for using BIM and VR to train construction workers safely.	(Getuli, Capone and Bruttini, 2021)
	4-D BIM enabled VR visualization improves information flow and knowledge exchange in a multilingual context when construction team cannot understand printed safety instructions, manuals, and documentation.	(Afzal and Shafiq, 2021)
	To assess the usage of virtual design construction (VDC) tools including BIM, VR, AR, and GIS to improve job-site safety in Gulf Cooperation Council (GCC) countries.	(Shafiq and Afzal, 2020)
	Through a case study of a 19-story hospital in China, a simulation framework of an indoor post-earthquake fire rescue scenario and a smoke visualization method integrating volume rendering and the particle system is provided.	(Lu et al., 2020)
	iInspect is a suggested app to gather collective intelligence for bulk inspections. BIM-based VR and an indoor real-time localization system are used to examine commercial and public buildings for fire safety.	(Zhang et al., 2019)
	A 3D BIM model with numerous features that may be viewed via VR video processing to ensure the safety of ancient buildings.	(Ji, Fang and Shim, 2018)

Table 8.3 Challenges for BIM-VR integration during O&M stage

S. No.	Challenges	Sub Challenges	Papers
1.	Capital/Cost Related	Hardware, software and networking costs for BIM and VR	(Mostafa and Alaqeeli, 2022)
		Staff training Costs	(Dixit et al., 2019; Eadie et al., 2013)
		Design to O&M model interoperability costs	(Becerik-Gerber et al., 2012; Dixit et al., 2019)
2.	Technical Related	Incompatibility between real and virtual models and devices	(Velev and Zlateva, 2017; Williams, Orooji and Aly, 2019)
		Provisions of special softwares to support VR, AR and MR	(Al-Kebsi;Shaima and Mostafa, 2021; Mostafa and Alaqeeli, 2022)
3.	User/Actor Related	Lack of knowledge/skills/ expertise among O&M team	(Cavka, Staub-French and Poirier, 2017; Jang and Lee, 2018; Brittany and A., 2016)
		Communication issues due to non-involvement of asset owners during design stage	(Dixit et al., 2019; Ghassemi and Becerik-Gerber, 2011)
4.	Information Related	Lack of/unnecessary information available in model during O&M stage	(Naghshbandi, 2016; Liu and Issa, 2013)
		Specialized agency employed during design phase results in lack of data understanding for O&M	(Thomas et al., 2017; Elmualim and Gilder, 2014)
		Non-availability of updated as built models	(Elmualim and Gilder, 2014; Dixit et al., 2019; Mostafa and Alaqeeli, 2022)

BIM with VR is expected to increase productive cooperation. BIM and VR tools make it simple to visualize HVAC devices or net systems (electrical, water, or gas supply, or elements of acoustic protection) in a real space that are otherwise inaccessible due to a fake ceiling. As a result, a 7D BIM model and VR environment might be helpful for management and upkeep of the facilities.

The use of VR in BIM has facilitated better collaboration, but a more realistic data retrieval platform still needs development. To properly do maintenance, the user must have unrestricted access to all of the BIM model's corresponding physical components. Several stages of the lifecycle can benefit from integrating BIM and VR, although O&M requires the most resources (both time and money). The literature on BIM-VR integration is limited to the design and construction stages of infrastructure projects.

As there exists a large number of benefits that can be reaped out of BIM-VR integration during O&M phase, there exist various challenges hindering this integration which are summarized in Table 8.3.

8.4 Conclusions

Findings from this study suggest that the infrastructure-building industry is a dynamic one that helps drive global economic growth and promotes enhancements in all areas of an infrastructure asset's life cycle. The study focuses on the two most effective technologies, BIM and VR, which help architects and designers improve design communication within the team members and with the client and bring interesting potentials in various lifecycle phases of a construction project. Using VR to showcase BIM for a project transforms how teams work together in the office and on-site. Many people assume that the only reason to use a BIM model with VR is to enhance the 3D model for visual aids, but a closer look at a BIM model reveals a wide range of advantages. With a view to better recognizing the need for a standardized BIM-VR framework at the O&M stage, this study set out to undertake an SLR on the topic of BIM and VR implementation. This literature review made use of the databases Scopus, Web of Science, and EBSCO host. After sorting and reviewing 68 articles by year and category, the researchers arrived at their final selection using a three-pronged strategy. The studies highlight the many applications of BIM-VR integration, including design and visualization, teaching, operations and maintenance, historic preservation, and construction safety. Currently, BIM and VR technologies are used in the construction industry for tasks such as worker safety planning, defect management, quality management, project scheduling, knowledge collection, safety management, logistics management, and progress assessment. Costs associated with building a mock-up can be mitigated with the help of BIM/VR software, as the costs are associated with correcting mistakes and disposing of excess materials on the job site. BIM and VR are the future of the construction industry because they are used in so many pivotal stages of projects. Finally, this study was able to identify a lack of scientific literature on the measured costs and benefits deriving from the application of BIM-VR integration in the O&M stage, thus highlighting the need for further development in this area of research and for a larger diffusion of such technologies in the AECO sector. Researchers and practitioners also need to examine the obstacles quantitatively and provide a standardized framework to successfully apply BIM-VR in O&M stages of an infrastructure asset. This study's findings can encourage policymakers, academics, and businesspeople to dig deeper and discover an efficient means of raising interest in using BIM-based VR in the AECO sector.

References

Adi, M.N. and Roberts, D.J. (2014) 'Using virtual environments to test the effects of lifelike architecture on people BT – technologies of inclusive well-being: serious games, alternative realities, and play therapy', in Brooks, A.L., Brahnam, S., and Jain, L.C. (eds.). Berlin, Heidelberg: Springer Berlin Heidelberg, pp. 261–285. 10.1007/978-3-642-45432-5_13.

Afzal, M. and Shafiq, M.T. (2021) 'Evaluating 4d-bim and vr for effective safety communication and training: A case study of multilingual construction job-site crew', *Buildings*. 10.3390/buildings11080319.

Afzal, M., Shafiq, M.T. and Al Jassmi, H. (2021) 'Improving construction safety with virtual-design construction technologies – A review', *Journal of Information Technology in Construction*, 26(July), pp. 319–340. 10.36680/j.itcon.2021.018.

Akofio-Sowah, M.-A. *et al.* (2014) 'Managing ancillary transportation assets: The state of the practice', *Journal of Infrastructure Systems*, 20(1), p. 04013010. 10.1061/(asce)is.1943-555x.0000162.

Al-Kasasbeh, M., Abudayyeh, O. and Liu, H. (2021) 'An integrated decision support system for building asset management based on BIM and Work Breakdown Structure', *Journal of Building Engineering*, 34(October 2020), p. 101959. 10.1016/j.jobe.2020.101959.

Al-Kebsi, S. and Mostafa, A. (2021) 'Trends and challenges of virtual reality in architectural design education', *Journal Architecture & Planning*, 33(2), pp. 191–215.

Aldowayan, A., Dweiri, F.T. and Venkatachalam, S. (2020) 'A review on current status of facility management practices in building industry and prospective bim intervention to manage the facilities effectively during its service life', *Proceedings of the International Conference on Industrial Engineering and Operations Management*, (August), pp. 831–846.

Aliu, J. and Aigbavboa, C. (2021) 'Reviewing the trends of construction education research in the last decade: a bibliometric analysis', *International Journal of Construction Management*, 0(0), pp. 1–10. 10.1080/15623599.2021.1985777.

Alizadehsalehi, S. and Yitmen, I. (2023) 'Digital twin-based progress monitoring management model through reality capture to extended reality technologies (DRX)', *Smart and Sustainable Built Environment*, 12(1), pp. 200–236. 10.1108/SASBE-01-2021-0016.

Alizadehsalehi, S., Hadavi, A. and Huang, J.C. (2020) 'From BIM to extended reality in AEC industry', *Automation in Construction*, 116(April), p. 103254. 10.1016/j.autcon.2020.103254.

Alizadehsalehi, S., Hadavi, A. and Huang, J.C. (2021) 'Assessment of AEC students' performance using BIM-into-VR', *Applied Sciences (Switzerland)*, 11(7). 10.3390/app11073225.

Alreshidi, E., Mourshed, M. and Rezgui, Y. (2017) 'Factors for effective BIM governance', *Journal of Building Engineering*, 10(July 2016), pp. 89–101. 10.1016/j.jobe.2017.02.006.

Azhar, S. (2011) 'Building information modeling (BIM): Trends, benefits, risks, and challenges for the AEC industry', *Leadership and Management in Engineering*, 11(3), pp. 241–252.

Banfi, F. (2020) 'HBIM, 3D drawing and virtual reality for archaeological sites and ancient ruins', *Virtual Archaeology Review*, 11(23), pp. 16–33. 10.4995/var.2020.12416.

Banfi, F. *et al.* (2023) 'Narrating ancient Roman heritage through drawings and digital architectural representation: From historical archives, UAV and lidar to virtual-visual storytelling and HBIM projects', *Drones*, 7(1), p. 51. 10.3390/drones7010051.

Bassanino, M. *et al.* (2010) 'The impact of immersive virtual reality on visualisation for a design review in construction', *Proceedings of the International Conference on Information Visualisation* (April 2014), pp. 585–589. 10.1109/IV.2010.85.

Becerik-Gerber, B. *et al.* (2012) 'Application areas and data requirements for BIM-enabled facilities management', *Journal of Construction Engineering and Management*, 138(3), pp. 431–442.

Behzadi, A. (2016) 'Using augmented and virtual reality technology in the construction industry', *American Journal of Engineering Research (AJER)*, 5(12), pp. 350–353.

Bhonde, D., Zadeh, P. and Staub-French, S. (2022) 'Evaluating the use of virtual reality for maintainability-focused design reviews', *Journal of Information Technology in Construction*, 27(June 2021), pp. 253–272. 10.36680/j.itcon.2022.013.

Biocca, F., Kim, T. and Levy, M.R. (1995) 'The vision of virtual reality', in *Communication in the age of virtual reality*. Hillsdale, NJ, US: Lawrence Erlbaum Associates, Inc (LEA's communication series.), pp. 3–14.

Bolpagni, M., Gavina, R. and Ribeiro, D. (2022) *Industry 4.0 for the Built Environment*. 1st edn. Springer Cham. 10.1007/978-3-030-82430-3.

Booth, A. (2008) 'Unpacking your literature search toolbox: On search styles and tactics', *Health Information and Libraries Journal*, 25(4), pp. 313–317.

Boton, C. (2018) 'Supporting constructability analysis meetings with immersive virtual reality-based collaborative BIM 4D simulation', *Automation in Construction*, 96(August), pp. 1–15. 10.1016/j.autcon.2018.08.020.

Boukamp, F. and Akinci, B. (2004) 'Towards automated defect detection: Object-oriented modeling of construction specifications', in *International Conference on Computing in Civil and Building Engineering*. Weimar, Germany, pp. 1–12. 10.25643/bauhaus-universitaet.131.

Brittany, G. and A., I.R.R. (2016) 'Framework for evaluating the BIM competencies of facility owners', *Journal of Management in Engineering*, 32(1), p. 4015024. 10.1061/(ASCE)ME.1943-5479.0000378.

Bruno, S. *et al.* (2022) 'Verbum – virtual enhanced reality for building modelling (virtual technical tour in digital twins for building conservation)', *Journal of Information Technology in Construction*, 27(January), pp. 20–47. 10.36680/j.itcon.2022.002.

Casini, M. (2022) 'Extended reality for smart building operation and maintenance: A review', *Energies*, 15(10), p. 3785. 10.3390/en15103785.

Cavka, H.B., Staub-French, S. and Poirier, E.A. (2017) 'Developing owner information requirements for BIM-enabled project delivery and asset management', *Automation in Construction*, 83(August), pp. 169–183. 10.1016/j.autcon.2017.08.006.

Chen, H. *et al.* (2021) 'Development of BIM, IoT and AR/VR technologies for fire safety and upskilling', *Automation in Construction*, 125(September 2020), p. 103631. 10.1016/j.autcon.2021. 103631.

Choi, J. and Lee, S. (2023) 'A suggestion of the alternatives evaluation method through IFC-based building energy performance analysis', *Sustainability*, 15(3), p. 1797. 10.3390/su15031797.

Cronin, P., Ryan, F. and Coughlan, M. (2008) 'Undertaking a literature review: A step-by-step approach', *British journal of nursing*, 17(1), pp. 38–43.

Davies, H.T. and Crombie, I.K. (1998) 'Getting to grips with systematic reviews and meta-analyses', *Hospital medicine (London, England: 1998)*, 59(12), pp. 955–958.

Dinis, F.M. *et al.* (2020) 'Improving project communication in the architecture, engineering and construction industry: Coupling virtual reality and laser scanning', *Journal of Building Engineering*, 30(December). 10.1016/j.jobe.2020.101287.

Dionisio, J.D.N. and Burns III, W.G. (2013) '3D virtual worlds and the metaverse: Current status and future possibilities', *ACM Computing Surveys 45*, 34(3), p. 38. 10.1145/2480741.2480751 This.

Dixit, M.K. *et al.* (2019) 'Integration of facility management and building information modeling (BIM): A review of key issues and challenges', *Facilities*, 37(7–8), pp. 455–483. 10.1108/F-03-201 8-0043.

Dunston, P.S. *et al.* (2011) 'An immersive virtual reality mock-up for design review of hospital patient rooms', *Collaborative Design in Virtual Environments* [Preprint], (January). 10.1007/978-94-007-0605-7.

Eadie, R. *et al.* (2013) 'BIM implementation throughout the UK construction project lifecycle: An analysis', *Automation in Construction*, 36, pp. 145–151. 10.1016/j.autcon.2013.09.001.

Elhakeem, A. and Hegazy, T. (2012) 'Building asset management with deficiency tracking and integrated life cycle optimisation', *Structure and Infrastructure Engineering*, 8(8), pp. 729–738. 10.1080/15732471003777071.

Elmualim, A. and Gilder, J. (2014) 'BIM: innovation in design management, influence and challenges of implementation', *Architectural Engineering and Design Management*, 10(3–4), pp. 183–199. 10.1080/17452007.2013.821399.

Fernando, T., Wu, K.C. and Bassanino, M. (2013) 'Designing a novel virtual collaborative environment to support collaboration in design review meetings', *Journal of Information Technology in Construction*, 18(November), pp. 372–396.

Flintsch, G.W. and Medina, A. (2001) 'Framework for a Web-based Intelligent Infrastructure Asset Management System', *5th International Conference on Managing Pavements* [Preprint].

Froese, T.M. (2010) 'The impact of emerging information technology on project management for construction', *Automation in Construction*, 19(5), pp. 531–538. 10.1016/j.autcon.2009.11.004.

Fu, M. and Liu, R. (2018) 'The Application of Virtual Reality and Augmented Reality in Dealing with Project Schedule Risks' (May 2020). 10.1061/9780784481264.042.

Gan, V.J.L., Liu, T. and Li, K. (2022) 'Integrated BIM and VR for interactive aerodynamic design and wind comfort analysis of modular buildings', *Buildings*, 12(3). 10.3390/buildings12030333.

Getuli, V., Capone, P. and Bruttini, A. (2021) 'Planning, management and administration of HS contents with BIM and VR in construction: An implementation protocol', *Engineering, Construction and Architectural Management*, 28(2), pp. 603–623. 10.1108/ECAM-11-2019-0647.

Getuli, V. *et al.* (2022) 'A smart objects library for BIM-based construction site and emergency management to support mobile VR safety training experiences', *Construction Innovation*, 22(3), pp. 504–530. 10.1108/CI-04-2021-0062.

Ghanem, S.Y. (2022) 'Implementing virtual reality – Building information modeling in the construction management curriculum', *Journal of Information Technology in Construction*, 27, pp. 48–69. 10.36680/j.itcon.2022.003.

Ghassemi, R. and Becerik-Gerber, B. (2011) 'Transitioning to integrated project delivery: Potential barriers and lessons learned', *Lean Construction Journal*, 2011, pp. 32–52.

Graham, K., Chow, L. and Fai, S. (2019) 'From BIM to VR: Defining a level of detail to guide virtual reality narratives', *Journal of Information Technology in Construction*, 24(December), pp. 553–568. 10.36680/J.ITCON.2019.031.

Hajirasouli, A. *et al.* (2023) 'BIM-enabled virtual reality (VR)-based pedagogical framework in architectural design studios', *Smart and Sustainable Built Environment* [Preprint]. 10.1108/SASBE-07-2022-0149.

Hao, Z., Zhang, W. and Zhao, Y. (2021) 'Integrated BIM and VR to implement IPD mode in transportation infrastructure projects: System design and case application', *PLoS ONE*, 16(11 November). 10.1371/journal.pone.0259046.

Hardin, B. and McCool, D. (2015) *BIM and Construction Management: Proven Tools, Methods, and Workflows*. 2nd edn. Wiley.

He, Z. and Liu, T. (2022) 'Exploring key elements and performance of BIM and VR technologies in the project management of assembled buildings', *Advances in Multimedia*, 2022, pp. 1–8. 10.1155/2022/1168953.

Heydarian, A. *et al.* (2015) 'Immersive virtual environments versus physical built environments: A benchmarking study for building design and user-built environment explorations', *Automation in Construction*, 54, pp. 116–126. 10.1016/j.autcon.2015.03.020.

Jang, S. and Lee, G. (2018) 'Impact of organizational factors on delays in BIM-based coordination from a decision-making view: A case study', *Journal of Civil Engineering and Management*, 24(1), pp. 19–30. 10.3846/jcem.2018.296.

Ji, X., Fang, X. and Shim, S.H. (2018) 'Design and development of a maintenance and virtual training system for ancient Chinese architecture', *Multimedia Tools and Applications*, 77(22), pp. 29367–29382. 10.1007/s11042-018-5979-4.

Kamari, A., Paari, A. and Torvund, H.Ø. (2021) 'Bim-enabled virtual reality (VR) for sustainability life cycle and cost assessment', *Sustainability (Switzerland)*, 13(1), pp. 1–24. 10.3390/su13010249.

Kim, S. *et al.* (2018) 'Investigating owner requirements for BIM-enabled design review', *Construction Research Congress 2018: Construction Information Technology - Selected Papers from the Construction Research Congress 2018*, 2018-April(June 2020), pp. 581–590. 10.1061/9780784481264.057.

Kivits, R.A. and Furneaux, C. (2013) 'BIM: Enabling sustainability and asset management through knowledge management', *The Scientific World Journal*, 2013. 10.1155/2013/983721.

Kreutzberg, A. (2015) 'Conveying architectural form and space with virtual', in *The 33rd eCAADe Conference*, pp. 117–124. 10.52842/conf.ecaade.2015.1.117.

Kuliga, S.F. *et al.* (2015) 'Virtual reality as an empirical research tool – Exploring user experience in a real building and a corresponding virtual model', *Computers, Environment and Urban Systems*, 54, pp. 363–375. 10.1016/j.compenvurbsys.2015.09.006.

Lewis, A., Riley, D. and Elmualim, A. (2010) 'Defining high performance buildings for operations and maintenance', *International Journal of Facility Management*, 1(2), p. 16.

Lin, Y.C. *et al.* (2018) 'Integrated BIM, game engine and VR technologies for healthcare design: A case study in cancer hospital', *Advanced Engineering Informatics*, 36(August 2017), pp. 130–145. 10.1016/j.aei.2018.03.005.

Liu, R. and Issa, R.R.A. (2013) 'Issues in BIM for facility management from industry practitioners' perspectives', *Computing in Civil Engineering – Proceedings of the 2013 ASCE International Workshop on Computing in Civil Engineering*, (June), pp. 411–418. 10.1061/9780784413029.052.

Love, P.E.D. *et al.* (2015) 'A systems information model for managing electrical, control, and instrumentation assets', *Built Environment Project and Asset Management*, 5(3), pp. 278–289.

Lu, X. *et al.* (2020) 'Scenario simulation of indoor post-earthquake fire rescue based on building information model and virtual reality', *Advances in Engineering Software*, 143(May 2019), p. 102792. 10.1016/j.advengsoft.2020.102792.

Mahalingam, A., Yadav, A.K. and Varaprasad, J. (2015) 'Investigating the role of lean practices in enabling BIM adoption: Evidence from two Indian cases', *Journal of Construction Engineering and Management*, 141(7), pp. 1–11.

Majumdar, T., Fischer, M.A. and Schwegler, B.R. (2006) 'Conceptual design review with a virtual reality mock-up model. Paper presented at the Building on IT', *Joint International Conference on Computing and Decision Making in Civil and Building Engineering*, pp. 2902–2911.

Manzoor, B. *et al.* (2021) 'A research framework of mitigating construction accidents in high-rise building projects via integrating building information modeling with emerging digital technologies', *Applied Sciences (Switzerland)*, 11(18). 10.3390/app11188359.

Molinari, C., Paganin, G. and Talamo, C. (2002) 'Information Systems for Real Estate Management', in *Proceedings of the CIB W070 2002 Global Symposium, CIB and CABER*, pp. 349–362.

Mora-Serrano, J., Muñoz-La Rivera, F. and Valero, I. (2021) 'Factors for the automation of the creation of virtual reality experiences to raise awareness of occupational hazards on construction sites', *Electronics (Switzerland)*, 10(11). 10.3390/electronics10111355.

Mostafa, A. and Alaqeeli, A. (2022) 'Benefits and challenges of integrating IoT, VR & AR in the BIM-based facility management process: Literature and case-based analysis', *Journal of Engineering Research*, 6(4), pp. 25–40. 10.21608/erjeng.2022.265269.

Naghshbandi, S.N. (2016) 'BIM for facility management: Challenges and research gaps', *Civil Engineering Journal*, 2(12), pp. 679–684.

Natephra, W. *et al.* (2017) 'Integrating building information modeling and virtual reality development engines for building indoor lighting design', *Visualization in Engineering*, 5(1). 10.1186/s40327-017-0058-x.

Noruwa, B.I., Arewa, A.O. and Merschbrock, C. (2020) 'Effects of emerging technologies in minimising variations in construction projects in the UK', *International Journal of Construction Management*, 0(0), pp. 1–8. 10.1080/15623599.2020.1772530.

Osello, A., Lucibello, G. and Morgagni, F. (2018) 'HBIM and virtual tools: A new chance to preserve architectural heritage', *Buildings*, 8(1), pp. 1–12. 10.3390/buildings8010012.

O'Grady, T. *et al.* (2021) 'Circular economy and virtual reality in advanced BIM-based prefabricated construction', *Energies*, 14, pp. 1–16. 10.3390/en14134065.

Paes, D. and Irizarry, J. (2018) 'A usability study of an immersive virtual reality platform for building design review: Considerations on human factors and user interface', *Construction Research Congress 2018: Construction Information Technology – Selected Papers from the Construction Research Congress 2018*, 2018-April(April), pp. 419–428. 10.1061/9780784481264.041.

Paes, D., Kim, S. and Irizarry, J. (2017) 'Human factors considerations of First Person View (FPV) operation of Unmanned Aircraft Systems (UAS) in infrastructure construction and inspection environments', *6th CSCE-CRC International Construction Specialty Conference 2017 – Held as Part of the Canadian Society for Civil Engineering Annual Conference and General Meeting 2017*, 2(May), pp. 822–831.

Rondeau, E.P., Brown, R.K. and Lapides, P.D. (2006) *Facility Management.* 2nd edn. NY, USA: Wiley: New York.

Safikhani, S. *et al.* (2022) 'Immersive virtual reality for extending the potential of building information modeling in architecture, engineering, and construction sector: Systematic review', *International Journal of Digital Earth*, 15(1), pp. 503–526. 10.1080/17538947.2022.2038291.

Salem, O., Samuel, I.J. and He, S. (2020) 'BIM and VR/AR technologies: From project development to lifecycle asset management', *Proceedings of International Structural Engineering and Construction*, 7(1). 10.14455/ISEC.res.2020.7(1).AAE-11.

Sampaio, A.Z. (2018) 'Enhancing BIM methodology with VR technology', *State of the Art Virtual Reality and Augmented Reality Knowhow* [Preprint]. 10.5772/intechopen.74070.

Sampaio, A.Z. and Simões, D. (2015) 'Maintenance of buildings using BIM methodology', *The Open Construction and Building Technology Journal*, 8(1), pp. 337–342. 10.2174/18748368014 08010337.

Sartor, M. *et al.* (2014) 'International purchasing offices: Literature review and research directions', *Journal of Purchasing and Supply Management*, 20(1), pp. 1–17.

Schiavi, B. *et al.* (2022) 'BIM data flow architecture with AR/VR technologies: Use cases in architecture, engineering and construction', *Automation in Construction*, 134(March 2021), p. 104054. 10.1016/j.autcon.2021.104054.

Schraven, D., Hartmann, A. and Dewulf, G. (2011) 'Effectiveness of infrastructure asset management: Challenges for public agencies', *Built Environment Project and Asset Management*, 1(1), pp. 61–74.

Seuring, S. and Gold, S. (2012) 'Conducting content-analysis based literature reviews in supply chain management', *Supply Chain Management*, 17(5), pp. 544–555.

Seyman Guray, T. and Kismet, B. (2022) 'VR and AR in construction management research: Bibliometric and descriptive analyses', *Smart and Sustainable Built Environment* [Preprint]. 10.1108/SASBE-01-2022-0015.

Seyman Guray, T. and Kismet, B. (2023) 'Applicability of a digitalization model based on augmented reality for building construction education in architecture', *Construction Innovation*, 23(1), pp. 193–212. 10.1108/CI-07-2021-0136.

Shafiq, M.T. and Afzal, M. (2020) 'Potential of virtual design construction technologies to improve job-site safety in gulf corporation council', *Sustainability (Switzerland)*, 12(9). 10.3390/ su12093826.

Shahinmoghadam, M., Natephra, W. and Motamedi, A. (2021) 'BIM- and IoT-based virtual reality tool for real-time thermal comfort assessment in building enclosures', *Building and Environment*, 199(May), p. 107905. 10.1016/j.buildenv.2021.107905.

Shringi, A. *et al.* (2023) 'Safety in off-site construction: Simulation of crane-lifting operations using VR and BIM', *Journal of Architectural Engineering*, 29(1), pp. 1–10. 10.1061/(asce)ae.1943-55 68.0000570.

Sood, R. and Laishram, B. (2021) 'Infrastructure Asset Management by Integrating BIM and Lean Tools', in *Proceedings of Indian Lean Construction Conference*. Ahmedabad, pp. 43–52.

Srivastava, A. *et al.* (2022) 'Imperative role of technology intervention and implementation for automation in the construction industry', *Advances in Civil Engineering*, 2022, pp. 1–19. 10.1155/2022/6716987.

Tallgren, M.V., Roupé, M. and Johansson, M. (2021) '4D modelling using virtual collaborative planning and scheduling', *Journal of Information Technology in Construction*, 26(April), pp. 763–782. 10.36680/J.ITCON.2021.042.

Thomas, B. *et al.* (2017) 'Management of collaborative BIM data by federating distributed BIM models', *Journal of Computing in Civil Engineering*, 31(4), p. 4017009. 10.1061/(ASCE)CP.1 943-5487.0000657.

Tjahjono, B. *et al.* (2010) 'Six sigma: A literature review', *International Journal of Lean Six Sigma*, 1(3), pp. 216–233.

Velev, D. and Zlateva, P. (2017) 'Virtual reality challenges in education and training', *International Journal of Learning*, 3(1), pp. 33–37. 10.18178/IJLT.3.1.33-37.

Wang, K.C. and Tung, S.H. (2023) 'Cost estimation model based on building information modeling and virtual reality for customizing presold homes', *KSCE Journal of Civil Engineering* [Preprint]. 10.1007/s12205-023-0825-2.

Wang, P. *et al.* (2018) 'A critical review of the use of virtual reality in construction engineering education and training', *International Journal of Environmental Research and Public Health*, 15(6). 10.3390/ijerph15061204.

Wang, W. *et al.* (2022) 'BIM information integration based VR modeling in digital twins in industry 5.0', *Journal of Industrial Information Integration*, 28(April), p. 100351. 10.1016/j.jii. 2022.100351.

Weber, B., Staub-Bisang, M. and Alfen, H.W. (2016) 'Infrastructure as an asset class: Investment strategy, sustainability, project finance and PPP', *Infrastructure as an Asset Class: Investment Strategy, Sustainability, Project Finance and PPP*, pp. 1–392. 10.1002/9781119226574.

Whyte, J. (2002) *Virtual Reality and the Built Environment*. Architectural Press, An imprint of Elsevier Science.

Williams, J.E., Orooji, F. and Aly, S.J. (2019) 'Integration of virtual reality (VR) in architectural design education: Exploring student experience', *American Society for Engineering Education*, p. 11. 10.18260/1-2--32999.

Wong, J.Y. *et al.* (2020) 'BIM-VR framework for building information modelling in engineering education', *International Journal of Interactive Mobile Technologies*, 14(6), pp. 15–39. 10.3991/IJIM.V14I06.13397.

Zaker, R. and Coloma, E. (2018) 'Virtual reality-integrated workflow in BIM-enabled projects collaboration and design review: A case study', *Visualization in Engineering*, 6(1). 10.1186/s40327-018-0065-6.

Zhang, D. *et al.* (2019) 'Taking advantage of collective intelligence and BIM-based virtual reality in fire safety inspection for commercial and public buildings', *Applied Sciences (Switzerland)*, 9(23). 10.3390/app9235068.

9 Verification and Validation of a Framework for Collaborative BIM Implementation, Measurement and Management (CIMM)

Andrew Pidgeon

School of Computing, Engineering and Digital Technologies,
Teesside University, UK

9.1 Introduction

Building Information Modelling (BIM) is an area that has significantly and progressively developed from theoretical underpinnings of enhanced coordination to a more data-centric, technological approach to project delivery (Yang and Mao, 2021). Further, this is also reinforced by the development of supportive technology ranging from the initial digitisation of 2D model production (Taylor, 2007; Whyte et al., 1999) utilising low latency network connectivity and a paperless storage functionality (McCuen, 2008) through the utilisation of a single source of truth via a Common Data Environment (CDE) (Eadie et al., 2013). This advancement facilitates near real-time connectively of teams who are geospatially separated yet conjoined through a coordinated approach to information management (Chahrour et al., 2020). Moreover, research undertaken by Oraee et al. (2021) shows that due to these complimentary and concatenated advances, project teams are able to collaborate, coordinate and communicate if not more effectively than if they were physically located in the same location (Ojo and Pye, 2020). This harbours the realisation that intra- and inter-organisational collaboration of project delivery can be undertaken regardless of location, time zone, specialism or affiliation (Grilo et al., 2013). Moreover, this increases the quality of outputs through joint working (Ashcraft, 2008) and unified decision-making via a transformative shift in collaborative culture. Gilligan and Kunz (2007) state that as a resultant factor of these positive experiences there is a prediction that an ever-increasing uptake of BIM is likely to continue to shape the project delivery environment. However, although a range of technological advancements alongside refinement of project delivery processes have assisted in the exponential growth of the application of digital tools, its underpinning foundation and future positioning stems from the application of collaborative theory (Colbry et al., 2014; Wood and Gray, 1991). This results in enhanced integration of interdisciplinary teams and stakeholders across the entire lifecycle of a project. Moreover, increased risks have been a consequence due to both the amount and variations in available technology (Yang and Mao, 2021), which has resulted in the trend of increased inherited complexity (Eastman et al., 2011) and thus an upsurge in potential risk of failure due to new processes and applied methodologies through the vast array of toolsets available to facilitate people utilising collaborative BIM. To counter these problematic areas and to determine clear objective focus on what the application of collaborative BIM is intended to be applied to deliver, research undertaken by Pidgeon and Dawood (2021; 2021a; 2021b) outlines that objective focus through the creation of plain language analysis and determination in both quantifiable and qualitative forms assist in bridging the gap between

DOI: 10.1201/9781003408949-9

theoretical understandings and practical achievements. BIM in its simplest form is a project delivery methodology intended to manage the digital flow of information accurately and informatively with the inclusion of coordinated 3D geometric modelling in order to develop an asset across multiple stages of its lifecycle (Succar, 2009; Barlish and Sullivan, 2012; Zuppa et al., 2009; Miettinen and Paavola, 2014). Further, collaboration is key to the success of BIM (Pihlak et al., 2010) with the Efficiency and Reform Group (2011) stating that BIM itself is a 'key driver for collaboration' across multiple disciplines, individual infrastructure projects and teams of people respectively.

Research undertaken by Hollenbeck et al. (2012) on the subject of collaborative teams states that in the 1980s '20% of work was team based', whereas three decades later this figure increased to circa 80% showing an exponential shift if working culture supported by advancements in the technology field and through affordability via economies of scale (O'Sullivan and Sheffrin, 2003). Benefits of collaboration include but are not limited to strategic coordination (Yukl et al., 2005), alongside the pull from the diversification and inclusion of multiple stakeholders across separate non-geolocated sectors (Abuelmaatti and Ahmed, 2014). According to Gilligan and Kunz (2007), the latter will result in a continual increase in the adoption of BIM alongside returned recompenses, so long as a more detailed understanding of operational 'inter-organisational' requirements are provided prior to implementation (Taylor and Bernstein, 2009).

The objectives of this research are therefore, in terms of process to (1) improve the implementation of collaborative BIM through more informed and agreed objectification of BIM; (2) enhance how quantifiable and qualitative inputs and outputs are available to manage and measure goals, risks and task assignment; and (3) develop a robust and novel implementation and collaborative BIM framework. Further, it is important that a diverse range of stakeholders with relevant and wide-ranging experience and expertise are included to ensure that the fundamental hypothesis is critiqued, adjusted and adapted, so that varying perspectives and inputs are instilled within.

Expanding on from the above, the sections which follow specifically explore further, via a systematic literature review, the areas of collaborative BIM and collaborative practice alongside validation methodologies, in order to support the validation through pre-existing rigorously sound theories and methodologies, whilst supporting and equally challenging collaborative BIM in order to achieve a greater perspective of underpinning knowledge within this field. Moreover, a two-phased approach to verifying and validating the framework hypothesis, as well as incorporating beneficial adaptions was derived from the inclusion of academic and industry-focused BIM practitioners, alongside an initiated workshop on a real-world building structures project in London. Furthermore, this better positions the latter part of the research in validating and developing the implementation framework methodology applied to industry participants via a participative case study, to positively expose more detailed benefits and to counter the inefficiencies of collaborative BIM through the framework development verification process.

9.2 Methodology

The aim of this research was to develop a two-stage verification and validation process, adapted from the Delphi model (Dalkey and Helmer, 1963) alongside the 'action research' methodology (Dick, 2014), building upon the theoretical underpinning literature as well as research findings developed by the authors. Whereby the proposal of a collaborative BIM implementation framework is analysed and progressively developed,

Table 9.1 Data input source and description

Data Input Source	Description of the Utilised Methodology
Literature Review	A systematic literature review of existing underpinning theories as well as reporting and methodologies spanning industrial application, including the authors progressive research findings (Ashcraft, 2008, Barlish and Sullivan, 2012; Colbry et al., 2014; Eastman et al., 2011; Grilo et al., 2013; Ojo and Pye, 2020; Pidgeon and Dawood, 2021; Pidgeon and Dawood, 2021a; Pidgeon and Dawood, 2021b)
Focus Group Questionnaire (verification stage)	The development and distribution of an electronic questionnaire consisting of 8 fundamental questions to 15 core BIM experts from academia, design and construction domains, focused on reviewing the framework hypothesis and providing qualitative and quantitative insights towards seeking adaptions (pre-workshop case study)
Project Case Study (validation stage)	Facilitating, as a participative observer, a collaborative workshop inclusive of 8 participants based on applying the implementation framework on a real-world building structures BIM project for validation purposes

taking into consideration evaluative feedback from subject matter participants i.e., 15 BIM experts at the questionnaire stage and eight at the project case study stage (separating each group as per the intended research design), whilst applying rigour and completeness towards seeking validation, and thus implementing adaptations of a novel collaborative BIM framework. Furthermore, there is additional benefit in analysing the practical use of a theory/methodology as part of a realistic application (workshop case study), rather than solely relying upon assumed and entirely theoretical perspective (Bromley, 1986).

Further, Table 9.1 summarises the three-phased data input sources which support the design of the research study following the survey research technique, as well as supplying further descriptive detail for the applied methods.

9.2.1 Design and Procedure

As part of the design of this research study a process flow methodology has been developed (presented in Figure 9.1) which outlines successively the progress from start to completion, building upon a systematic literature review (Denyer and Tranfield, 2009) as the outlined steps above in order to present a sound methodology that encourages a feedback loop alongside a summarised evaluation post review of attained discoveries (Rosnow and Rosnow, 2011).

As aforementioned within the introduction of the methodology section, a two-stage process was used as part of the design following the Delphi model (Dalkey and Helmer, 1963), which has been adapted and utilised within the structure in order to present a method which facilitates consistency and rigour in order to incorporate a wider range of perspectives and viewpoints. Ultimately this approach through reshaping via examination presents an opportunity for improved development through participant input and observation (Wondolleck and Yaffee, 2000). Figure 9.2 highlights the adapted process and presents the two-stage structure incorporating a feedback loop and thusly, applied validation through the collection and analysis of data from the focus group survey (stage 1), as well as project workshop case study (stage 2). Moreover, Buur (1990) states that the

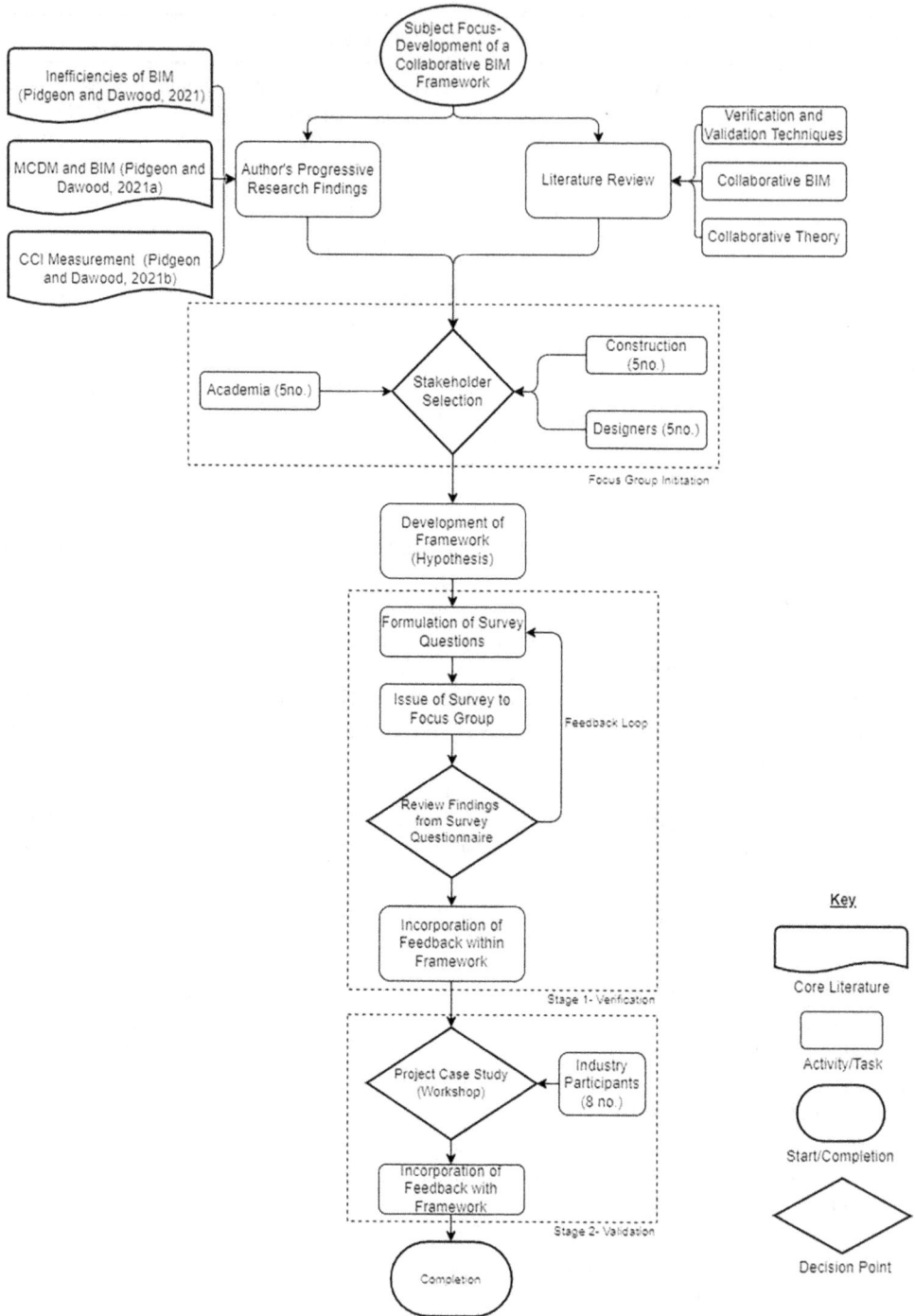

Figure 9.1 Process flow methodology.

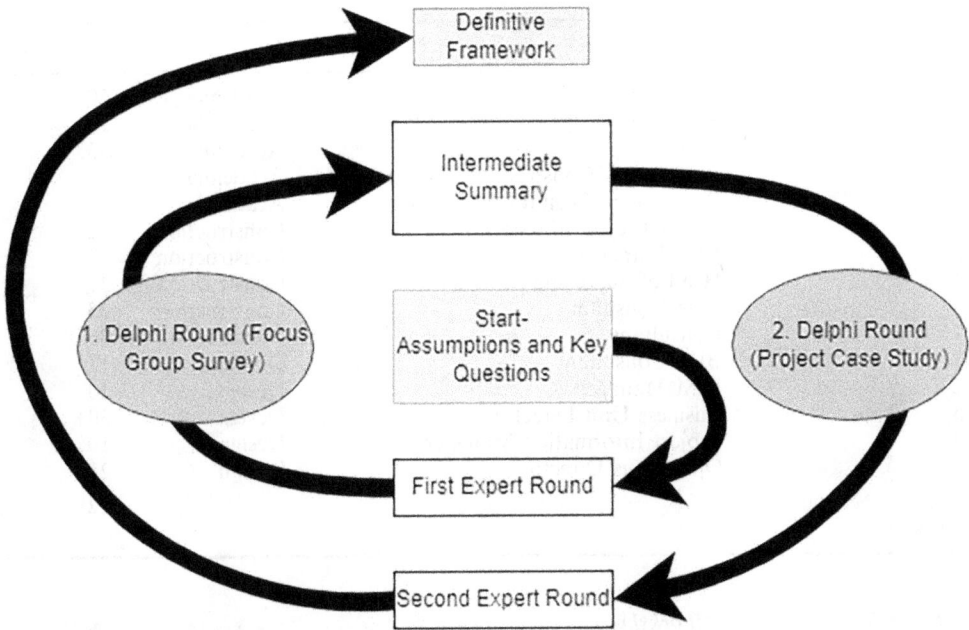

Figure 9.2 The adapted Delphi methodology (Dalkey and Helmer, 1963).

benefit to adopting this approach creates a valued acceptance of such a proposal, with Blessing and Chakrabarti (2009) additionally advising that the structure of a complimentary system attracts incremental observations and benefits in the form of improved recommendations to the original hypothesis.

9.2.2 Stakeholder Selection (Focus Group)

A dynamic group of participants spanning academia, construction and design with a bias towards BIM and experience were selected as part of the design of this research for stage 1, forming the focus group members who would participate in verifying the framework hypothesis. A more detailed breakdown of the participant's diverse background, experience and function is presented in Table 9.2 to highlight their skills, domain and relevance in terms of achieving a consistent and wide range of inclusive opinions and therefore feedback attained.

9.2.3 Survey Questionnaire

The semi-structured interview data was collected electronically via a developed questionnaire. This technique was focused on inviting the 15 focus group members spanning academia, design and construction domains, as part of the design of this research to review and respond to eight qualitative questions focused on the areas of novelty, transparency, benefit, ease of use and importance.

Further, these questions as part of the design were structured in order to achieve direct feedback from a diverse range of BIM competent stakeholders, which feeds into the verification process and thus strength of the framework hypothesis.

Table 9.2 Stakeholder inclusion summary

Participant Reference	Function	Sector	Experience
A1	Professor	Academia	20
A2	Professor	Academia	28
A3	Professor	Academia	33
A4	Professor of Construction Management	Academia	40
A5	Professor of Digital Construction	Academia	12
C1	Senior Project Manager	Construction	22
C2	BIM Manager	Construction	22
C3	Head of Surveying	Construction	37
C4	BIM Consultant	Construction	35
C5	Consultant	Construction	35
D1	BIM Consultant	Design	12
D2	BIM Manager	Design	10
D3	Business Unit Director	Design	30+
D4	Project Information Manager	Design	10
D5	Operations Director	Design	21
Average experience (in years)			22

9.2.4   Project Case Study (Workshop)

The purpose of an applied project case study is to expand upon the hypothesised framework via a rigorized grounded approach following the stage 1 verification, and to gain wider insights into the benefits and opportunities of the framework that a focus group of participants can support in developing through adaptive observations, as part of stage 2 validation (Shavelson and Towne, 2002). The utilisation and implementation of the framework was initiated on an active c. £70m building infrastructure BIM project (design and build) structures project in London involving eight project team members including engineers, project managers, surveyors and contractors, who have adopted BIM within the structural engineering design elements at RIBA (Royal Institute of British Architects) stage 2. Further, a core focus, due to the complexity of the clients' requirements i.e., the project is upgrading an existing asset towards a new purpose, meant that information management and the delivery mechanism of applied BIM adds only to the depth of the research field and validation thereof again through independent perspective, cross-cutting academic and industry investigative research study. Moreover, the main objective, and thus the high-level pre-determined focal area for the workshop case study, is towards clarifying and achieving better 'Information Requirements' in coordinating, collaborating and executing the complex project to the highest possible standard. Information requirements were specifically determined defined as an area of utmost importance when aspiring to establish and achieve collaborative BIM (Pidgeon and Dawood 2021; Pidgeon and Dawood, 2021a) and is justified in terms of building upon this area further through the practical workshop validation and thus implementation of the framework. Further, this approach enables the benefits of the system to be applied, holistically reviewed and adjusted in order to create a truly robust and valued methodology, so that the two-stage Delphi method (Dalkey and Helmer, 1963) is utilised adding weighting (prioritisations) and to support in addressing any limitations, whilst taking into consideration any further positive changes that can be exposed within the hypothesised framework. Further, although

Table 9.3 Case study participants

Case Study Focus (Objective)	BIM Project Type	Delivery Team Sector	Specialism	Role	Years' Experience	Ref.
		Lead Appointed Party (Tier 1)	Design	Apprentice Engineer	3	PA1
Utilisation of the	Building	Structural	Design	Engineer	6	PA2
collaborative	Structures-	Engineering	Design	Principal Engineer	13	PA3
BIM implementation	Large		Information Management	BIM Manager	20+	PA4
framework towards	Scale		Contractor	Consultant	40	PA5
Information	Redevelop-		Project Management	Project Director	30+	PA6
Requirements	ment		Project Management	Technical Director	33	PA7
			Contractor	Head of Surveying	30	PA8
Total Workshop Participants					c. 22 years	

there is no 'minimum sample size' for the qualitative survey element of research investigation (Patton, 2002; Yin, 2003), it is important that a sufficient range of people with perspectives and expertise is included to add to the validation via the case study technique (Lincoln and Guba, 1985). In light of this latter statement, Table 9.3 summaries for visibility purposes the eight participants (excluding the author who facilitated the workshop acting as a 'participative observer'), who formed part of the workshop case study group that were engaged in the project, as well as their specialisms, roles and the overarching area of application (Building Structures collaborative BIM project).

In regard to the main objective of the workshop, therein applying the framework as a methodology that supports the delivery of collaborative BIM and from an implementation perspective, is beneficial to validate that it is wholly inclusive, dynamic and adjustable for a range of BIM activities such as but not limited to information exchange management, 'nD' coordination, quality control and assurance, clash detection and model validation, etc. However, for the purposes of the case study, the subject and targeted focus was on supporting the developed transparency of the information requirements from a collaborative BIM perspective of a building structures project, with direct inclusion of the structural design task team members including early contractor involvement.

Further and in more detail, determination of prioritizations of the agreed criteria and alternatives alongside the risk mitigation and approach to achievements were also sub-elements that were focal entities of the framework validation. This is due to its presented importance and centrality in the delivery of complex projects to the information manager (project manager) via BIM as delivery mechanism as aforementioned.

9.3 Results

9.3.1 Data Analysis

This next section extrapolates further from the general summary section above with additional qualitative and quantifiable evidence ascertained through the semi-structured survey questionnaire technique. Table 9.4 concludes the key responses in terms of times

Table 9.4 Qualitative response contributions to core themes ('X' marks a reference to counted returns and therefore their inclusion within respondents' feedback)

Ref.	Clarity	Benefit	Novelty	Ease of Use	Importance	Transparency	Recommended Adaptions
A1	X	X	X			X	N/A
A2	X	X	X	X	X	X	
A3	X	X	X	X	X	X	
A4		X	X		X	X	
A5	X		X			X	Visual flow, stage clarity and project application
C1	X	X	X		X		Simplistic workflows
C2	X	X	X		X	X	N/A
C3	X	X	X		X		Reword framework flow so it's understandable to non-BIM experts
C4	X	X	X	X	X		Make processes straightforward/ simplistic
C5	X	X	X	X	X	X	N/A
D1	X	X	X		X		Apply to the wider audience
D2	X	X	X	X	X	X	Project case study to widen adaptions
D3	X	X	X		X		More guidance for usability
D4	X		X	X	X		N/A
D5		X	X		X		
Total Counts	13	13	15	6	13	8	

they were emphasised, utilising a general grouping methodology, with a dedicated column outlining where more detailed recommendations for adaptions to the framework would benefit and bolster the framework hypothesis.

Reflecting upon the feedback consolidated within Table 9.4, the lowest scoring i.e., least weighted item, referenced theme was the 'ease of use' (6 out of 15) followed closely by 'transparency' (8 out of 15) in appliable terms. Upon review of the recommendations of the observations and adaptions to the framework hypothesis, this was due to the presentation to the focus group of a framework and process flow as opposed to a tool or solution which guides them through the steps towards execution.

9.3.2   *Objectifying and Improving the Way BIM Goals Are Achieved*

In regard to feedback gained on how the framework assists in achieving and advancing on how collaborative BIM can return positive outcomes, one respondent (ref. D1) stated that there is heightened assistance to 'categorise goals and criteria', with another (ref. D2) reaffirming this by adding that the language and structure aids in creating a 'clear way of expression' to the interested goals and objectives.

However, one remark was made (ref. D3) in that they believe in its 'current form' the framework would add limited benefit if not directly aimed at 'BIM people', with another (ref. C3) agreeing that it must be agnostic of 'BIM professionals' with the language to suit so that it does not exclude certain stakeholders who want to deliver collaborative BIM. However, in terms of generalising the feedback, the consensus was relatively positive with comments focused on the framework attractiveness such as 'drawing out a discussion on the importance and priorities would help' (ref. A2), 'better and transparent communications' (ref. A3), 'makes every process and project more transparent' (ref. C5) and 'enable benchmarking of performance' (ref. C1). These responses add to the collective weighting that the hypothesis is a strong foundation to support the implementation of collaborative BIM.

9.3.3   Identification of Core Elements for an Inclusivity and Measurability

Generalising on the core elements that the framework supports and addresses, one respondent (ref. D2) stated the framework is 'arranged in a way that the is no margin for the uncertainty' with another (ref. D3) similarly adding 'the intent is reasonably clear, solid brief' which is also in agreement with one respondent from academia (ref. A2) who stated, 'By using the framework and identifying factors but also having a greater understanding of the risk factors through doing this would allow for clearer mitigation on the project for all parties'.

However, one respondent (ref. C1) contributed with the comment 'the identification of objectives at the start is key', with another adding an err of caution (ref. D5) stating that the 'framework is difficult to understand without the need of a more detailed discussion and worked examples'.

Summarising on these comments, data shows that the methodology itself is robust in collating multiple factors of both qualitative and quantitative assertions, in order to gain clarity of the applied benefits which 'allows stakeholders better understand the expectations of each party' (ref. A3).

9.3.4   Framework Understandability and Usability

In terms of the understandability and useability of the framework, each participant responded to the questionnaire by providing a numerical value i.e., 1 star = poor, 5 stars = excellent, based on their own interpretation and perspective. These values were then converted to a percentage factor i.e., 0% = poor, 100% = excellent, which is collated in Figure 9.3.

Extrapolating further on Figure 9.5, the average in terms of agreement that the framework was sufficiently understandable and usable was c. 74.6%. Further, responses show that those participants from academia felt more strongly (84%) that the framework was clear and useable, which is correlated with their familiarity with high-level outlined methodologies, which was then followed by construction at 76% and design at 64%, respectively.

Table 9.5 summarises these average responses from the perspective of each participant group (domain) and then from the perspective of the combined focus group averaged total.

9.3.5   Novelty of Hypothesis

Additionally, participants were asked whether they were aware of any other similar implementation framework that utilises qualitative and quantitative objectives alongside prioritised weightings across the lifecycle of collaborative BIM. In response to this question, the consensus was unanimous as no, with 100% of respondents spanning academia, design and construction agreeing with the novelty of the proposal.

Furthermore, and specifically, one respondent (ref. A2) added that a significant novelty was due to the 'inclusion of risk factors' with another (ref. C5) stating one reason for this,

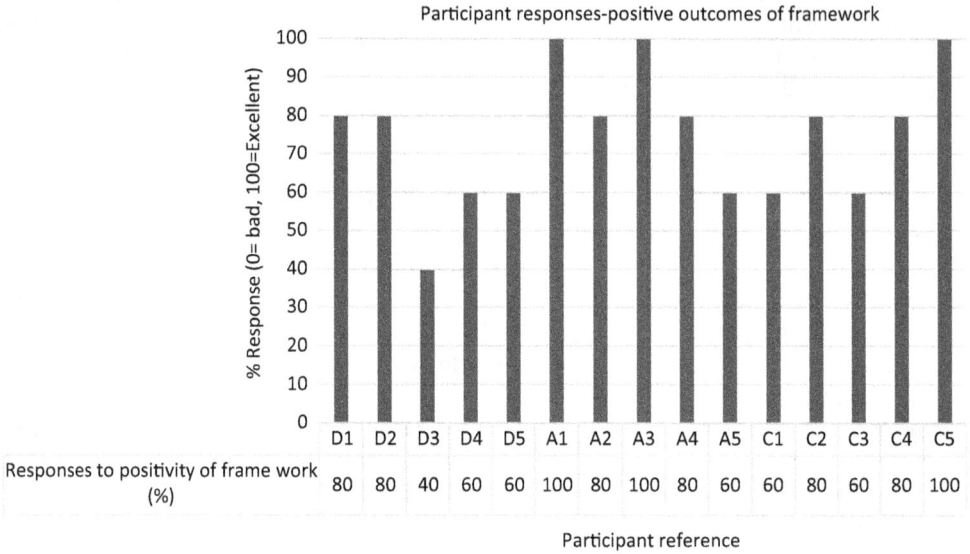

Figure 9.3 Responses towards positive outcomes and usability of framework.

Table 9.5 Average agreed consensus to framework understandability (0% = negative, 100% = positive)

Participant Domain	Domain Average	Combined Total Average
Design	64%	
Academia	84%	74.6%
Construction	76%	

taken from their 35 years in industry, was due to lack of pre-existing 'weighting values', which through this framework now enables both quantifiable and qualitative applied data input, extraction and analysis.

9.3.6   Likelihood of Improving Collaboration BIM

In response to the question of how likely the framework is in its current state to achieving positive outcomes driven via collaborative BIM, the average response rating was that 69.3%, with 100% being highly likely to achieve success and zero being the exact opposite i.e., unusable and restrictive. Further, academia and construction average responses to this question were both tied at 72%, with designers' responses averaging slightly lower at 64%. However, this was explained in greater detail as to why these weren't higher as one respondent (ref. D3) stated that 'more supporting materials' were required in order to make it more transparent and understandable, with another (ref. A5) expressing that 'better visualisation' of the process would improve and increase the intended purpose.

Further, one respondent (ref. C4) asserted that a key factor to the success is ensuring that 'the full team buy into the process' with another (ref. C2) stating that currently there is nothing else currently available to support project execution, with a direct benefit being its ability to 'focus on specifics' that existing methodologies fail to transfer, which opens project teams up to 'denial of understanding'.

9.3.7   *Process Improvements and Clarity*

The responses to the question on immediate observable process improvements were mixed, however, they were largely more positive than negative. For example, one respondent (ref. D2) stated 'the framework structure is well designed, and depth thought' with another (ref. D3) adding that due to their experience and more so confidence they found it 'confusing in its language', with another (ref. A3) stating that the process methodology 'makes good sense' and is 'solid'.

Finally, and in a positive remark, one respondent (ref. C2) believed the process is helpful due to 'the steps and the process are succinct and to the point'. Expanding on the clarity of the framework, a similar approach to visualising the previous weighting responses was used to capture quantitative data for the understanding of the framework, as shown within Table 9.5. The dominance on positive feedback for the framework's observable clarity was 72% from both academia and construction, with design biased participants averaging at 64% respectively, which is shown below as a total combined average as 69.3% in Table 9.6.

As shown in Figure 9.4, more than 50% of participants rated the hypothesis as 80% effective in terms of clarity, with one from construction in fact responding with a value of 100%. In addition, three respondents provided feedback with values of 40%, which appears to be focused on their perception in terms of 'simplicity', reduced clarity on whether the framework is aimed at specific 'BIM people' and/or whether industry would work hand in hand with 'industry experts implicit knowledge'.

Table 9.6 Average agreed consensus to framework clarity (0% = negative, 100% = positive)

Participant Domain	Domain Average	Combined Total Average
Design	64%	69.3%
Academia	72%	
Construction	72%	

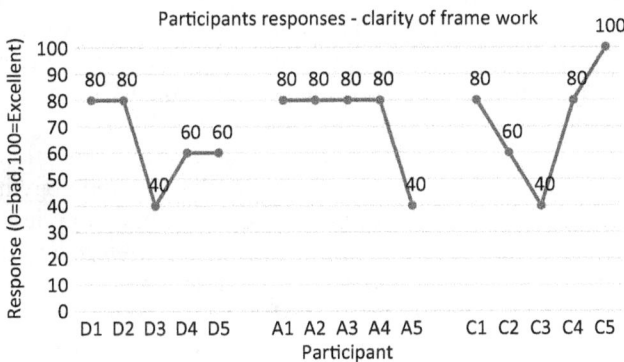

Figure 9.4 Responses towards clarity of framework.

9.3.8 *Summary*

In summary, the headline items and findings regarding the observations, comments and thus responses received from the focus group participants were grouped around the following items:

- The inclusion of a diverse range of stakeholders was achieved, which is relayed in the depth of responses;
- It is important and key that simplicity remains throughout the framework's philosophy;
- The designed framework methodology was determined as 100% novel by all participants;
- The framework has the ability to positively affect change towards the implementation of collaboration BIM, due to its objective focus, namely its specific and targeted relevance;
- People are the core focus and keeping the process clear and understandable for BIM experts as well as non-BIM experts is of necessary importance;
- Measurability in both quantitative and qualitative aspects for objectives, prioritisation, risk mitigation and wider visibility in determining the early required outcomes of collaborative BIM is original and advantageous;
- Adjustments to the hypothesis are required (see Table 9.4) in order to increase the understanding and thus the benefit of the applied framework; and
- Implementation on a real project case study workshop would add value and definitive advantage to be gained.

Due to the received observations, post analysis of the focus groups responses to the hypothetical framework proposal via the survey data collection process, there are opportunities as outlined within this section for core areas to be adapted prior to validation on an applied project (workshop). Finally, these adaptions, prior to undertaking a workshop case study, are presented within the next section in terms of physical amendments to the framework process flow.

9.4 Framework Hypothesis – Adaptions Post-Verification (Stage 1)

Following the review and analysis section of the data collated as part of the survey questionnaire, a targeted approach towards seeking a verification exercise following the two-stage Delphi methodology (Dalkey and Helmer, 1963) was gained. The core findings following the summary above show that there are viable areas for adjustment which will improve the understanding and implementation of the framework, from hypothesis to executable delivery. Therefore, the revisions made to the collaborative BIM implementation hypothesis have been incorporated as shown in Figure 9.5.

Further, one raised item captured within the 'transparency' and 'ease of use' sections of the data collection exercise, and thus initial hypothesis, was that the methodology itself took people some time to digest and interpret, which exposes the methodology itself to subjectivism and in all likelihood, proneness to attract errors in use. Therefore and prior to the project workshop as part of the validation stage, an interactive and collaborative tool (solution) was created in order to facilitate the workshop; automating the process flow step by step by automatically calculating prioritisations (weighted values) as part of the utilisation of complex areas such as AHP, as well as streamlining the process between

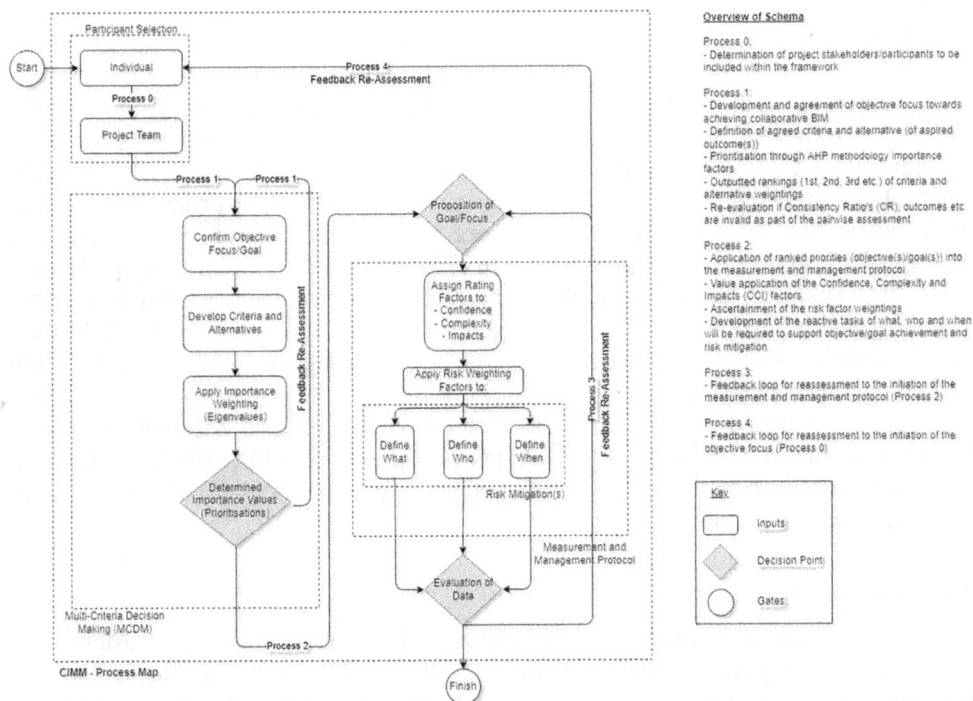

Figure 9.5 Framework – post-verification stage (adjustments made to process map).

objectifying the goal, weighting the priorities and defining the risks, opportunities and stage gates to increase the likelihood of positive outcomes as part of the collaborative BIM process.

Additionally, three distinct and complimentary supportive items were developed following review of participant feedback deduced from the completion of the questionnaire (framework verification), in order to reduce confusion and increase likelihood of successful outcomes in the application of the methodology. These were

• The development of the high-level framework (workflow map);
• The development of the process map (block steps to success); and
• The development of the schema flow (implementation flow at an executable level).

The rationale behind these three differing levels of supportive outputs, all of which are consolidated as part of the design of the technical solution, is to provide firstly a documented mechanism in which the methodology is understandable, appliable and implementable from feedback gained in terms of improving clarity at the first stage verification. In terms of expanding in detail upon each of the bulleted items above, the first is a high-level insight into the development methodology (workflow map), the second is a process map of how the methodology is to be practically applied (block steps) and the third is a granular flow of what is undertaken when and by whom at which stage (schema flow).

9.5   Project Case Study (Workshop): Validation Phase

As outlined within the methodology of the research and following the Delphi method-ology (Dalkey and Helmer, 1963), the second phase (validation) was to undertake a project workshop and implement the developed process onto a real-world case study, utilising the 'grounded theory' approach to absorbing a range of examined perspectives (Marttunen et al., 2015). The focus of the workshop project was to practically apply the methodology on a high value c. £70m building infrastructure BIM project (design and build), integrated into the structure engineering team including early contractor involvement, in support of the implementation of BIM at RIBA Stage 2 (concept design). This further enabled clarification and insight on the objective, prioritisations of goals, supporting factors, risk mitigation and heightened collaboration across all stakeholders with the author acting as a 'participative observer' throughout, utilising the 'action research' methodology (Dick, 2014).

Further, a schema flow was developed focused on the outlined requirements of the workshop in order to provide pre-planned approach in support of the validation process, taking into consideration the required amendments and factors gained from the first stage verification process as shown in Figure 9.6.

9.5.1   Core Findings

The workshop case study was initialised by stating the intent of utilising the framework to focus on defining the supporting elements (criteria and alternatives) in light of the main objective of the information requirements in scope of the project requirements. Ultimately, the aim of the workshop was to assist in defining the objectification of BIM on the specific project inclusive of the core stakeholders. Further, this was to pro-mote clarity of qualitative and quantitative outputs and mitigation of risks potentially challenging the ability to achieve successful collaboration.

The sections which follow outline the data that the participants provided as a unified, collaborative team with the pre-set objective focus of defining the core supporting elements and requirements for 'information requirements' at the project delivery level.

9.5.1.1   Identifying the Criteria and Alternatives

The first activity was for the collective participants to agree, with focus on the outlined project requirements and application of collaborative BIM, what the core criteria and alternatives are in support of achieving the objective through discussive dialogue. Further, criteria and alternatives were described as being the following, in order to aid clarity and awareness pre-assignment:

- Criteria: the upmost important elements in relation to the objective; and
- Alternatives: important supporting sub-elements which assist in achieving the objective, but feed into the criteria elements.

Firstly, PA5 (see Table 9.3 for anonymised reference list) made a valid suggestion from the outset that information requirements should be 'based on a vision, based on a common language' and were not typically defined at a project level. Further, 'key points' such as agreeing these from the outset would ensure that the criteria and alternatives were

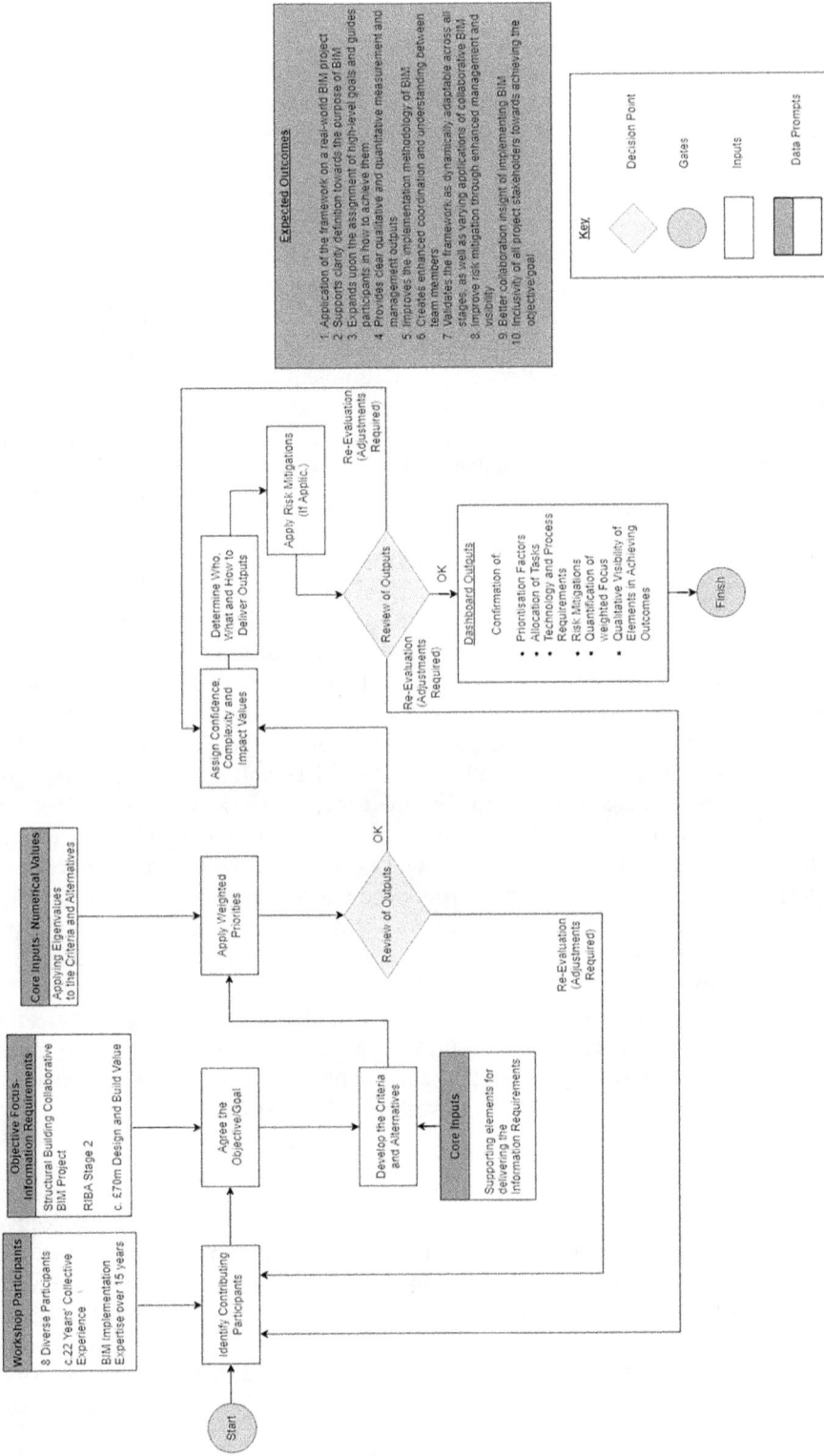

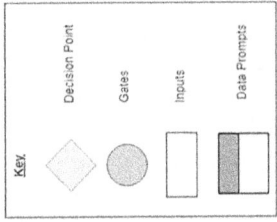

Figure 9.6 Schema flow supporting the framework validation workshop.

collectively and correctly defined, in order to support the positive delivery 'with a same understanding about the vision, the topics and the purpose'. Participant PA4 initiated the definition of the qualitative group inputs for the criteria, stating the clients 'end vision' was important to have in mind and therefore the 'Exchange Requirements' were collectively agreed that these should be a core criteria item. Participant PA2 followed this discussion by reaffirming that 'coordination' was a core element and 'not just coordination with the other disciplines on the project, but with the existing structure' due to the complexity surrounding the location (central London) as well as being a redevelopment of an existing asset with multiple teams with varying interests. Next, PA5 built upon the last response with the input of 'Stakeholders Requirements' are very important due to a range of perspectives, interests and priorities, with PA8 adding to this that the 'Constraints' were similarly important and intrinsically linked to planning, on-site activities as well as any other general activity required. PA4 added a comment that 'Level of Information Need' was key in determining not just for 'model data but also general information' such as reports, documents, schedules as well as geometric models, designs and calculations respectively. The group started to also challenge their five criteria for the matrix on topics such as 'budget and spend' with PA5 and PA3 stating that this could be part of the constraints item.

The concluding phase of the initial group activity was to determine of the secondary elements through agreements of the alternatives. PA2 initiated the first alternative with 'Budget' due to its detail and impact on each of the previously determined criteria. Next, PA5 added that 'Time' was a granular linked activity and important to every stakeholder, with PA4 adding that 'Responsibility' was required so that determination of 'who is doing what' is important for coordination of tasks teams in complex environments such as this project. Further, PA3 added that the 'intended aim' of the application of BIM needed to be outlined and understood by all stakeholders and therefore 'Purpose of Application' was another core element. Finally, the group was asked to reflect upon on defined criteria and alternatives that were collectively defined and object or challenge if they didn't think they were appropriate for the next stage of weighting the prioritisations – all team members agreed with their inputs and added that you could expand past the 5 × 5 matrix. Figure 9.7 is a visual summary of the objective, criteria and alternatives data inputs following the previous section, which is organised within a 5 × 5 matrix (5 criteria, 5 alternatives).

In the sub-sections which follow an exercise was undertaken to cross analyse the criteria against each of the other criteria in a quantitative scaled manner i.e., how much more important are the 'Exchange Requirements' versus each of the other elements, which was then also repeated for the alternatives in relation to the criteria.

9.5.1.2   *Applying the Prioritisation Factors to Criteria (Including CI and CR Values)*

Following the determination of the criteria and alternatives in support of the objective and a similar approach to collective agreement, eigenvalues (prioritised scaling factors) were discussed, agreed and assigned to each criterion in respect to their importance with each of the criterion towards the focus of achieving the goal. Furthermore, the eigenvalues, which were reduced to odd values only for reduction in range and thus an improvement in understanding and appliable speed, were presented in a tabular form throughout the workshop. This was important to make it clear to all participants the

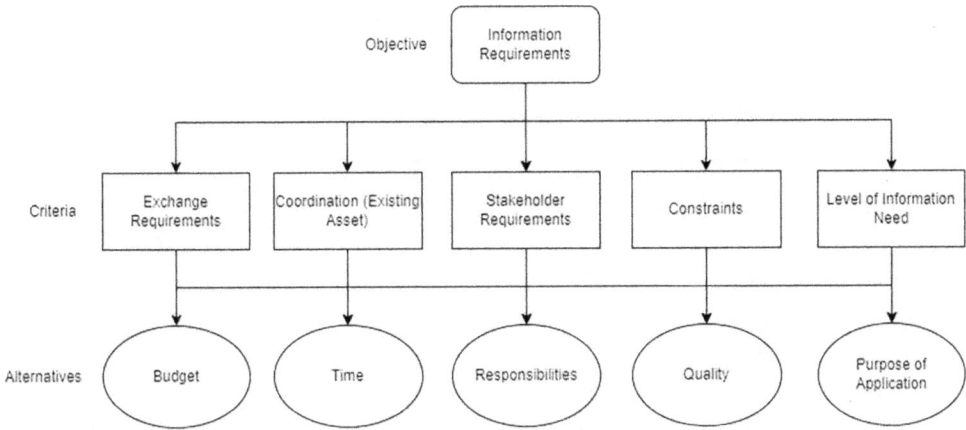

Figure 9.7 Analytical Hierarchy Process (AHP) matrix visual.

Table 9.7 Scaling factors (Eigenvalues)

Scaling Factors	
*Value**	*Description*
1	Equal to
3	Moderately important
5	Strongly important
7	Very important
9	Extremely important

*If the item assessed is actually less important, then the reciprocal value shall be used, i.e., 1/3, 1/5, 1/7, 1/9 respectively
**If CR value is >10% then it is deemed as inconsistent following the rationale of Saaty (1987), and values should be reapplied to gain below 10%. However, as this is an inherently inconsistent space

value, description and weighting of the scaling factors, as shown below in Table 9.7 (BIM) with multiple stakeholders' inputs it reaffirms the need for applying the method to prioritise focus and objectify through a unified team, as found by Pidgeon and Dawood (2021a) (Table 9.8).

In addition, a prioritisation summary was produced which was automatically calculated utilising the AHP-rooted metrics to emphasize and clarify the core findings, from this exercise. Moreover, it facilitates in providing increased clarity on which of the criteria was of impost importance, and thus, should be a focal point in achieving information requirements when delivering via collaborative BIM.

Furthermore, and upon reflection of Table 9.9, 'Constraints' were determined as having the upmost importance when thinking about prioritisations, followed by 'Stakeholder Requirements' and then 'Coordination of the existing asset' when directly focused on achieving good information requirements to satisfy the needs of the project, utilising collaborative BIM as a delivery mechanism.

Table 9.8 Pairwise comparison – Information Requirements (objective focus)

Criteria Description	*1. Exchange Requirements*	*2. Coordination (Existing Asset)*	*3. Stakeholder Requirements*	*4. Constraints*	*5. Level of Information Need*
	Focus – Information Requirements				
1. Exchange Requirements	1.00	1.00	0.11	0.14	1.00
2. Coordination (Existing Asset)	1.00	1.00	0.14	1.00	5.00
3. Stakeholder Requirements	9.00	7.00	1.00	0.20	9.00
4. Constraints	7.00	1.00	5.00	1.00	7.00
5. Level of Information Need	1.00	0.20	0.11	0.14	1.00
Total	19.00	10.20	6.37	2.49	23.00
Consistency Ratio (CR) Value (Inconsistent >10%)**				0.302 (30.2%)	
Consistency Index (CI)				0.339	
Lambda Max				6.354	

Table 9.9 Summarised prioritisation findings

Prioritisation Summary		*Ranking*
1. Exchange Requirements	5.4%	4th
2. Coordination (Existing Asset)	15.9%	3rd
3. Stakeholder Requirements	35.8%	2nd
4. Constraints	39.2%	1st
5. Level of Information Need	3.8%	5th

The section which follows is a repeat process of the above, however, it is targeted at the collective alternatives focused on each of the criteria respectively.

9.5.1.3   *Applying the Prioritisation Factors to Alternatives (Including CI and CR Values)*

The participants proceeded as part of the next stage of the workshop in analysing and assigning numerical weightings to the grouped alternatives, versus each of the dedicated individual criteria. This was in order to cross-examine each of the criteria and bring to the front where the emphasis and focus should be in a quantitative perspective.

9.5.2   *Exchange Requirements*

The tables which follow contain the key weighted values agreed and applied by the collective group members, with focus on the importance of each of the criteria versus the objective towards attaining exemplar information requirements via collaborative BIM execution, utilising the adapted AHP principles and weighting scaling factors (Tables 9.10–9.14).

Table 9.10 Pairwise comparison – Exchange requirements

	Focus – Exchange Requirements				
Criteria Description	*1. Budget*	*2. Time*	*3. Responsibilities*	*4. Quality*	*5. Purpose of Application*
1. Budget	1.00	5.00	9.00	3.00	3.00
2. Time	0.20	1.00	9.00	0.20	0.20
3. Responsibilities	0.11	0.11	1.00	0.14	0.14
4. Quality	0.33	5.00	7.00	1.00	7.00
5. Purpose of Application	0.33	5.00	7.00	0.14	1.00
Total	1.98	16.11	33.00	4.49	11.34
Consistency Ratio (CR) Value (Inconsistent >10%)**				0.258 (25.8%)	
Consistency Index (CI)				0.289	
Lambda Max				6.158	

9.5.3 Coordination (Existing Asset)

Table 9.11 Pairwise comparison – Coordination (existing asset)

	Focus – Coordination (Existing Asset)				
Alternative Description	*1. Budget*	*2. Time*	*3. Responsibilities*	*4. Quality*	*5. Purpose of Application*
1. Budget	1.00	1.00	1.00	0.14	0.14
2. Time	1.00	1.00	1.00	0.14	0.14
3. Responsibilities	1.00	7.00	1.00	0.14	0.11
4. Quality	7.00	7.00	7.00	1.00	0.14
5. Purpose of Application	7.00	7.00	9.00	7.00	1.00
Total	17.00	17.00	19.00	8.43	1.54
Consistency Ratio (CR) Value (Inconsistent >10%)**				0.107 (10.7%)	
Consistency Index (CI)				0.120	
Lambda Max				5.480	

9.5.4 Stakeholder Requirements

Table 9.12 Pairwise comparison – Stakeholder requirements

	Focus – Stakeholder Requirements				
Criteria Description	*1. Budget*	*2. Time*	*3. Responsibilities*	*4. Quality*	*5. Purpose of Application*
1. Budget	1.00	1.00	7.00	1.00	1.00
2. Time	1.00	1.00	3.00	0.33	0.11
3. Responsibilities	0.14	0.33	1.00	1.00	5.00
4. Quality	1.00	3.00	1.00	1.00	0.14
5. Purpose of Application	1.00	9.00	0.20	7.00	1.00
Total	4.14	14.33	12.20	10.33	7.25
Consistency Ratio (CR) Value (Inconsistent >10%)**				0.838 (83.8%)	
Consistency Index (CI)				0.939	
Lambda Max				8.756	

9.5.5   Constraints

Table 9.13 Pairwise comparison – Constraints

Criteria Description	1. Budget	2. Time	3. Responsibilities	4. Quality	5. Purpose of Application
			Focus – Constraints		
1. Budget	1.00	1.00	7.00	0.14	0.20
2. Time	1.00	1.00	1.00	0.14	0.14
3. Responsibilities	0.14	1.00	1.00	0.14	0.14
4. Quality	7.00	7.00	7.00	1.00	1.00
5. Purpose of Application	5.00	7.00	7.00	1.00	1.00
Total	14.14	17.00	23.00	2.43	2.49
Consistency Ratio (CR) Value (Inconsistent >10%)**				0.109 (10.9%)	
Consistency Index (CI)				0.122	
Lambda Max				5.490	

9.5.6   Level of Information Need

Table 9.14 Pairwise comparison – Constraints

Criteria Description	1. Budget	2. Time	3. Responsibilities	4. Quality	5. Purpose of Application
			Focus – Level of Information Need		
1. Budget	1.00	5.00	9.00	1.00	0.33
2. Time	0.20	1.00	9.00	1.00	0.33
3. Responsibilities	0.11	0.11	1.00	0.33	0.14
4. Quality	1.00	1.00	3.00	1.00	5.00
5. Purpose of Application	3.00	3.00	7.00	0.20	1.00
Total	5.31	10.11	29.00	3.53	6.81
Consistency Ratio (CR) Value (Inconsistent >10%)**				0.349 (34.9%)	
Consistency Index (CI)				0.390	
Lambda Max				6.562	

The section which follows provides an exclusive summary of the core findings determined by the workshop participants, taking data presented within the former sections.

9.5.6.1   *Summary of Criteria and Alternative Prioritisation*

Within Table 9.15 which follows, the data from the previous section is summarised with the ranked criteria alongside their supporting alternatives i.e., prioritised sub-element.

Participants collectively observed that these ranked findings made complete sense due to the complexity and requirements of the project within a 'closed complex environment, in central London, involving multiple disciplines and perspectives' with constraints being intrinsic to the challenges, stakeholder requirements being at the forefront of importance and coordination is integral to awareness when facing such a complex existing asset.

Table 9.15 Ranking table alongside supporting items

Ranking	Criteria	Supporting Alternatives	Prioritisation Percentage
1st	4. Constraints	4. Quality	39.17%
2nd	3. Stakeholder Requirements	5. Purpose of Application	35.78%
3rd	2. Coordination (Existing Asset)	5. Purpose of Application	15.86%
4th	1. Exchange Requirements	1. Budget	5.38%
5th	5. Level of Information Need	5. Quality	3.81%
Total Percentage			100.00%

9.5.6.2 Determining Confidence, Complexity and Impacts (CCI) Plus Risk Mitigation Factors

This CCI element took the top three priorities derived by the execution of the framework solution alongside the participants' inputs and facilitated a discussion to assign 'who, what and when' is required in order to increase the likelihood of successfully achieving or mitigating the risks and constrictions associated with these expectations. Further, these items were not isolated to qualitative steer but were supported by numerical assignment and detailed examination, by creating a simple ranked system (Likert scale) with further specification by the workshop participants on the CCI items linked to their connected focal input i.e., first, second, third ranked criteria.

Moreover, Table 9.16 contains those collaborative inputs gained from the participants utilising this section of the framework, with focus on applying the CCI principles to the progressive findings within the scope of the project requirements, alongside mitigations and targeted focus on how to better deliver against the required outcomes.

9.6 Conclusive Remarks

This research study's aim was to verify (stage 1 – survey questionnaire) and validate (stage 2 – workshop case study) a novel approach to executing collaborative BIM through progress research findings stemming from outlining the inefficiencies through to the development of an implementable framework methodology, that can be applied across a range of stakeholders and project focus, in order to better define both quantitatively and qualitatively what the objective purpose is of collaborative BIM for a range of stakeholder inputs. Moreover, the appropriation of how to implement collaborative BIM was refined by utilising the Delphi method through verification of the hypothesis including a range of 15 BIM experts spanning academia, construction and design, as well as validation of the developed framework on a real-world, high value and complex building bound by BIM as a delivery mechanism as part of the workshop case study, involving eight core project participants. Throughout this research it was determined that there was a novel and positive result in having a targeted implementation strategy, and thus framework, that doesn't just imply that collaborative BIM is the requirement but rather assists project teams spanning multiple stakeholders in outlining what they want to achieve, what their expectations are in terms of core underpinning supportive criteria and alternatives alongside numerical logic that provides percentage prioritisation factors

Table 9.16 Confidence, Complexity and Impact (CCI) workshop input

Confidence, Complexity and Impact (CCI) Measurement and Management Framework

Focal Point (Objective)	Confidence (in Being Able to Deliver the Task) 1=High 2=Medium 3=Low	Complexity (of the Task/Goal) 1=Non-complex 2=Complex 3=Highly Complex	Impact (on Wider Objectives if Not Achieved) 1=Low 2=Medium 3=High	Total Sum (C+C+I)
4. Constraints	1	3	3	7
3. Stakeholder Requirements	1	3	3	7
2. Coordination (Existing Asset)	2	3	3	8

Risk Factor (from total sum above)

Low	1–3
Medium*	4–6
High*	7–9

What (needs to be undertaken or utilised to improve the likelihood of success?)

4. Constraints — Identification, diligence, review of data input (awareness of asset data)
3. Stakeholder Requirements — Gateway review, consistent dialogue (early), key client contact
2. Coordination (Existing Asset) — Diligence and coordination, data formation, tool selection (appropriate)

Who (needs to support the task for effective delivery?)

4. Constraints — Design team
3. Stakeholder Requirements — All stakeholders
2. Coordination (Existing Asset) — All project stakeholders, designers, owners/operator

When (will this start, be re-evaluated i.e., monthly, and finish?)

4. Constraints — Early and throughout (key milestones)
3. Stakeholder Requirements — Permanent process, tender
2. Coordination (Existing Asset) — Permanent process but early through to as built

*Mitigations (if risks are 'medium' and/or 'high')

4. Constraints — Surveys, stakeholder engagement, brief review
3. Stakeholder Requirements — Brief review (iterative), milestones, skilled personnel (consistency)
2. Coordination (Existing Asset) — Surveys, regular reviews, historical info, risk budget

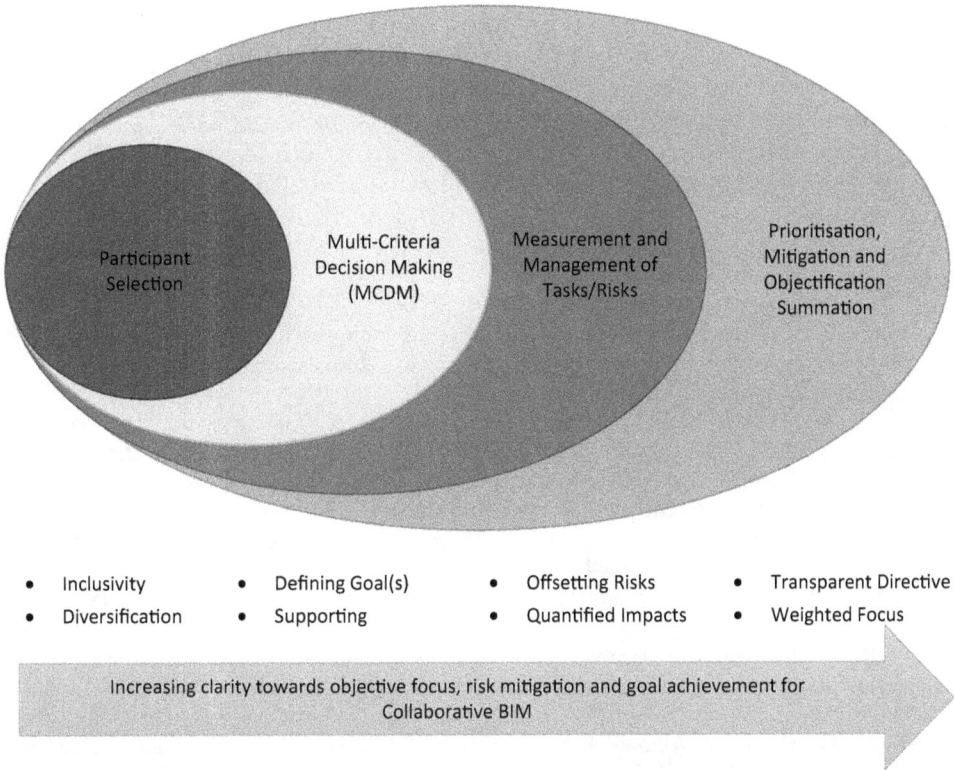

Figure 9.8 Collaborative implementation, measurement and management (CIMM) BIM framework.

based on dialogue and reasoning through multiple stakeholder inputs. Furthermore, risk
mitigation and targeted focus on what is required in order to increase the likelihood of
successfully implementation collaborative BIM alongside the typical project complexities
was also confirmed as unique and highly valuable, which was reflected in both the survey
data collection spanning multiple sectors as aforementioned, as well as being realised at
the project execution level.

Finally, post verification and validation, Figure 9.8 visually represents the finalised
Collaborative Implementation, Measurement and Management (CIMM) BIM Framework
in terms of summarized validation.

9.7 Limitations and Implications

The limitations of this research were that only 15 BIM expert practitioners were utilised at
stage 1 (verification). Further, a limitation of, eight industry-focused BIM expert practitioners were utilised at stage 2 (validation). Further, validation of the framework proposal may benefit from exposure to a wider range of projects and practitioners to
ascertain further the advantageous elements as well as restrictive factors in order to
additionally develop the framework. Additionally, as found at stage 2 (validation) by
hosting the platform as a web-based interactive solution enhanced data analytic

methodologies could be applied using intelligence factors such as Machine Learning (ML). This would de-silo the findings within the common areas of application and infer better lessons learnt, best practices and detection of trends and themes. However, this could also expose the solution to complexities such as requiring projects to be literature with these applied techniques. Finally, due to the ongoing impacts of COVID-19 both the survey questionnaire and project workshop case study were held virtually utilising an interactive web-hosted questionnaire tool, as well as online conference video technologies respectively.

References

Abuelmaatti, A, & Ahmed, V. (2014). Collaborative technologies for small and medium-sized architecture, engineering and construction enterprises: Implementation survey, *Itcon*, 19, 210–224, https://www.itcon.org/2014/12 (accessed 17 April 2022)

Ashcraft, H. W. (2008). Building information modelling: A framework for collaboration. *Construction Law*, 28, 5., http://www.scl.org.uk/files/101-ashcraft.pdf (accessed 17 April 2022)

Barlish, K., & Sullivan, K. (2012) How to measure the benefits of BIM—A case study approach. *Automation in Construction*, 24, 149–159, 10.1016/j.autcon.2012.02.008

Blessing, L. T., & Chakrabarti, A. (2009), *DRM, a Design Research Methodology*, Springer, London, 2009, ISBN 978-1848825864

Bromley, D. B. (1986), *The Case-Study Method in Psychology and Related Disciplines*, ISBN 978-0471908531

Buur, J. (1990), *A Theoretical Approach to Mechatronics Design [Ph.D. Thesis]*, Institute for Engineering Design, Technical University of Denmark, Lyngby, Denmark, 10.1080/15710880802 098099

Chahrour, R., Hafeez, M. A., Ahmad, A. M., Sulieman, H. I., Dawood, H., Rodriguez, S., Kassem, M., Naji, K. K., & Dawood, N. (2020). Cost-benefit analysis of BIM-enabled design clash detection and resolution. *Construction Management and Economics*, 39(1), 55–72, 10.1080/014461 93.2020.1802768

Colbry, S., College, C., & Adair, R. (2014). Collaboration theory. *The Journal of Leadership Education*. 13, 63–75, 10.12806/V13/I4/C8.

Dalkey, N., & Helmer, O. (1963), An experimental application of the Delphi method to the user of experts. *Management Science*, 9(3), 458–467, https://www.jstor.org/stable/2627117?seq=1 (accessed 17 April 2022)

Denyer, D., & Tranfield, D. (2009). Producing a systematic review. In P. D. Buchanan & P. A. Bryman (Eds.), The Sage Handbook of Organizational Research Methods (pp. 671–689). London: Sage Publications, ISBN 978-1-4462-0064-3

Dick, B. (2014), Action research. In Qualitative Methodology (pp. 50–66). SAGE Publications, Inc., 10.4135/9781473920163

Eadie, R., Browne, M., Odeyinka, H., Mckeown, C., & McNiff, S. (2013). BIM implementation throughout the UK construction project lifecycle: An analysis. *Automation in Construction*. 36, 145–151, 10.1016/j.autcon.2013.09.001.

Eastman, C., Eastman, C. M., Teicholz, P. et al. (2011), BIM Handbook: A Guide to Building Information Modeling for Owner, Manager, Designer, Engineers and Contractors, Third Edition, ISBN: 978-1-119-28753-7

Efficiency and Reform Group (2011), Government Construction Strategy, Cabinet Office, London, UK, https://www.gov.uk/government/publications/government-construction-strategy (accessed 17 April 2022)

Gilligan, B., & Kunz, J. (2007), VDC Use in 2007: Significant Value, Dramatic Growth, and Apparent Business Opportunity, Center for Integrated Facility Engineering, ReportTR171 (2007), http://cife.stanford.edu/ (accessed 17 April 2022)

Grilo, A., Zutshi, A., Jardim-Goncalves, R., & Steiger-Garcao, A. (2013), Construction collaborative networks: The case study of a building information modelling-based office building project, *International Journal of Computer Integrated Manufacturing*, 26(1–2), 152–165, 10.1080/0951192x. 2012.681918

Hollenbeck, J. R., Beersma, B., & Schouten, M. E. (2012). Beyond team types and taxonomies: A dimensional scaling conceptualization for team description. *Academy of Management Review*, 37(1), 82–106, https://www.jstor.org/stable/23218853 (accessed 17 April 2022)

Lincoln, Y., & Guba, E (1985), *Naturalistic Inquiry*, Sage Publications, USA, 1985, ISBN 978-0-803-92431-4

Marttunen, M., Mustajoki, J., Dufva, M., & Karjalainen, T. P. (2015), How to design and realize participation of stakeholders in MCDA processes? A framework for selecting an appropriate approach. *EURO Journal on Decision Processes*, 3(1–2), 2015, 187–214, ISSN 2193-9438 10.1007/s40070-013-0016-3.

McCuen, T. L. (2008), Scheduling, Estimating, and BIM: A Profitable Combination. *AACE International Transactions*, 1–8, http://ascpro0.ascweb.org/archives/cd/2008/paper/CPRT276002008.pdf (accessed 17 April 2022)

Miettinen, R., & Paavola, S., (2014), Beyond the BIM utopia: Approaches to the development and implementation of building information modelling. *Automation in Construction*, 43, 84–91, 10.1016/j.autcon.2014.03.009

O'Sullivan, A., & Sheffrin, S. M. (2003), *Economics: Principles in Action*, Pearson Prentice Hall, Upper Saddle River, NJ, pp. 157. ISBN 978-0-13-063085-8 .

Ojo, A. & Pye, C. (2020), BIM implementation practices of construction organisations in the UK AEC Industry. *Project Management Journal*, IX, http://usir.salford.ac.uk/id/eprint/61801/1/pmwj98-Oct2020-Ojo-Pye-BIM-implementation-in-construction-orgs-in-UK-AEC-industry.pdf (accessed 17 April 2022)

Oraee, M., Hosseini, M. R., Edwards, D., & Papadonikolaki, E. (2021). Collaboration in BIM-based construction networks: A qualitative model of influential factors. *Engineering Construction & Architectural Management*. ahead-of-print. 10.1108/ECAM-10-2020-0865.

Patton, M. (2002), *Qualitative Research and Evaluation Methods*, Sage Publications, USA, 2002, pp. 244–245, ISBN 978-0-7619197-1-1

Pidgeon, A., & Dawood, N. (2021). BIM adoption issues in infrastructure construction projects: Analysis and solutions. *Journal of Information Technology in Construction (ITcon)*, 26, 263–285, 10.36680/j.itcon.2021.015

Pidgeon, A., & Dawood, N. (2021a). Bridging the gap between theory and practice for adopting meaningful collaborative BIM processes in infrastructure projects, utilising multi-criteria decision making (MCDM). *Journal of Information Technology in Construction (ITcon), Special issue: 'Construction 4.0: Established and Emerging Digital Technologies within the Construction Industry (ConVR 2020)'*, 26, 783–811, 10.36680/j.itcon.2021.043

Pidgeon, A., & Dawood, N. (2021b), The development of an adaptable outcome-focused measurement and management methodology, incorporating risk mitigation assignment in support of collaborative BIM. In N. Dawood, F. Rahimian, & M. Sheikhkhoshkar (Eds.), *Proceedings of 21st International Conference on Construction Application of Virtual Reality* (pp. 152–165). Middlesbrough, UK: Teesside University Press, ISBN: 978-0-9927161-3-4

Pihlak, M., Poerschke, U., Holland, R. J., & Messner, J. I. (2010). BIM collaboration across six disciplines. Proceedings of the International Conference on Computing in Civil and Building Engineering, Pennsylvania State University, PA, USA, ISBN 978–1907284601

Rosnow, R., & Rosnow, M. (2011). *Writing Papers in Psychology*, Cengage Learning, Belmont, CA, ISBN 9780534243784

Shavelson, R. J., & Towne, L. (Eds.). (2002). *Scientific research in education*, National Research Council, National Academy Press, Washington, DC, 10.17226/10236.

Succar, B. (2009), Building information modelling framework: A research and delivery foundation for industry stakeholders, *Automation in Construction*, 18 (3), 357–375, 10.1016/j.autcon.2008. 10.003

Taylor, J. (2007). Antecedents of successful three-dimensional computer-aided design implementation in design and construction networks. *Journal of Construction Engineering and Management*, 133, 10.1061/(asce)0733-9364(2007)133:12(993)

Taylor, J. E., & Bernstein P. G. (2009). Paradigm trajectories of building information modelling practice in project networks. *Journal of Management in Engineering*, 25(2), 69–76, http://ascelibrary.org/doi/abs/10.1061/%28ASCE%290742-597X%282009%2925%3A2%2869%29 (accessed 17 April 2022)

Whyte, J., Bouchlaghem, N., Thorpe, A., & McCaffer, R. (1999). A survey of CAD and virtual reality within the house building industry. *Engineering, Construction, and Architectural Management*, 6(4): 371–379, 10.1108/eb021125

Wondolleck, J. M., & Yaffee, S. L. (2000), *Making Collaboration Work: Lessons from Innovation in Natural Resource Management*, Island Press, Washington, DC, 10.5070/G311610473

Wood, D. J., & Gray, B. (1991). Toward a comprehensive theory of collaboration, *The Journal of Applied Behavioral Science*, 27(2), 139–162, 10.1177/0021886391272001

Yang, C., & Mao, L. (2021). Analysis on risk factors of BIM application in construction project operation and maintenance phase. *Journal of Service Science and Management*, 14, 213–227. 10.4236/jssm.2021.142013.

Yin, R. K. (2003), *Case Study Research: Design and Methods*, 5th Edition, Sage Publications, Thousand Oaks, CA, USA, ISBN 978-1506336169

Yukl, G., Chavez, C., & Seifert, C. F. (2005), Assessing the construct validity and utility of two new influence tactics. *Journal of Organizational Behaviour*, 26, 705–725, 10.1002/job.335

Zuppa, D., Issa, R., & Suermann, P. (2009). BIM's impact on the success measures of construction projects. *Journal of Computing in Civil Engineering*, 346, 503–512, 10.1061/41052(346)50.

10 BIM Adoption Issues in Infrastructure Construction Projects

Analysis and Solutions

Andrew Pidgeon and Nashwan Dawood

School of Computing, Engineering and Digital Technologies,
Teesside University

10.1 Introduction

In 2011, the UK Government set out in prescribing a UK Building Information Modelling (BIM) Mandate within its Government Construction Strategy (HM Government, 2012). The intention by UK Government was to forge together and strengthen project stakeholders at the project delivery stage with a requirement that all centrally procured Construction contracts shall require fully "collaborative 3D BIM" from 2016 onwards. Furthermore, this mandate sought to achieve a "20% saving" through efficiencies via adoption and advancements realised through anticipated value, as well as improving the quality, timeliness and productivity of Project Delivery of Collaborative BIM via Digital methods. This intention here is to also incorporate new digital ways of working as part of the UK BIM Framework Roadmap (UK BIM Alliance, 2020).

The expectation was that over time, digitisation across projects, from initial feasibility studies through to the execution and delivery stages would result in efficiencies gained alongside the development of specifications and standards, with focus on Digital Engineering and BIM advancements. The standardised processes were and are namely the PAS1192 suite of specifications (BIM Level 2) and the more recent ISO19650-1:2018 and ISO19650-2:2018 (UK BIM Framework), with inclusion of Digital best practice, seamless sharing of information and a collaborative approach to Information Management and delivery. Further, ISO19650-1:2018 (BSI, 2018) states that BIM is defined as being "The use of a shared digital representation of a built asset to facilitate design, construction and operation processes to form a reliable basis for decisions".

A timeline of these standardisation developments alongside the BIM Mandate, frameworks and focused groups is shown in Figure 10.1, sourced from the UK BIM Framework (UK BIM Alliance, 2020). Whilst all of these standards and UK government initiatives have impacted construction processes, there are still issues with implementation of BIM resulting in inefficiencies, gaps in proper implementation and inability of the industry to cultivate the value of BIM. In this context, the aim of this study is to identify and provide solutions of the issues that preventing the industry from embracing BIM.

In order to achieve the aim of this research project, four main objectives have been set. These are to

- Undertake a systematic literature review of academic and a critical review of industrial literature – This involves academic and industrial literature and reports reviews;

DOI: 10.1201/9781003408949-10

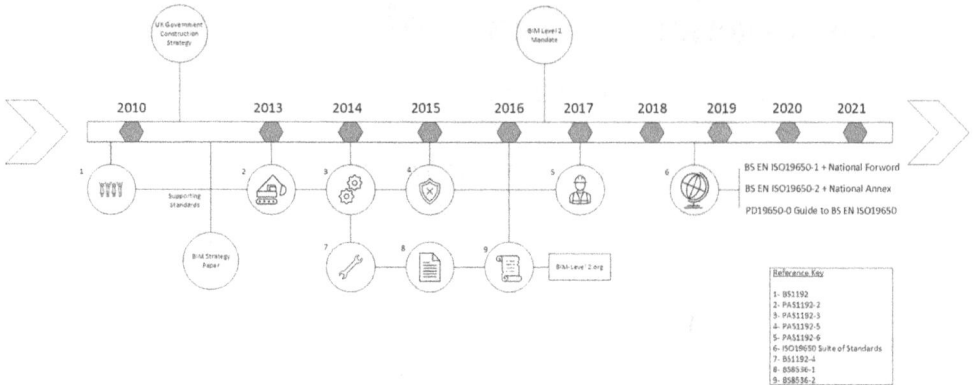

Figure 10.1 UK BIM framework timeline (UK BIM Alliance, 2020).

- Analyse, contest and support literature by undertaking a data collection exercise with a range of industry experts who are delivering construction projects and in particular infrastructure via BIM;
- Provide a comparative analysis between data collection, academic and industrial literature in a holistic format; and
- Propose a solution focused on countering the disadvantages and inefficiencies in the adoption of BIM to the benefit of future projects.

Following on from the above, each section within this research paper is intended to build upon the knowledge attained, from both a theoretical and practical perspective respectively. A comparative analysis and summary outline a proposal for a solution framework, which is focused on increasing opportunities and efficiencies and primarily aimed at enhancing the delivery of projects via BIM. This is by reducing the preventative and restrictive gaps and inefficiencies towards gaining applied advantage.

10.2 Theoretical Development of BIM Adoption in Infrastructure Projects

10.2.1 *Systematic Literature Review*

A systematic literature review (Borenstein et al., 2009) which has been undertaken as part of the design procedure, providing a "Grounding Theory" (Glaser and Strauss, 1967), whilst also underpinning and including research methodologies, observations and conclusions ascertained from previous research (Paré et al., 2015). The Preferred Reporting Items for Systematic Reviews and Meta-Analyses (PRISMA, 2009) approach was utilised in this research to coordinate and construct the main elemental structure of the research methodology (Cohen, 2017), as well as providing a focus on the keyword string searches to filter onto appropriate literature to support the research topics and its findings (outlined within Tables 10.1 and 10.2).

Further to the latter, exploration of relevant literature subject titles from online academic libraries (e.g. ResearchGate, Elsevier, University Library, etc.) and areas by research experts was undertaken to add credibility and targeted focus to the methods

Table 10.1 Systematic literature review – Academic literature subject focus

Subject Focus/Searched Topics	Citation
Benefits of BIM Application	Singh et al, 2011; Vernikos, 2016; Da Silva et al., 2019
Return on Investment	Eadie et al, 2014; Doloi, 2013; Azhar et al., 2008; Akponeware et al., 2017
BIM v. Non- BIM Project Review	Barlish and Sullivan (2012)
Benefits of BIM	Young et al., 2009; Chahrour et al, 2021; Bargstädt, 2015
Lessons Learned and Case Studies	Ustinovičius et al., 2018; Hardin & McCool, 2015; Highways England, 2019
Cultural and Behavioural Changes	Latham, 2003; Bargstädt, 2015
Training and Development Needs	Hore et al, 2011
Adoption and Implementation of BIM	Zuppa et al., 2009; Doloi, 2013; Vernikos, 2016

Table 10.2 Systematic review – Industrial literature subject focus

Subject Focus/Searched Topics	Citation
BIM Case Studies	Constructing Excellence, 2014
Measuring Collaboration	Greenwood Consultants, 2020
Industry Performance Reports	CITB, 2018
National Infrastructure Strategy	HM Government, 2020
BIM Implementation	UK BIM Alliance, 2020
Construction Reporting	McKinsey, 2017

presented (Hart, 2001). A selection of some of these academic areas alongside their subject focus can be found in Table 10.1, with Table 10.2 outlining similar approaches but with focus on Industrial literature.

To date there have been extensive academic and literature research techniques applied, highlighting some of the benefits of BIM such as but not limited to application within 4D construction scheduling (Da Silva et al, 2019), off-site construction techniques (Vernikos, 2016) and Clash detection and avoidance (Akponeware et al, 2017; Chahrour et al, 2021).

In addition, literature from academic research teaches us that there is a benefit to applying BIM, with a "33% saving in cost reduction" as affirmed by Azhar et al. (2008) and thus financial benefit being realised through the utilisation of BIM for activities such as Clash Detection (Chahrour et al, 2021). Further, a key success factor at the Project Implementation Stage (from Design to Construction) stems from creating a collaborative, engaged and motivated Teams (Vernikos, 2016). This allows projects and programmes to shift their delivery teams' focus on thinking more broadly about how they will deliver the Projects, with Bargstädt (2015) stating that the selection of appropriate toolsets is key to achieving better outputs via the BIM process(es).

However, there is also counter-evidence to suggest that benefits are very much limited in specific areas such that there is a "lack of coherent collaborative working" (Kirby et al, 2018) with Shennan (2016) stating one reason for this is that procurement routes are "rooted in a pre-BIM world" and have an of absence modernisation and appropriation. Moreover, Aljumaiah (2020) believes that compliance of the processes and procedures specifically for BIM and "not upheld well", and due to these reasons "only one-third of projects are delivered on time and budget".

Furthermore, Ojo and Pye (2020) confirm that the Construction industry is a "complex" and dynamic in nature, which appears to be a probable reason why construction projects are difficult endeavours to deliver successfully and instil innovative methods. This is not purely in terms of or due to their volumetric quantum, but a resulting factor due to their entangled elements such as plant, people, processes, high risk and cost (Baloi and Price, 2003). There is however a fragmentation within the AEC industry as well as reluctance to adopt and adapt to change, roles and responsibilities plus awareness and wider benefit appreciation (Singh et al., 2011).

Eadie et al. (2014) also reaffirm this in agreement in that the Return On Investment (ROI) and benefits of adopting and implementing BIM are typically less likely to be realised, which may result in poor future adoption by Delivery Teams. Moreover, Barlish and Sullivan (2012) state that there is actual very little to no data available to compare "non-BIM projects against BIM projects" which makes the former statement difficult to validate (both positively and negatively).

Moreover, Young et al. (2009) have undertaken research which outlines some of the benefits of BIM, which include a Return on Investment (ROI) that can be achieved via application of BIM via experts, which can add "three to four times" more value in terms of internal benefits in comparison to traditional production of and delivery methods.

Hardin and McCool (2015) affirm that this latter statement then results overall in "better outcomes" for all stakeholders; be they directly or indirectly involved in the project itself and/or from lessons learned and best practice feeding back from project delivery examples. Moreover, whilst harnessing and utilising beneficial and novel approaches to Project Delivery moving away from the inherent and slow-to-change inefficiencies within Construction projects. Latham (2003) outlines in the "Five Step Approach to Behavioural Change" that it is of utmost importance that systems of measuring outcomes "must be aligned with goals". Further to this Latham positions himself further by stating if outcomes aren't aligned with goals then "dysfunctional behaviours" are guaranteed. However, Zuppa et al. (2009) have also found that the adoption and more so successful implementation of BIM "is slow due to ambiguity surrounding its definition purpose and value".

Hore et al. (2011) provide a "people driven" theory behind why such assurance values aren't realised more positively by BIM, stating that the majority of the time "Training and Education" is not "subsidised by the Client" (and thus a cost to the Project), which typically and usually isn't delivered resulting in poor awareness and experience of how to apply theory at a delivery level. Moreover, Doloi (2013) states that another of the fundamental reason as to why time, cost and poor management of projects isn't realised is due to "poor reporting" as well as poor planning and scheduling, which have a more direct impact on cost performance, delivery due to immeasurability.

There is in fact further evidence from Doloi (2013) towards suggesting why Construction projects are inherently late, overbudget and are one of the most underproductive areas of work is due to "vagueness in scope", with a further study suggesting that innovative Contractors are "50–60%" more productive (McKinsey, 2017).

Khosrowshahi and Arayici (2012) also reaffirm that although there are standards, the barriers to adoption are "driven by culture", which is people led and includes humanistic behaviours and attitudes. However, Forcael et al. (2020) outline this as an "over complex nature" but appreciate that although the Technical and Functional Requirements are important, the behavioural elements alongside BIM in terms of training people to further understand the collateral in a succinct manner who are non-specialists must also be a priority.

Moreover, Da Silva et al. (2019) state that although a desire can be assigned as a goal or objective, "lack of integration" results in a fundamental incompatibility between aspiration and realisation at the delivery phase.

10.2.2   Conclusions from the Literature Review

In summary following the systematic literature review, it appears that there are clearly stated advantages from previous research undertaken such as benefits in cost realisation and "value of BIM" (Chahrour et al., 2021), success driven by effective collaboration and "motivated teams" (Vernikos, 2016), as well as the selection of toolsets to align with the "likelihood of achieving success" (Bargstädt, 2015). In addition, the value of implementation was also affirmed by Young et al. (2009) with benefits gained are "three to four times more likely" by having skilled and qualified staff supporting project delivery via BIM.

However, equally, there are also issues restricting adoption such as legacy procurement routes restricting the level of adoption (Shennan, 2016), innovation being typically hard to apply and complex (Forcael et al., 2020) and a "lack of integration" (Da Silva et al., 2019) restricting the uptake and understanding of how BIM is to be applied practically. Moreover, common trends such as poor reporting (Doloi, 2013) are a "contributing factor" in reducing successful delivery, with Khosrowshahi and Arayici (2012) stating this is also a factor which is "driven by culture" and thus human-centric. Further, Zuppa et al. (2009) state that there is a reduction in projects being able to gain foresight of what the realised benefits could be, leading to poor planning and a reduced ROI.

Therefore, and in conclusion to this section, a deeper analysis of both literature and data collection through the means of semi-structured interviews directly linked to infrastructure projects is needed to gain insight, improve clarity and provide wider opportunities to support and develop knowledge gained from the research findings.

10.3   Analysing the Infrastructure Project Authority (IPA) Annual Report

The IPA is required by a Government mandate originally set in 2011 (HM Government, 2011) and revised in 2021 (HM Government, 2020), to produce annual findings for all projects that form part of the Government Major Projects Portfolio (GMPP); outlining their performance, comparison against previously reports, confidence factors in terms of delivering to scope and areas for future development respectively.

These annual IPA reports essentially split UK Major Projects into four Departmental areas which include Infrastructure and Construction, Military Capability, Government Transformation and Service Delivery and Information and Communications Technology. The GMPP areas are summarised and further outlined in Figure 10.2 for 2019–2020, which amount to £448bn of value aligning to Whole Life Costs (WLC).

Another key function of the IPA is to provide governance, assurance and guidance to better inform Project Management and Delivery across their portfolio, as projects progress throughout their lifecycle, but to also stakeholders and interested parties of best practice, guidance and lessons learned. Through the use of tools such as a Delivery Confidence Assessments (DCA), the likelihood of success factors are evaluated on a Red, Amber, Green (RAG) traffic light system methodology, with appropriate steer and

Government Major Projects Portfolio	125 projects	£448bn Whole Life Cost
Infrastructure and Construction	34 projects	£214bn Whole Life Cost
Government Transformation and Service Delivery	34 projects	£64bn Whole Life Cost
Military Capability	30 projects	£162bn Whole Life Cost
Information and Communications Technology (ICT)	27 projects	£8bn Whole Life Cost

This comprises:

1   The Government's GMPP Data Transparency Policy covers Whole Life Cost, Delivery Confidence Assessment, and project end dates. It does not include project benefits. See Annex for further information.

Figure 10.2 Extract from the IPA 2019–2020 major projects review report (IPA, 2020).

support to inform so that turnaround is more likely if negative elements are evidenced, and more positively best practice shared to the wider GMPP.

Focusing back to the evaluation of the report itself, one of the fundamental roles of the IPA to its "commitment towards to improving Project Delivery across Government" with its "three P's". These founding principles are:

1 People (need for the skills, competence and behaviours);
2 Principles (the clarity and basic understanding); and
3 Performance (driving a step change in performance together with industry).

The IPA's mantra is to focus efforts on improving the successful delivery of Major Projects, reporting directly into Government at a portfolio level. However, despite their being a consensus of a proven track record of delivering projects against the Cabinets priorities and requirements, Infrastructure and Construction projects are shown within as being still inherently late and overbudget, with complexities having wider impacts on Social, Economic and Environmental areas respectively. Projects themselves receive independent scrutiny, assurance support/services plus guidance from IPA however as stated there are significant fundamental problems with Project Delivery that typically result in negative outcomes, such as overspend, late decision-making, and programme overrun.

10.3.1   *Benchmarking System*

Since 2013 at the inception of the mandate and thus the analysis of projects via the production of annual reporting, the IPA has undertaken DCAs for all projects which are part of the GMPP and measured via a simple yet effective Red, Amber, Green (RAG) rating system.

The IPA reports are evaluated and summarised within this section of the paper. Coincidentally, these findings, ranging back seven years sit alongside the implementation and transition of the UK BIM Mandate (HM Government, 2012) as part of the UK BIM Framework (UK BIM Alliance, 2020).

Table 10.3 Explanation of DCA colour ratings (redrawn from IPA, 2020)

IPA Colour Rating	Description
Green	Successful delivery of the project on time, budget and quality appears highly likely and there are no major outstanding issues that at this stage appear to threaten delivery significantly.
Amber/Green	Successful delivery appears probable; however, constant attention will be needed to ensure risks do not materialise into major issues threatening delivery.
Amber	Successful delivery appears feasible but significant issues already exist, requiring management attention. These appear resolvable at this stage and, if addressed promptly, should not present a cost/schedule overrun.
Amber/Red	Successful delivery of the project is in doubt, with major risks or issues apparent in a number of key areas. Urgent action is needed to address these problems and/or assess whether resolution is feasible.
Red	Successful delivery of the project appears to be unachievable. There are major issues with project definition, schedule, budget, quality and/or benefits delivery, which at this stage do not appear to be manageable or resolvable. The project may need re-scoping and/or its overall viability reassessed.
Reset	A significant change to a project's baseline which involves a business case refresh or change.
Exempt	Data can be exempt from publication under exceptional circumstances and in accordance with Freedom of Information requirements – i.e. national security

Each project has an evaluation undertaken at key decision points or "gates" which focus at specific intervals on the likelihood of a Project being able to achieve as a minimum timely delivery, benchmarked costings and the least disruption possible away from the Project scope. Moreover, the ratings are based on the qualitative RAG conclusion with interim points i.e. Amber/Red and Amber/Green amounting to five areas of measurement across the 125 projects for the year 2019–2020.

To better inform and describe the benchmarking criterion behind the RAG system, Annex A within the appendices of the IPA report explains further in plain language the rationale in that "The DCA is the IPA's evaluation of a project's likelihood of achieving its aims and objectives and doing so on time and on budget" (IPA, 2020). This is further summarised in Table 10.3 and is used alongside the evaluation of each project scoring metric.

10.3.2 *Analysis of Project Confidence Data*

Figure 10.3 provides background and historic data outlining the trends and typical themes of Project measurements over the yearly periods from 2013 to 2019 inclusively. It combines and summarises the DCA's measured confidence for all projects undertaken on the GMPP list, spanning back over seven years, whilst following the 2012 discussion on and introduction to a proposed BIM mandate (UK Government Construction Strategy, 2011).

Further to the historic reflection/year-on-year breakdown of DCA evaluations, Figure 10.4 below focuses solely on the 2019–2020 "confidence ratings" for the total number of projects across the four categories namely Infrastructure and Construction, Government Transformation and Service Delivery, Military Capability and ICT. Projects within these categories total 125.

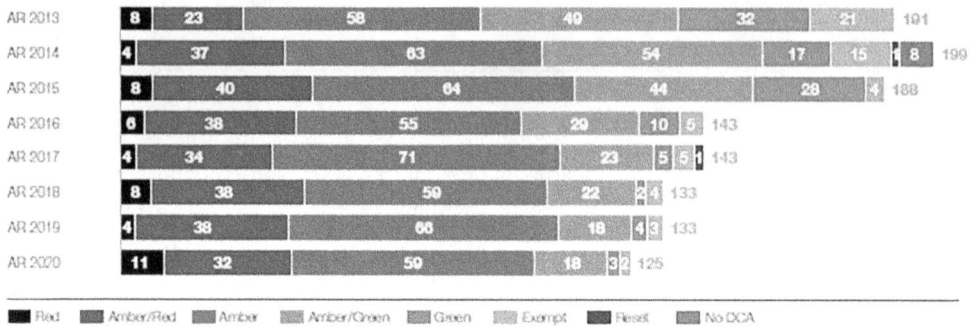

Figure 10.3 DC analysis 2013–2020 (IPA Major Projects Review, 2020).

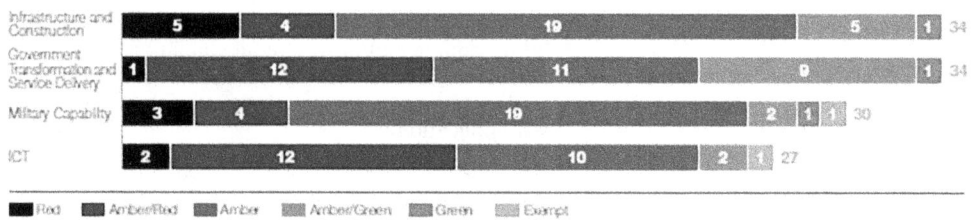

Figure 10.4 DCA's by project category (IPA Major Projects Review Report, 2020).

Analysing the data further, the Infrastructure and Construction Department is joint top in terms of the total number of projects (a total of 34) alongside Government Transformation and Service Delivery. However, in terms of Whole Life Cost (WLC), Infrastructure and Construction has a much larger average value of £6.3bn, only being topped by Military Capability which is measured at £6.5bn and has four fewer projects (30 projects). Furthermore, the Infrastructure and Construction Department has direct synergies and links into the requirement to deliver against a set of defined criteria instructed as Collaborative BIM via the UK Government Construction Strategy (HM Government, 2012) BIM Mandate. Naturally, this is of particular interest in spotting trends linked to the adoption and implementation of BIM and thus benefits gained in quantitative and qualitative measures.

The total WLC in 2020 for the 125 projects analysed and evaluated is stated as £448bn, which is an increase of £6bn from the previous 2019 appraisal report. Further, the Infrastructure and Construction Department has an average Project length of 12 years, which is five years behind in respect to Military Capability sector, but is relatively long in duration compared to others such as the Information and Communication Technology sector.

10.3.3   *Infrastructure and Construction Sector – Analysis*

Looking specifically at, as set out at the beginning of this chapter, the project category "Infrastructure and Construction" for the year 2019–2020, there is a disproportionately high percentage of projects that are either in the Red, Amber/Red or Amber.

Table 10.4 Summary of DCA for infrastructure and construction; Project confidence

DCA Status	No. of Projects	Percentage (%)
Red	5	14.7
Amber/Red	4	11.8
Amber	19	55.9
Amber/Green	5	14.7
Green	1	2.9
Total	34	100

Furthermore, 45% of Infrastructure and Construction projects fall directly into the Red area when averaged across the four categories. Extrapolating and investing further into the "Project Confidence" metrics across the five DCA ratings for the year 2019–2020, percentages are tabled alongside the categories of confidence in Table 10.4.

Although there is evidence presented within the report that shows that 27% of projects have improved in their DCA measure between 2019 and 2020 as well as being better aligned for successful delivery, there is a significant risk that the rather top-heavy measurement of 73% of projects will fail to deliver in confidence against their committed scope (time, budget, etc). This negative appraisal is also reflected as an increasing trend when looking backwards in time at the number of "Green" confidence ratings, which are decreasing in number since the reports were initially produced in 2013 (refer to Figure 10.3).

10.4 Methodology of Data Collection from Construction Experts

10.4.1 Design and Procedure

To gain both a wider and an enhanced understanding regarding the adoption of BIM in Design and Construction from both academic and industry perspectives combined, a framework was developed. This framework has a strategic approach towards data collection, understanding core themes and evidence attainment to essentially the bridge gap between the theory and real-world application, which was investigated further via semi-structured interviews.

Following the above, several methodologies were selected for use within this paper to create a comparative analysis alongside "experiential knowledge" to gain a more holistic view across various rich datasets (Maxwell, 2005), forming part of the data-gathering exercise.

The core underpinning methodologies selected and applied as part of the study design were as follows:

- Systematic Literature Review (Borenstein et al., 2009);
- Semi Structured Interviews (Bernard, 1988); and
- Critical Review (Baumeister and Leary, 1997).

These methods were practically applied to assist in analysing, comparing and drawing on conclusions from a range of theories, opinions and observations, with no disproportionate bias towards one method over an alternative (Paré et al., 2015). Further,

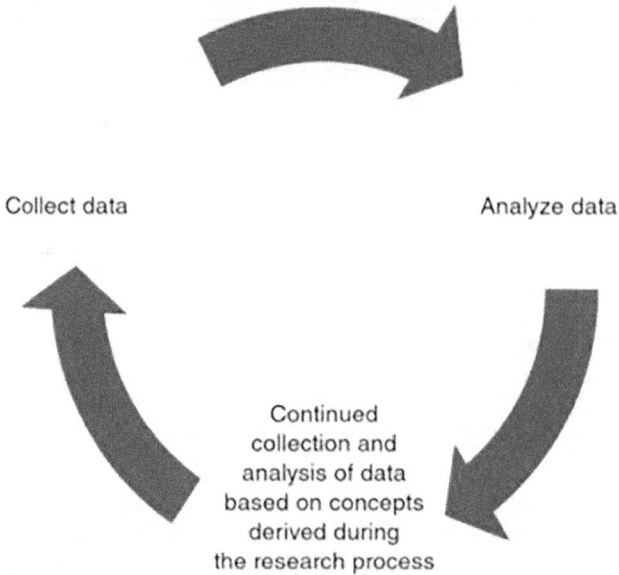

Collect data   Analyze data

Continued
collection and
analysis of data
based on concepts
derived during
the research process

Figure 10.5 Interrelationship between data collection and analysis (Corbin and Strauss, 2008).

this methodology allowed an in-depth exploration as part of the "Grounded Theory" (Glaser and Strauss, 1967) of academic and industrial literature by systematically focusing and selecting targeted data. The former was used alongside semi-structured interviews (Bernard, 1988), as well as the reported project measures of BIM implementation via critically reviewing supplementary datasets (Figure 10.5).

Further, "Thematic Analysis" (Holton, 1973) was used throughout the structure of the methodology as an overarching approach to identifying themes and patterns and progressing to review and distinguish data collection from a qualitative perspective. Braun and Clarke (2013) state that this approach encompasses and instils a "distinctive method with a clearly outlined set of procedures".

Figure 10.6 visually outlines the described process and procedure discussed above, whilst also highlighting the feedback loop between evidence found at the data collection stage linking back into possible addition of exploration at the literature review stage, as well as forwards towards consolidating and capturing results and concluding remarks from evidence ascertained.

Finally, the conclusion section of this chapter synthesises the data, provides a conclusive insight into the emerged findings, trends and relationships, post hoc analysis, opportunities and observations (Braun and Clarke, 2013), as well as outlining a framework solution aimed at identifying and countering the issues faced.

10.4.1.1 *Semi-Structured Interviews*

The semi-structured interviews' core purpose was primarily aimed at adding a further insight through data collection of applied BIM at the execution stage, and to also compare and analyse data collected against academic and industrial literature, and thus draw more rounded conclusions.

Figure 10.6 Block diagram of data collection incorporation.

Following on from the above, a range of "descriptive, exploratory and interpretive" (Gray, 2009) questions were designed as part of the qualitative research study to enable data capture from multiple sources and methods. Denzin (1989) states that this is to bolster the feedback received from a range of "core" respondents and to apply focus on "reliability" of the answers received. This means that despite who the interviewer is, answers have higher likelihood in being consistent if designed using the "triangulation" methodology, adding to, and increasing the validity of the data collection (Denzin, 1989).

These questions which can be found with Appendix A of this report, encompass and add to the framework of issues and opportunities, utilising the "pragmatist" methodology (Rossman and Wilson, 1985). The questions developed focus on understanding further areas such as the benefits of BIM at the application stage, the impact of standardisation (and subsequent updates), how success is measured, how goals and objectives are set and how people are impacted by the adoption and implemented of BIM. In addition, this is also alongside responder feedback on the UK Government reporting of Project confidence via industrial literature and a free input question with the interviewer playing an "Observer" role (Hart, 2001).

Individuals were purposefully selected who were industry facing with a minimum of ten years experience across project delivery and construction via BIM, with a bias to both practical application and fundamental understanding of theory but also with "critical similarities" of the subject focal point (DiCicco-Bloom and Crabtree, 2006); in this case the adoption and implementation of BIM. This was to ensure that the design of the applied methodology complimented and added to the "Grounded Theory" (Glaser and Strauss, 1967), which Strauss and Corbin (1994) state uses "the interplay between analysis and data collection to produce theory".

Table 10.5 Summary of DCA for infrastructure and construction; Project confidence

Function	Role	Date of the Interview
Construction	BIM Infrastructure Manager	09-02-21
Construction	Regional Director	18-01-21
Construction	BIM Lead	19-01-21
Construction	Project Director	11-02-21
Construction	Regional Chief Engineer	23-02-21
Design	BIM Coordinator	23-02-21
Construction	Senior Project Manager	26-01-21
Design	BIM Designer	12-02-21
Design	Operations Director	10-02-21
Function	Role	Date of the Interview
Design	BIM Manager	10-02-21
Construction	Construction Director	16-02-21

Furthermore, to ensure that an "underpowered study" was not inadvertently designed and reflected in the data collection and analysis (Fraley and Vazire, 2014), the sample size was slightly dominant towards construction, but also included a connection with design experts too (36.3% Design, 63.6% Construction). This was intended to not limit the data collection and so that the evaluations made reduced the likelihood of bias towards a specific stage or stakeholder as part of the study.

In addition, Table 10.5 has been provided to show the diverse range of interviewee's who were invited as part of the semi-structured interviews, focused on gaining further insight into areas of success, negative impacts and highlighting potentially new areas for reaffirming evidence and/or providing a platform to explore new areas (complimenting the literature review). Diversity is typically difficult to incorporate into such data-collection activities (Gray, 2009). However, due to the strategy and approach taken the group of individuals invited to interview purposefully have diversity both in terms of their experience, accountability and within their operation roles.

The interviews were undertaken over a duration between 30 and 45 minutes, against ten predetermined questions, with one final "open question" for participants to respond to (adding further evidence if they think it will add value) – this enables a more in-depth account of feelings, evidence and responses (Chilban, 1996), as well as a facilitating an "observational" role in regard to the open response final question.

10.4.1.2   Critical Review

A critical review of core industrial literature, in both qualitative and quantitative forms was undertaken as part of the design of the analysis to "provide a critical evaluation and interpretive analysis" (Baumeister and Leary, 1997). This method facilitates and enables dissection, interpretation and critical evaluations of literature which can be undertaken without any intentional bias (Paré et al., 2015).

This review, which was focused directly on the Infrastructure Projects Authority (IPA) Major Project Annual Report 2019–2020 (IPA, 2020), was undertaken to extract core project Delivery "Confidence" data from the industry-focused report to further understand and analyse current figures and data comparatively against other methods, in critical yet systematic manner (Hart, 2001). This analysis and results are found in the "Results" section.

This critical analysis approach of the subject allows further determination and dissemination of the success factors and benefits as well as the inefficiencies, disadvantages and ultimately how the application of BIM is being reflected quantitatively in the Key Performance Indicators (KPIs), as reported within the UK Government report (IPA, 2020).

10.5   Results

The sections which follow are divided into the two predominant areas of analysis, with their intended focus on understanding further the inherent issues of BIM adoption within infrastructure construction projects.

1 IPA Annual Report Examination; and
2 Data Collection via Semi-Structured Interviews.

10.5.1   Data Collection – Semi-Structured Interviews

As defined in the methodology section, to gain a holistic view of the applied status of BIM implementation across academic and industrial literature, thought-provoking questions were designed as outlined within Appendix A and themes summarised in Table 10.6 as part of the data collection process.

The questions, which form part of the research methodology, were designed to be placed directly to industry experts who are practically applying BIM at the Design and Construction stages, to gain their thoughts, comments, knowledge and insights on the success, failures, advantages and misconceptions of BIM in terms of adoption and implementation.

10.5.1.1   BIM Standards, Methods and Procedures

All eleven of the respondents agreed that the standards, methods and procedures were important in creating a common approach to delivery of projects via BIM methodologies, and specifically over the past five years they have added a "huge benefit" and more are "very important in methodology" in alignment of the requirements of BIM. However, due to the vast amount of literature and transfer from BS to PAS to ISO several

Table 10.6 Interview themes and descriptions

Themes Identified in the Interviews	Rationale
BIM Standards, Methods and Procedures	Insight into standardised methodology benefits
Lessons Learned and Benefits	Exploration of extending opportunities
Productivity and Project Confidence Factors	Discovering the benefits and impacts of implementation
Benefits, Desires, and Inefficiencies of BIM Implementation	Ascertaining the drivers and blockers of BIM
Role Specific Gain and Benefit from BIM	Understanding benefits to people
Measuring Successful Adoption	Learning from data capture
Reviewing Goal Achievements	Exploration of how objectives are realised
Open Question (Free Input)	Opportunity to act as an observer

respondents added that some benefit has been lost and a "lack of understanding of the requirements" at a practical level has reduced the attraction and advantage that BIM can offer. Furthermore, one person responded with a comment stating that although they understand the rationale behind the suite of standards they did feel that most of the responses at the tender and even delivery stage were "lip service to clients", that clients "didn't really know what they were looking for" with BIM adoption and implementation other than "nice fluffy 3D models" which was also reaffirmed by others.

In addition, another commented that implementation is "Definitely going in the right direction, but I think that there's problems lies in the skills" adding a further statement that they think the success of standardisation is limited by "people understanding technology and how to them".

10.5.1.2   *Lessons Learned and Benefits*

In regard to measuring and sharing lessons learned and benefits of BIM, general agreement from all respondents was that this was not done sufficiently, that it was "difficult to put a number of the benefits" and is generally more because of this it's "an issue, and I don't think we have good measurement tools". In addition, it was typically relayed by several statements that people in industry could "tell you a story about the benefits, but not measure them quantifiably" which meant communicating either through conversations or visual outputs of case studies. One respondent advised that a difficulty was due to the benefits being harder to measure unless "every function buys into it" and went further stating that "not all parties are delivering via the BIM requirements", which was also a common theme reproduced by respondents.

10.5.1.3   *Productivity and Project Confidence Factors*

Directly related to the IPA report evidencing year on year a reduction in project confidence of being delivered on time, on budget and with reduced disruption one respondent believed this was because the Client "wants BIM because it's mandated" but what they actually want is to simply "deliver to programme, within budget and not have any defects". It was stated by five respondents that "objective focus" which was described as reviewing, agreeing and positioning stakeholders towards what is required from BIM alongside a reduction of specific targeted goals via BIM was a possible reason why there was inefficiency and lack of return and stunt in advancements made (as evidenced within the IPA reporting). Insightfully, a respondent stated that "scatter brain at the last 10% of a project" is where the inefficiencies come from at the Construction stage, regardless is the method for delivery is BIM or not with another, adding that "the tangible benefits were you make the cost savings at the end are not realised, because at the end [BIM, Data Exchanged and Digital Methodologies] are not being used". Another added that a potential and fundamental reason was that the focus on reducing cost, improving punctuality and adding value was "a fantastic dream that they tried to sell to everyone" via the BIM mandate, but there are "many external factors that impact the lifecycle of a project".

10.5.1.4   *Benefits, Desires and Inefficiencies of BIM Implementation*

When faced with delivering and/or implementing BIM on a project, a common theme in terms of desires was that they wanted to "add value" to their delivery via BIM processes and digital engineering but felt like the understanding of "what BIM is" and lack of clarifying what BIM is for the project "despite the BEP, core documents, etc" was a reason for the inefficiencies. One respondent passionately stated that "BIM is not just a 3D model", and others believed that "the most important thing is all about data" when actually it should be about focus and using these tools and methods to achieve the goal or objective. Further, it was advised by another respondent that a pain point was getting "data from the design model across to the construction model", which at the moment required manual intervention and 'manufacturing models' were different to the former, which is an ongoing "pain and inefficiency", with missing elements included within the "RIBA" (Royal Institute of British Architects frameworks).

Another respondent advised that applying BIM is a "massive step change onto their project(s)" adding further that it" takes time and when you are on a project with tight timescales the last thing anything wants to do is take a step back and try to learn how to use something".

In terms of understanding what BIM is fundamentally, the consensus was that there is still a lack of understanding by most stakeholders at the application stage ranging from "Clients who want 100% BIM" but then ask "what's an EIR", to sub-contractors who say "they are just supplying what you require". Further, those who have a greater awareness and understanding are the people who use specific elements to "benefit" their outputs and roles. Benefits are realised so that it "becomes normal" with one respondent confirming this by stating that adoption "of course takes time and you'd almost consider this as part of a transition period".

Expanding further on the latter paragraph, benefits from respondents were linked to "avoidance of rework and defects" with one respondent stating they believe defects have reduced in buildings from "5% to 3%" because of targeted BIM application. However, a counter statement was that "other benefits can come, but it's a big investment" as well as "a lot of the competition aren't making the same investment", which meant some of the supply chain weren't disincentivised by investment needed (time, cost and productivity impacts).

Further, another individual replied that they believed the key to successful understanding was by putting "Trust in the people and training them I the same way", so that "awareness is set" and therefore much more likely to succeed.

10.5.1.5 Role Specific Gain and Benefit from BIM

Regarding whom gains most benefit from BIM adoption the consensus was split, however, one comment was that "Designers and Contractors benefit most", as better accuracy will result in additional projects and improved collaboration between Contractors. Another stated that if the goals and approach to why BIM is being applied is agreed then "all can benefit from building safely and efficiently" as well as another proposing that the end user and client will have most benefit as they know "what he's getting" linked to "what he's going to use it for" which links into downstream use of "3D models that can be constructed and maintained from". Other respondents however countered this opinion, stating that the "client will have most benefit" but adding further this can be true "but it's got to be an educated client". Further, another stated that "at the moment 90% of clients can't see and don't get the benefits, and most are forced into BIM" due to a lack of understanding of "what it means". This translates into "poor governance and adoption" and is more often misunderstood by "middle to senior to client" management levels.

In addition, there was consensus between 50% of respondents that those who do understand and "get BIM" are the "younger engineers and designers" and more so the "guys on the ground", but there was a "real detachment the more you go higher" adding to the previous respondent's thoughts on hierarchy disconnect. It was believed this is the case due to "management thinking that it's all about technology, models and software".

Another respondent added that effectively the gain should be initially for the "Designers and Contractors" but more so the ultimate gain should be "to the UK Taxpayer" via the appropriate delivery of projects via BIM. Further, this person added that they would "hope that projects don't cost as much [if BIM is applied correctly], so it creates enough efficiencies that the public purse is the one that benefits".

10.5.1.6   *Measuring Successful Adoption*

A very common and consistent theme in relation to measuring the success of BIM adoption and implementation was that it is "difficult to record it" in a tangible and quantitative way, with one respondent advising that "we are definitely seeing less hours against rework" but this is coupled with more effort needed in BEP production, modelling inputs and some "general inefficiencies" (due to lack of clarity). Furthermore, a "BIM v Non-BIM" assessment and measurement of the benefits in relation to traditional methods has "not been undertaken across sectors", to confirm advantages and disadvantages across all areas of implementation.

In addition, a respondent stated that the "quantifiable element that PM's really care about is hours spent and I don't think we really go into detail previously how many hours we spend currently doing something" with BIM certainly demanding "more time for management of tasks, CDEs and so on". This is also linked to difficulties in having less of a "justification" for opting for and implementing BIM and more so being able to "track the benefits".

However, one Highways and Civil Engineering interviewee stated that a reason for this was that "we are not doing the same thing day in day out" therefore making it difficult to measure the variables, with another stating "that is the problem, it isn't measured is it". Moreover, another input in response to the question of measuring the success only one respondent commented that they are "measuring the time spent on model production, issue resolution, change mechanisms, KPIs, etc." on a "weekly basis", which then looks at comparative data of similar tasks on previous occasions to spot trends and efficiencies from "taking ownership of the model through to the handover at the end".

Another added that they "aren't aware of qualitative or quantitative data being produced" including "appropriate KPIs" to support measures and management, with another stating that although this is largely true "large scale frameworks and alliances should be doing that across 10 year commissions having all those metrics an assessing them".

Furthermore, one comment was made that success is simply measured when BIM is "done in the way that when the Client is happy" the company is awarded additional work. Others similarly added to this point separately and stated that they didn't "think its measured at all, and that's why there is so much confusion around BIM" and that there possibly is "something missing in the middle", to gain clarity of the required measures.

10.5.1.7 Reviewing Goal Achievements

In terms of reviewing goals for the application of BIM alongside measuring the success on achieving these objectives, one respondent commented that these have been "done on a central level [for commercial reasons] and not at a project level" with another stating that at this stage "we're still in a fact-finding element" of measuring success and achieving objectives. Moreover, another respondent commented that although the goals for them were "quite simply to get projects delivering digitally" with a reference to an "Engineering Excellence" program across a business which focuses on sharing "best practice and innovations" across an intranet with success being shared to other colleagues most often via "a conversation as it always is about people".

Commonalities are made from other respondents with responses such as although goals are set and achieved, they could be perceived as "empty goals" in that "they are not defined" and are typically an "opened ended empty list of 'things". Additional responses were simply put, in that measuring, reviewing and acting upon goals set "isn't measured", and that this area was lacking maturity and that there wasn't "any measure in designer organisations" but this can only be done on at a similar sector based project level due to the fact across sectors and disciplines "it is so very different".

10.5.1.8 Open Question (Free Input)

In terms of the free input question, a comment was made that there is heightened complexity now with additional systems and it "just seems at times they're so complicated, difficult to use and we are losing value in them". It was also stated that the development of BIM requirements has meant that teams have to "bring in a load of specialists" to digest and translate when all that the project is trying to do is "make the lives easier and more efficient". This is countered by the initial intent of BIM by "reducing the value".

Further, the responder added that once the above happens staff "don't use tools correctly and it [BIM] goes back to lip service".

Another respondent commented that a problem with "parts of the BIM process" is when people don't "put documentation on place" linked to creating clarity, which means you have to "unstitch" on ad-hoc basis which can lead to inefficiencies and rework.

In addition, similar comments were made by several respondents stating that "The gap between the Construction and Design is a must to address and to make sure everyone is and are in the loop" which is linked to "education and awareness" and the appropriation at the application stage of BIM.

Finally, a respondent added a final statement with some thought-provoking questions such as "How do we stop it [BIM] being a computer game pull it back to the reality of what we actually do in the ground and make it efficient", because "You're probably getting 80/20; 80% benefit for 20% of the aspect and to get the last 20% you're probably footing 90% of the cost ... is that really where we want to be ... maybe not?".

10.5.1.9 Additional Analysis

The figures which follow highlight and summarise repetitive words that each respondent used throughout the interview process, with those linked topics and words shown within

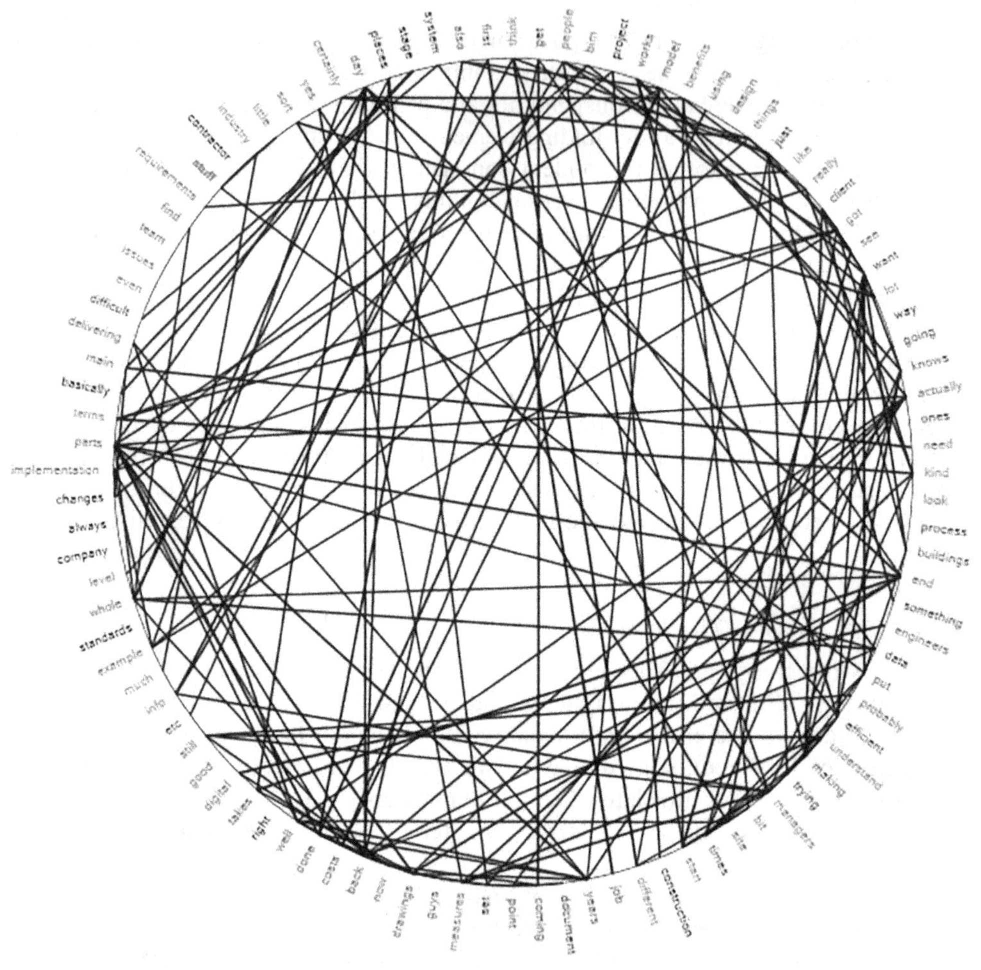

Figure 10.7 Linked words and topics post-interview analysis (100 most frequently used words).

Figure 10.7, and those that were replicated more boldly dominant as shown within Figure 10.8.

It is evidenced visually above that common words such as "project, think and client" were used most often, with dominantly yet equal weighted trends of "design, people and construction" received most at the data collection stage from industry experts.

What this shows is that the issues with adoption from data collection from infrastructure construction facing experts at the interview stage appear to be people-centric and not directly restricted because of or by the technologies available to BIM projects. In addition, the proportionate collation of words used shows in Figure 10.8 that "requirements" and "everyone" are also linked and common to gaining benefits from BIM such as collaboration, implementing digital strategies and delivering core outputs to the client's information requirements. This is also reaffirmed within the former sections in that these specific points are very important in removing the obstacles and are relevant in respect to issues facing poor adoption techniques.

Figure 10.8 Dominant words and topics post-interview analysis.

10.6 Summary and Conclusion

10.6.1 Discussion

Evidence and thus findings within this research show that although there is a suite of developing standards from BS to PAS to ISO, positioning projects compliantly alongside the UK BIM Framework requirements, there are inherent inefficiencies to the successful adoption and implementation of BIM. These inefficiencies are directly reflected in industrial reporting, notably within the IPA annual reports stating 14.7% of projects have a red (low) and 2.9% have a green (high) confidence rating for 2019–2020.

In addition, these difficulties and inefficiencies have been extracted from the semi-structured interviews which highlight a reduction in detailed goal setting and objective focus on what BIM could facilitate and enable, through clear line of sight to seek objective focus. Further, a lack of qualitative and quantitative measures available to stakeholders to assist in the measurement and management of BIM delivery as the project is being developed and delivered is not typically available for BIM i.e. none of the respondents were aware of a system or convention available. Typically, when projects are measured this is undertaken either through generic KPIs and usually close out stage, when it is often too late to gain a benefit, make sufficient changes if objectives are heading off track or educate staff in some of the advantages that may increase the value of BIM and elements therein. Again, evidence from interviewees stated that they weren't aware of a commonly available system or method for capturing BIM advancements, goals or outputs at the application stages across the project lifecycle.

All respondents from industry believe the standardisation and protocols are required and increase benefit, which have largely helped in advancing from pre-BIM to current

BIM but at the application stages. However, it is also believed because of the "complex" languages and reduced focus on the applicable intent, there are some poor methodologies and fragmentation incurred. These are associated specifically to being able to utilise BIM as a mechanism to deliver against clearly defined outcomes and delivering goals with a finite objective focus.

10.6.2 Synthesis of Results

This section provides a cross-examination and draws a comparative analysis and synthesis on the findings and observations from the three main areas within the research methodology (see Figure 10.2), focused on the adoption issues within infrastructure construction projects.

Digital tools, methodologies and processes have been regarded as bringing a "20% saving" at the Construction stage via "collaborative 3D BIM" (HM Government, 2011), which has been realised at a case study level by Chahrour et al. (2021) focusing on elements such as clash detection and achievable financial benefits through appropriate capturing and measurement of issues faced, pre-realisation. However, industry experts responded unanimously that although standardisation paired with technological advancements was a core requirement and an enabler to creating efficiencies via BIM, they feel the "lack of objective focus" means that the application doesn't solely reap benefits when applied wholesale across the project lifecycle. Moreover, nor is it measured sufficiently between "BIM v Non-BIM" and a lack of ROI justification means BIM is perceived as "just a cost" and thus reflected in update.

Kirby-Turner and Whittington (2018) state that a reason for this is gained by "potential confusion" stemming from a vast range of standards, protocols and guidance respectively. Further, education targeted towards more senior levels within industry detailing what BIM is at a project level is compacted, which results in a reduced understanding at a management level and lack of prioritisation on specific goals towards the application of BIM (and thus not being agreed and set, measured and reviewed accordingly). This reduces the value through "lack of appropriation", with consensus believing for data collection that BIM is simply perceived in most cases as a "flashy 3D modelling" and similar outputs. Zuppa et al. (2009) state that for the successful adoption and to "enable the successes" education, role and awareness must be a priority and there was evidence from the semi-structured interviews that this is missing again at the execution stage of BIM projects.

Further, data collection from the interview stage suggests in actual fact that the ROI is very difficult to apply to all projects, not just in terms of the subject diversification, but because of the poor understanding of "what to measure", when and how to measure this quantifiably (for example, there is no framework in place). Adding further to this point, one respondent commented that it was typical that BIM application was measured in "staff Professional Development Reviews (PDRs) but not recorded at a project level" which adds to the difficulties of measuring the benefits and lessons learned at the delivery stage of the system and not just peoples achievements. This said some projects do receive benefits which were stated as not being "directly measured" when comparing one BIM project to another, as well as industry typically not specifying what the "benefits of BIM" adoption should be, and then measuring and managing these specifics in a targeted manner (Kim et al., 2013). The IPA report defines projects as reducing in confidence in them being delivered on time and budget as shown in Table 10.5. Taylor (2007) affirms this in that a reason for the lack of positive return and success is due to "a poor understanding of how technology is implemented" at the delivery stages of Design and Construction projects.

Moreover, fragmentation between developed standards and people being able to firstly understand them and then apply at a local level is clear both from data collected at the interview stage. This is also visible in quantifiable figures measured by the IPA and the "lack of agreed value" of BIM from academia (Chinowsky and Rojas, 2003). Data from the IPA report in 2019–2020 shows that 45% of the "red" rated project confidence is directly associated to the Infrastructure and Construction sector, which outlines again issues with practical application of BIM across a diverse range of projects. Barlish and Sullivan (2012) argue that one of the issues of poor measurement is that there "are no definite measures to the benefits of BIM in use", which shines focus on a disconnect between theoretical application and real-world situations.

Further, feedback from industry following data collection and analysis was that there is a perception that Client understanding is limited on how to adopt, practically administer BIM and govern BIM, and the "further that you go up the hierarchy", a disconnect appears from the value and an appreciation of the upfront effort that BIM requires. It is also believed that this is the case because Clients see BIM as a "technology tool", when in actual fact it is an all-encompassing "set of processes" and methodologies positioned to improve quality, whilst advancing the measurement, management and proactive governance of project delivery (across all stages of the lifecycle).

Reoccurrences and trends linked to people being overwhelmed by the nature of BIM (standards, toolsets and conventions) alongside the theoretical methodology and mandatory requirements have resulted in reduced effectiveness and compacted understanding from staff of why they are implementing BIM and other Digital Engineering processes, procedures and tools.

Quantifiable measurement as progress is made in terms of goal achievement, alignment with requirements for the application of BIM alongside measuring the success of goals and outcomes is perceived as "too difficult" from industry experts, with evidence from Barlish and Sullivan (2012) stating there is 'no universal benefit of BIM' from either an academic or industrial perspective, without an objective focus. This is also reaffirmed in the IPA report findings in the form of disproportionately low confidence ratings specifically within the Infrastructure and Construction sector. In plain language, BIM out of the Box simply doesn't add a benefit by itself, and a lack of project comparators (BIM v Non-BIM), coupled with no quantifiable measurement or approaches with objective focus measurements appear to reaffirm this, as mirrored by Ibbs et al. (2007).

Further, the IPA report(s) shows that across the diverse category of Infrastructure and Construction, data within shows a common trend that the success of projects being delivered on time and budget is actually reducing year on year, despite the UK Government Strategy (HM Government, 2011) statement of achieving "20% savings" plus additional benefits and advantages via collaborative BIM adoption and implementation. Further to this, interviewee data shows that the benefit and gain to be had from BIM at the implementation stage relied upon "data appropriation" and by removing the "specialisms" required as they were "confusing and prohibitive", which results in a need for a clearer way of removing clutter, creating clarity and advancing towards benefit realisation.

Despite the development of UK BIM Standards and Frameworks (UK BIM Alliance, 2020) agreements from several industry experts outlined that the deliverable process hasn't radically changed to accommodate BIM requirements, but rather a "different language" and "more toolsets" were needed to deliver Design and Construction via BIM. Eadie et al. (2014) state that "managers are reluctant to add new processes and technology" which explains alongside industry feedback why there is a competing set of old and new procedures will little ROI due to BIM. Effectively, these competing processes are doubling workloads (Hardin and McCool, 2015), exacerbating errors and requiring staff to have a heightened understanding of the methodologies, which ultimately results in inefficiencies gained alongside an increased cost in software, training and of course, time delays.

Further comparative conclusions are evident and common throughout in that new processes (such as BIM) are more often 'bolted' onto existing frameworks with a reduction in assertiveness in terms of the application and governance despite them being mandated from the UK Government on all publicly awarded contracts from 2016 onwards. Further and due to this, industry experts felt within the interviews undertaken most of the benefits gained from adopting BIM were merely "talked about" qualitatively, and were difficult to quantitatively measure, primarily due to the lack of a measurement and assurance frameworks/system available at a project delivery level.

Furthermore, evidence confirms that there is a poor connection between the design, construction and operational stages in terms of an agreed and seamless transfer of information, as well as other benefits that are being lost, such as unifying and collaborating systems to improve efficiencies across jurisdictions, using BIM for a defined purpose and as previous, measuring success. Adding to this, other than utilising "3D models" for time sequencing and planning, a missing part of the value of BIM was seen to be the "data management" and more so the ability to manage, measure and deliver projects by utilising information in a more coordinated and considered way. Again, this is where objective focus is missing, which would have a high likelihood of enabling them to be delivered on time, on budget and to an improved standard, respectively.

It is also evident that in 2010 the IUK Infrastructure Cost Review (ICE, 2015) outlined that the typical costs of project delivery are "40% higher" specifically due to poor production management and pressure to deliver projects which can be likened to BIM (a new applied methodology). Fortune (2006) reaffirms this by stating this is partly due to "no whole project delivery review of project performance".

Finally, it is very important to remember that although BIM is an information delivery process that utilises advancements in technology for quality assurance/improvement, as well as enabling advancements and collaboration in modelling techniques (3D, 4D, 5D), it is in fact mostly intrinsically cultural and driven by people, who are supported and assisted by Digital technologies, toolsets and processes.

10.6.3   *Conclusive Remarks*

The purpose of the research methodology herein was to broadly investigate the adoption issues of BIM across academic literature and industrial reports, alongside interviewing implementation stakeholders who were responsible for delivering BIM on construction infrastructure projects. Findings within this chapter show that although there are clear opportunities for the benefits to be achieved via the adoption of BIM across the entire lifecycle of a project, there is a gap between the theory and realised benefits at the application stage by industry.

For example, proportionate application is key in ensuring that goals are aligned, agreed and measured via a methodology throughout the entire delivery cycle and not limited to a single stage or isolated task. This is also alongside setting appropriate goals and adjusting them accordioning, which provides a consistent focus on what the intention of the outcome should be, because of BIM. Further, a clear and understandable strategy of what to apply and when to do so for those who aren't necessarily BIM specialists is required to break down the barriers of confusion and see existing tasks such as BIM Execution Plans (BEPs) and model coordination meetings a hindrance, not a help.

In response to this, a solution framework focused on improving current practices and enhancing adoption and transparency to teams across all project delivery has been proposed in Figure 10.8. The framework aims to deconstruct BIM into its constituent parts of objective focus which will better enable stakeholders in (a) determining their problems and desires; (2) reviewing the impacts, dependencies and core requirements of the objective(s); and (3) selecting how the objective(s) will be delivered, who will be responsible and creating measurement criteria to drive success or counter negative impacts stifling adoption.

This then enables stakeholders to holistically address their use of BIM and thus more likely to remove the barriers of confusion whilst improving productivity, increasing in quality of outputs, enhancing collaboration and transparency of object focus. This also creates an opportunity to develop quantitative and qualitative measurements alongside the concentrated areas, which allow increased visibility of the ROI of applied BIM.

In addition, the solution framework positions the objective setting of targets directly towards the project requirements, in order to break down cultural and hierarchical barriers, whilst increasing the understanding of desires and opportunities centrally for discussion and agreement. This is rather than the current method of assigning a goal simply to the core BIM team, as this is one of the areas that drives reduced understanding and poor governance/acceptance of BIM at a delivery level.

Finally, an assurance feedback loop is added to the framework to further ensure that the objectives, the impacts and the outputs are sufficiently clear and proportionate. Further, KPI's development is proposed with a review of staff suitability (skills and qualifications) to improve BIM adoption and gain benefits measured both qualitatively and quantitively.

10.6.4 Limitations

Some of the limitations faced within this research are as follows:

- Although there is a diverse range of industry experts across design and construction sectors, 10 out of 11 questions were predetermined which may have focused the study too directly;
- Industrial research topics were centralised around the 2019–2020 IPA Annual Report and could potentially limit the data analysis (IPA reports however are produced centrally by the UK Government for all publicly awarded contracts (BIM requirement as standard)); and
- The latter statement said, there is limited data available at a project and/or client level in terms of quantitative reporting of progress, advantages and benefits of BIM from industry at the application stage.

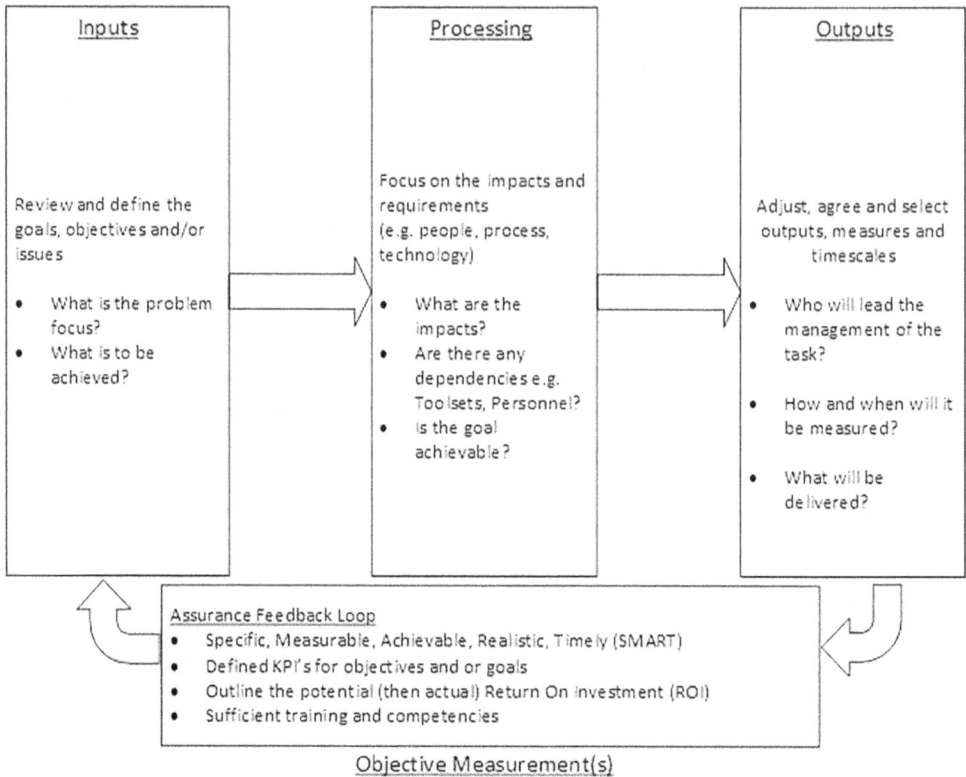

Figure 10.9 Proposed solution framework for targeting objective focus alongside the measurement of BIM goals and objectives.

10.6.5 *Future Works*

Following the findings within this paper and the framework solution outlined within Figure 10.9, there is an advantage and benefit in exploring further the impacts of objective/goal outcome focus of BIM, as well as developing the framework further by integrating across both qualitative and quantitative measurement formats. The benefit is in order to align the application and implementation of BIM towards objective goal achievement and thus increase the benefit at the execution and deliver stages respectively via people, processes and technologies.

Acknowledgements

The authors would like to thank Bjarnhedinn Gudlaugsson, a PhD research student of energy systems dynamics at the School of Computing, Engineering and Digital Technologies, Teesside University, for his guidance in the initial structure of this research.

References

Akponeware et al. (2017), *Clash Detection or Clash Avoidance – An Investigation into Coordination Problems in 3D BIM*, School of Civil and Building Engineering, Loughborough University

Aljumaiah B. (2020), *Building Information Modeling to City Information Modeling (BIM to CIM)*, Eastern Province Municipality

Azhar, S., Nadeem, A., Mok, J. Y. N., & Leung, B. H. Y. (2008). Building information modeling (BIM): A new paradigm for visual interactive modeling and simulation for construction projects. *Proceedings of First International Conference on Construction in Developing Countries*, Karachi, Pakistan, 435–446.

Baloi, D., & Price, A. D. F. (2003), Modelling global risk factors affecting construction cost performance. *International Journal of Project Management*, 21(4), 261–269.

Bargstädt, H. J. (2015), Challenges of BIM for construction site operations, *Procedia Engineering*, 117, 52–59. ISSN 1877-7058, 10.1016/j.proeng.2015.08.123

Barlish, K., & Sullivan, K (2012), How to measure the benefits of BIM – A case study approach, *Automation in Construction* 24 (2012) 149–159

Bernard, H. (1988), *Research Methods in Cultural Anthropology*. Newbury Park, California: Sage.

Baumeister R. F., Leary M.R. (1997), Writing narrative literature reviews. Review of General Psychology.

Borenstein M., Hedges L., Higgins J., & Rothstein H. (2009), *Introduction to Meta-analysis*. Hoboken, NJ: John Wiley & Sons Inc.

Braun, V., & Clarke, V. (2013). *Successful Qualitative Research: A Practical Guide for Beginners*. London, UK: Sage.

BSI (2018), BS EN ISO19650-2:2018+Revision, https://shop.bsigroup.com/ProductDetail/?pid= 000000000030420198. (accessed 3 August 2021).

Chahrour, R., Hafeez, M. A., Ahmad, A. M., Sulieman, H. I., Dawood, H., Rodriguez-Trejo, S., Kassem, M., Naji, K. K., & Dawood, N. (2021), Cost- benefit analysis of BIM-enabled design clash detection and resolution, *Construction Management and Economics*, 39:1, 55–72, 10.1080/01446193.2020.1802768

Chilban J. (1996), *Interviewing in Depth: The Interactive Relational Approach*. Thousand Oaks, California: Sage.

Chinowsky, P.S., & Rojas, E.M. (2003). Virtual teams: Guide to successful implementation. *Journal of Management in Engineering*, 19(3) 98–106.

CITB (2018), Industry Performance Reports, https://www.greenwoodconsultants.com/knowledge/industry-performance-reports/ (accessed 8 January 2021).

Cohen, M. (2017). A systematic review of urban sustainability assessment literature. *In Sustainability (Switzerland)*, 9(11). MDPI AG. 10.3390/su9112048

Constructing Excellence (2014), BIM Case Studies, https://constructingexcellence.org.uk/digital/bim-case- studies/ (accessed 2 March 2021).

Corbin, J, & Strauss, A (2008), *Basics of Qualitative Research: Techniques and Procedures for Developing Grounded Theory*, 3rd ed. Thousand Oaks, California: Sage Publications; 2008.

Da Silva, P. H., & Crippa, J, & Scheer, S. (2019), BIM 4D in Construction Scheduling; Details, Benefits and Difficulties, *PARC Pesquisa em Arquitetura e Construção* 10:19010

Denzin, N. (1989), *Interpretive Biography, Volume 17 of Qualitative Research Methods*, SAGE

DiCicco-Bloom, B., & Crabtree, B. F. (2006). The qualitative research interview, *Medical Education*, 40, 314–321. 10.1111/j.1365-2929.2006.02418.x

Doloi, H (2013), Cost Overruns and Failure in Project Management: Understanding the Roles of Key Stakeholders in Construction Projects, *Journal of Construction Engineering and Management,* 139(3), 267–279.

Eadie, R., Odeyinka, H., Browne, M., McKeown, C., & Yohanis, M. (2014). Building information modelling adoption: an analysis of the barriers to implementation. *Journal of Engineering and Architecture*, 2(1), 77–101.

Forcael, E., Martínez-Rocamora, A., Sepúlveda-Morales, J., García-Alvarado, R., Nope-Bernal, A., & Leighton, F. (2020). Behavior and performance of BIM users in a collaborative work environment. *Applied Sciences*, 10, 2199. 10.3390/app10062199

Fortune, C. (2006), Process standardization and the impact of professional judgement on the formulation of building project budget price advice, *Construction Management and Economics*, 24(10), 1091–1098.

Fraley, R. C., & Vazire, S. (2014), The N-pact factor: Evaluating the quality of empirical journals with respect to sample size and statistical power. *PLoS One*, 9(10), e109019. 10.1371/journal.pone. 0109019 (accessed 15 March 2021).

Glaser, B., & Strauss, A. (1967). *The Discovery of Grounded Theory: Strategies for Qualitative Research*. Mill Valley, CA: Sociology Press

Gray, D. E. (2009), *Doing Research in the Real World*, 2nd ed. Thousand Oaks, California: Sage Publications.

Greenwood Consultants (2020), Measuring Collaboration, https://constructingexcellence.org.uk/wp-content/uploads/2020/10/GCR-CEPMF002_MeasuringCollaboration_20201014_Rev2.pdf (accessed 2 March 2021).

Hardin, B., & McCool, D. (2015), *BIM and Construction Management; Proven Tools, Methods and Workflows*. Wiley 2nd Edition.

Hart, C. (2001). *Doing a Literature Search*. London: Sage.

Highways England (2019), A14 What we've delivered, https://highwaysengland.co.uk/our-work/a14-cambridge-to-huntingdon/what-we-ve-delivered/ (accessed 8 January 2021).

HM Government (2012), Government Construction Strategy, https://assets.publishing.service.gov.uk/government/uploads/system/uploads/attachment_data/file/61157/Procurement-and-Lean-Client-Group-Final-Report-v2.pdf (accessed 12 July 2020).

HM Government (2011), Prime Ministers Mandate for the Major Projects Authority, https://assets.publishing.service.gov.uk/government/uploads/system/uploads/attachment_data/file/378509/PM_mandate_for_MPA_2011.pdf (accessed 8 January 2021).

HM Government (2020), National Infrastructure Strategy, https://www.gov.uk/government/publications/national-infrastructure-strategy (accessed 2 March 2021).

Holton, G. J. (1973), *Thematic origins of scientific thought: Kepler to Einstein*. Cambridge, MA: Harvard University Press.

Hore, A., Montague, R., Thomas, K., & Cullen, F. (2011), Advancing the use of BIM through a government funded construction industry competency centre in Ireland, CIB W78 2011: 28th International Conference, Paris, 26–28 October 2011.

Ibbs, W., Nguyen, L. D., & Seulkee, L. (2007), Quantified impacts of project change, *Journal of Professional Issues in Engineering Education and Practice* (2007) 45–52.

Institute of Civil Engineers (ICE) (2015), Production Management in Design and Construction, ICG Guideline, 2015

Infrastructure and Projects Authority (2020), Annual Report on Major Projects 2019–20, https://assets.publishing.service.gov.uk/government/uploads/system/uploads/attachment_data/file/899401/IPA_AR_MajorProjects2019-20.pdf (accessed 11 July 2020).

Khosrowshahi, F., & Arayici, Y. (2012), Roadmap for implementation of BIM in the UK construction industry, Engineering, *Construction and Architectural Management*, 19(6), 610–635.

Kim, H., Anderson, K., Lee, S. H., & Hildreth, J. (2013), Generating construction schedules through automatic data extraction using open BIM (building information modeling) technology. *Automation in Construction*, 35, 285–295, 2013. https://doi.org/10.1016/j.autcon.2013.05.020.

Kirby-Turner, C, & Whittington, C. (2018), BIM, Collaboration and NEC4, https://www.constructionmanagermagazine.com/bim-collaboration-and-nec4/ (accessed 20 July 2020).

Latham, G. (2003), Goal setting: A five-step approach to behavior change. *Organizational Dynamics*, 32(3), 309–318. ISSN 0090-2616. 10.1016/s0090-2616(03)00028-7

Maxwell, J. A. (Ed.). (2005), *Qualitative Research Design: An Interactive Approach* (2nd ed.). Thousand Oaks, CA: Sage.

Mckinsey (2017), Reinventing Construction through a Productivity Revolution, https://www.mckinsey.com/industries/capital-projects-and-infrastructure/our-insights/reinventing-construction-through-a-productivity-revolution (accessed 14 July 2020).

Ojo, A., & Pye, C. (2020), BIM Implementation Practices of Construction Organisation in the UK AEC Industry. *Project Management Journal*, IX.

Paré, G., Trudel, M.-C., Jaana, M., & Kitsiou, S. (2015), Synthesizing information systems knowledge: A typology of literature reviews. *Information & Management. 2015*, 52(2), 183–199.

Rossman, G. B., & Wilson, B. L. (1985), Numbers and words: Combining quantitative and qualitative methods in a single large-scale evaluation study. *Evaluation Review*, 9(5), 627–643.

Shennan, R. (2016), BIM Level 2: What are the implications on procurement? https://www.mottmac.com/views/bim-level-2-what-are-the-implications-on-procurement (accessed 22 March 2022).

Singh, V., Gu, N., & Wang, X. (2011), A theoretical framework of a BIM-based multi-disciplinary collaboration platform. *Automation in Construction*, 20(2), 134–144. ISSN 0926-5805, 10.1016/j.autcon.2010.09.011

Strauss, A., & Corbin, J. (1994), Grounded theory methodology: An overview. In N. K. Denzin & Y. S. Lincoln (Eds.), *Handbook of Qualitative Research* (pp. 273–285). Sage Publications, Inc.

Taylor, J. E. (2007), Antecedents of Successful Three-Dimensional Computer-Aided Design Implementation in Design and Construction Networks. *Journal of Construction Engineering and Management*, 133(12), 993–1002.

The PRISMA Group (2009), Preferred reporting items for systematic reviews and meta-analyses: The PRISMA Statement. *Open Med 2009*, 3(3), 123–130.

Ustinovičius et al., (2018), Challenges of BIM technology application in project planning.

UK BIM Alliance (2020), The UK BIM Framework and the UK BIM Alliance, https://www.ukbimalliance.org/wp-content/uploads/2020/03/Paul_Wilkinson_UKBIMA.pdfi. (accessed 18 December 2020).

UKCO (2011), *Government Construction Strategy*, London: United Kingdom Cabinet Office.

Vernikos, V. K. (2016), Realising offsite construction in the civil engineering and infrastructure sector, *Loughborough University Institutional Repository*, https://dspace.lboro.ac.uk/2134/23521.

Young, N. W., Jones, S. A., Bernstein, H. M., & Gudgel, J. (2009). *The Business Value of BIM—Getting Building Information Modeling to the Bottom Line*. SmartMarket Rep., Bedford, MA: McGraw Hill Construction.

Zuppa, D., Issa, R., & Suermann, C. (2009), BIM's Impact on the Success Measures of Construction Projects, Conference Paper in Journal of Computing in Civil Engineering, June 2009

Appendices

Appendix A – Interview Questions

The semi-structured interview questions placed to the respondents as part of the data collection process were as follows:

1 In your opinion, has the development of the BIM Standards, Methods and Procedures helped with the adoption and implementation across industry? If yes how and if not, then why?

2 How do you/we measure and share the qualitative and quantitative benefits/lessons learned of BIM?

3 Why do you believe that reporting from the Infrastructure Projects Authority (IPA) and Constructing Excellence show no significant increase in productivity, cost and time improvement and general lack of return on investment since 2011 despite government mandating BIM?

4 When implementing, procuring or supporting BIM what are your three desires (in terms of outputs) and conversely what are your three pain points?

5 Adding to question above, what are the benefits and inefficiencies resulting from adopting BIM?
6 More generally, who on your Project (in terms of role/function) do you feel "gets" BIM and conversely who struggles and why?
7 When instructed to deliver projects to or accept into operations assets that have BIM requirements, what are your initial thoughts?
8 In your opinion, who gets the most benefit from adopting BIM i.e., clients, designers, contractors, maintainers, supply chain or other parties like local authorities and why?
9 When measuring the success of BIM adoption and Implemented on projects across Design, Construction and Operations perspectives when and how is it measured and who measures this?
10 Following Q9, in terms of goals and objectives for the use of BIM and Digital tools/ methodologies, when are these reviewed and how is success measured, reported and if required alternatives put in place to bring the goal/objective back on track? [qualitatively and quantitatively]
11 Anything else to add from Interviewee?

11 A Review of Barriers and Enablers of the BIM Adoption in Quality Management System

Nandini Sharma[1] and Boeing Laishram[2]
[1]*Department of Civil Engineering, IIT Guwahati, India;* [2]*Department of Civil Engineering, IIT Guwahati, India*

11.1 Introduction

The construction industry (CI) is a dynamic and complex field that involves the integration of multiple disciplines and stakeholders in lifecycle of buildings and infrastructure projects (Lee et al., 2005). The industry faces many challenges, including meeting project schedules, controlling costs, and ensuring quality outcomes (Duttenhoeffer, 1992). As a result, the use of Quality Management (QM) systems has become increasingly important in the CI to improve project performance and meet client expectations. QM plans for construction projects are unique to each individual project and linked with all phases of project life cycle comprising providing quality checklists, inspecting and testing sites, reporting non-conformances, and corrective measures (Lukichev & Romanovich, 2016). In addition, traditional QM in construction can be excessively bureaucratic and time-consuming, with a significant amount of paperwork and documentation required (Ringena et al., 2014). This can cause frustration among workers and project teams and slow down the construction process.

In response to the challenges associated with traditional QM in the CI, many organizations are exploring alternative approaches. One such approach is the utilization of digital tools and technologies to improve quality control and documentation (Hwang et al., 2019; Zhang et al., 2014). According to a study by several authors, BIM is one of the digital technologies as it facilitates real-time collaboration and information sharing between project stakeholders. Other digital tools like drones, sensors, and cameras can also be utilized to monitor construction progress and detect potential quality issues (Ding et al., 2015; Liu, Xie, & Tivendal, 2019; Zhang et al., 2014). By adopting these alternative approaches to QM, construction firms can improve their processes, reduce costs, and enhance overall project outcomes through collaboration and communication between project teams, starting in the planning and design phases. However, there may be challenges with implementing these approaches, including the need for training, resources, and infrastructure. Recently, several authors attempted to build a 4D BIM which is both functional and informative model for regulating quality of the project (Muhammad et al., 2019). BIM's 3D visualization capabilities enhance communication and coordination among teams, while its centralized platform can be used to manage project information, including quality control and documentation (Hwang et al., 2019). Additionally, BIM has the potential to improve overall project outcomes by integrating quality management (QM) into the planning and design phases of a project (Enegbuma et al., 2014). This approach can help reduce the likelihood of quality issues and minimize the need for rework or repairs later in the construction process, saving time and cost for all stakeholders involved.

DOI: 10.1201/9781003408949-11

BIM has emerged as an innovative approach that promises to revolutionize the CI, enabling it to keep pace with the changing times. As a rapidly growing digital technology, BIM has the ability to improve the way information is managed in construction projects, making it easier for stakeholders to collaborate and share information in real time. BIM has the potential to revolutionize the processes involved in designing, constructing, and maintaining construction projects. This technological advancement offers substantial advantages in terms of enhancing the overall quality, efficiency, and sustainability of such projects (Cheng, 2018). However, it was argued that BIM adoption has been hampered by challenges associated with BIM usage by architectural, engineering, and construction (AEC) firms in both developed and developing nations (L. Ding et al., 2017; Wang et al., 2016). It has been argued that there are multiple factors that restrict the progress of BIM, as previously asserted. Despite the extensive research on different aspects of BIM, only a few publications directly address digitalization in QM. Therefore, the primary objective of this study is to enhance the current understanding of QM by doing a systematic literature review (SLR) to explore the use of BIM in the CI. Additionally, this research seeks to identify and analyse the key problems that hinder the effective implementation of BIM for enhancing quality standards.

11.2   Literature Review

11.2.1   *What Is BIM?*

BIM is a word that has garnered significant attention and has been extensively discussed in academic journals. Essentially, BIM involves an intelligent 3D model-based process that provides relevant information to professionals at the appropriate time and to the appropriate individual. Its aim is to facilitate the effective planning, design, construction, and management of facilities (Enegbuma, W. I., Dodo, Y. A. & Ali, 2014; Paneru et al., 2023). Additionally, BIM is often viewed as a substitute for the traditional drawing design approach. The implementation of BIM offers industry professionals with more significant information, which aids in the efficient planning, designing, construction, and management of a project. Moreover, utilizing BIM can result in cost reduction, leading to better project affordability. BIM also improves overall project quality by enhancing information accuracy and quality during cost estimation (Babatunde et al., 2021). The use of BIM facilitates early clash detection and minimizes conflicts among project team members through powerful visualization and superior information quality. Additionally, BIM helps in the efficient operation of planning and construction procedures and enables project team members to preserve data for future maintenance and operation tasks. Furthermore, BIM reduces the time needed to access relevant information as softcopy data is more accessible than traditional hardcopy data. These benefits of BIM undoubtedly provide architects, engineers, and contractors with an effective approach to tackle the common challenges encountered in construction projects via efficient planning. However, in the CI, projects are managed collaboratively across various disciplines and team members, resulting in trust issues and a lack of clarity regarding responsibilities, duties, and interoperability (Chien et al., 2014). Therefore, when working on a BIM-based construction project, several factors must be addressed, and team participation is critical. BIM is a complex construction method, but it offers advantages over traditional construction estimation methods. Instead of using traditional CAD software, quantity surveys, material take-offs, and other measurement methods, BIM models save time and are more accurate. Cost is a critical factor in all phases of the construction lifecycle, and traditional methods like quantity take-off are time-consuming, less precise, and lack detail in some areas. These methods also make it difficult to discuss with stakeholders areas (Babatunde & Perera, 2020; Darwish et al., 2020). BIM has become increasingly popular in the CI due to its smart problem-solving capabilities and ability to facilitate knowledge sharing among all participants.

11.2.2 BIM and Its Application in QM

BIM has been gaining attention as a tool for QM in construction projects. By creating a digital representation of the building and all its components, BIM provides a platform for a more efficient and effective QM process (Darwish et al., 2017). According to Shah and Varghese (2008), BIM offers many benefits for QM, such as improved communication and collaboration between project stakeholders, enhanced visualization of project components, and the ability to identify and resolve design conflicts before construction begins. BIM also allows for better coordination and integration of project data, which can lead to more informed decision-making and a higher level of quality control. Through the use of BIM, construction teams can simulate and visualize the construction process, identify potential issues and resolve them before construction begins, and streamline project workflows. This results in better project outcomes, including reduced rework, increased efficiency, and improved quality of the final product (Azhar, 2011). Overall, the application of BIM in QM for construction projects is a promising approach to improving project outcomes. However, its successful implementation requires a shift in project workflows and a change in organizational culture to fully embrace the use of BIM (Ding et al., 2015). As such, the adoption of BIM for QM should be supported by training and education programs, clear guidelines for its use, and a commitment to ongoing technological advancement in CI.

11.3 Methodology

The aim of this study was accomplished through the utilization of systematic literature review (SLR) (David Tranfield, 2003). Unlike the traditional review, an SLR is a method that locates existing studies, selects and analyses in such a manner that reasonably clear judgments about what is and is not known may be obtained. A systematic literature search ensures transparency and reduces the potential for bias when identifying relevant publications for the research (Khan et al., 2003). As a result, to ensure a comprehensive and well-defined review of the literature, this study followed a three-stage procedure (David Tranfield, 2003). This rigorous approach ensured a systematic and transparent review, allowing for the identification of critical issues, concepts, and knowledge gaps in the field of study. Therefore, in this study, a three-stage procedure was adopted which includes planning the review, conducting the review, and analyzing the review (Figure 11.1).

11.3.1 Planning the Review

In conducting an SLR, the first step is to clearly define the research problem and formulate it. This initial stage is crucial as it sets the direction of the entire review process. Subsequently, a comprehensive search for relevant articles is carried out by identifying specific keywords and databases that are known for their extensive coverage of peer-reviewed journals in the relevant field. For instance, the Ebscohost, Scopus and web of

Identification

Selection of Database

Ebscohost, Scopus & WOS

Keyword search string in the database
N=1860

Initial search by year 1992-2023

Records after duplicates removed
N=303

Duplicates, non-peer reviewed,
non-English

Screening

Screening stage 1
N=103

Records only related to BIM
QM

Screening stage 1
N=78

Records only related to CI

Eligibility &
Analysis

Full text analysis
N=67

Excluded due to unavailability
of full articles on integrated
BIM & QM

Studies included
in the search
N=63

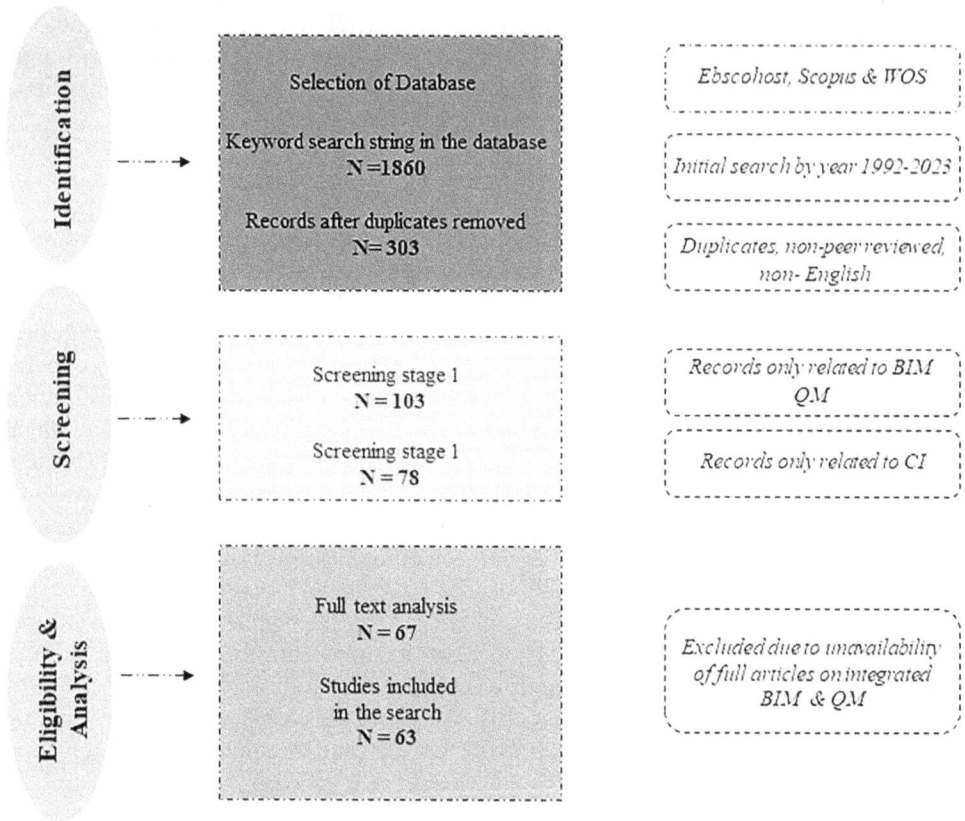

Figure 11.1 Literature search methodology.

science (WOS) databases were selected for its broad scope and inclusion of studies on project management and megaprojects, as mentioned by previous studies (de Araújo et al., 2017; L. Ding et al., 2017). Prior to conducting the database search, it is essential to develop a list of keywords and related terms that are associated with the research topic, which can be achieved through the use of building blocks as a search strategy. This method involves dividing the research problem into specific phrases and linking them together using Boolean operators like "AND" and "OR." In the current study, probable phrases such as "Quality Management," "Quality Improvement," "BIM," "Barriers," and "Implementation" were identified as relevant to the research problem through a combination of snowball and trial-and-error search method.

11.3.2 *Conducting the Review*

In the previous section, the methodology for searching and selecting relevant papers for the study is presented. The search process comprises three distinct steps, beginning with a keyword search in three databases, with no restriction on the initial publication date to ensure a comprehensive search. The search yielded 1860 publications until March 2023, which were then filtered by language, with only English language publications being retained. Duplicate papers were also removed, resulting in a reduced list of 303 papers. Secondly, we critically evaluated the titles and abstracts of the publications based on pre-defined filtering and selection criteria. These criteria involved including studies that focused on BIM and QM in the CI while excluding articles that did not meet these criteria. Using this process, 78 papers were selected for further consideration. Finally, a full-text manual analysis was undertaken, resulting in the selection of 63 papers for the systematic review analysis, as discussed in subsequent sections of this chapter.

11.3.3 Analysis and Reporting of the Review

In this step, a descriptive analysis of the publications was undertaken to obtain year-wise and infrastructure sector-wise publications. These results are shown using charts and tables in the following section. Furthermore, inductive content analysis was undertaken to determine the barriers in implementing BIM and how these barriers are tackled to improve quality. Each of the articles was read numerous times, back and forth, to extract and group codes, and eventually, classify them (Seuring et al., 2012).

11.4 Analysis and Result

11.4.1 Descriptive Analysis

As illustrated in Figure 11.2, the descriptive analysis includes the publication distributions in various years starting from 2000. The number of publications was initially limited as digital technology for infrastructure operations and maintenance was still in its early stages in the Indian CI. However, as technology has progressed, the number of publications has also increased. Figure 11.3 categorizes the articles according to the fields in which the majority of the research has been conducted. It is noticeable that since 2014–2015, there has been a significant surge in publications, indicating an increase in the usage and further research on BIM technology for QM in the Indian CI.

Moreover, the list of articles is classified according to the methods used to acquire the data. There were four distinct methods for data collection: conceptual research, case study, survey/questionnaire, and literature review. The literature review was the most

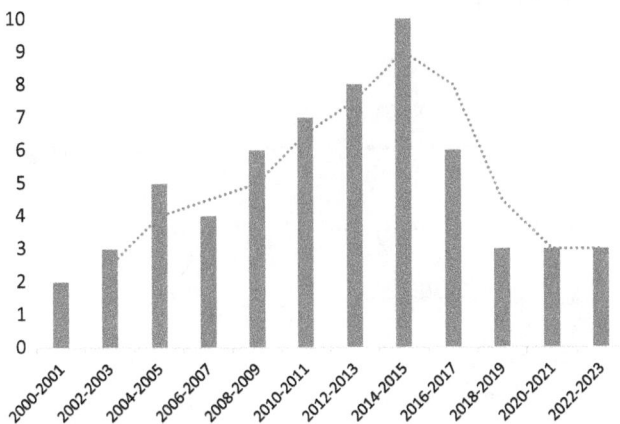

Figure 11.2 Number of published papers within the selected period.

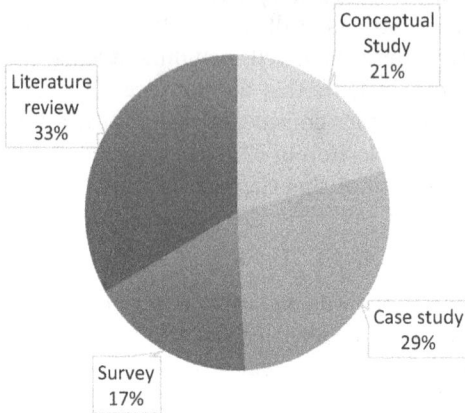

Figure 11.3 Overview of research methodologies on data collection.

frequently used method for data collection (21 articles). A significant number of re-searchers used the case study (18 articles) to collect data. The other techniques of data collecting include conceptual study (13 articles) and survey/questionnaire (11 articles).

11.4.2   Content Analysis

In the construction projects, the introduction of BIM to facilitate QM has been limited by a number of barriers. The insufficient support from top management, absence of suffi-cient technology and infrastructure, low level of BIM technical awareness, and staff re-sistance are the four most significant barriers. Each of these barriers can be further divided into two or three categories, as shown in fishbone diagram of barriers and its root causes (Figure 11.4).

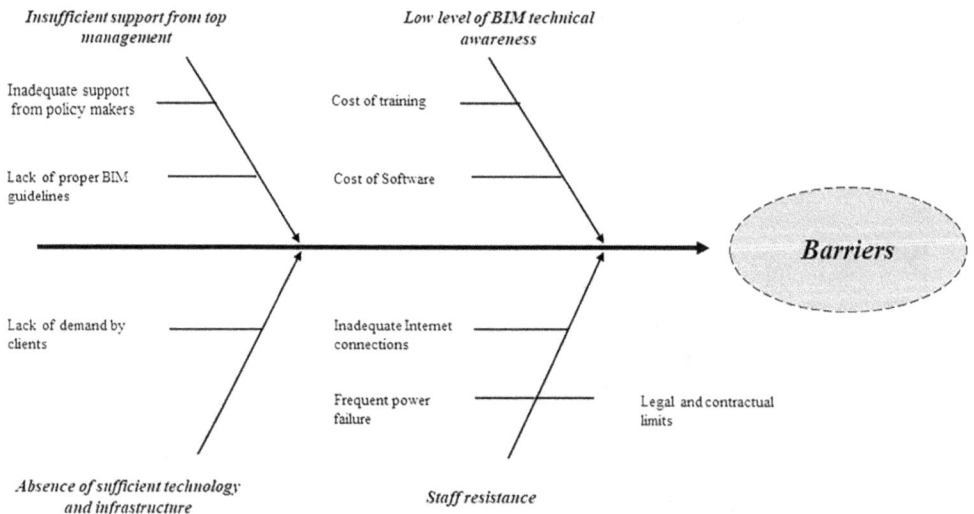

Figure 11.4 Fishbone diagram of barriers and its root causes.

11.4.2.1 Insufficient Support from Top Management

The successful implementation of BIM in the CI is hindered by several interrelated barriers, including inadequate support from top management, policymakers, and a lack of proper BIM guidelines. One of the main causes of inadequate support from top management is the absence of clear policies and guidelines for BIM adoption and implementation (Azhar, 2011; Chan et al., 2019). This lack of clarity can be attributed to inadequate support from policymakers who are responsible for creating such policies and guidelines. A lack of proper BIM guidelines can create confusion and uncertainty about the adoption and implementation of BIM, leading to a lack of support from top management. Studies have shown that inadequate support from top management can lead to a lack of commitment, resources, and funding for BIM projects, which can hinder the adoption and implementation of BIM in the CI (Coates et al., 2019; Wong et al., 2010). Additionally, inadequate support from policymakers can result in a lack of funding and resources for the development and implementation of BIM guidelines, which can further hinder the adoption and implementation of BIM in the CI (Babatunde et al., 2021; Enegbuma, Dodo, & Ali, 2014; Farouk et al., 2023). The absence of proper BIM guidelines can create a lack of standardization and consistency in BIM practices across different organizations, leading to inefficiencies and reduced benefits (Arayici et al., 2011). Therefore, policymakers must provide adequate support for the development and implementation of BIM guidelines. This includes funding and resources for technical experts to develop and maintain BIM guidelines, as well as collaboration with industry stakeholders to ensure the guidelines are relevant and effective. In addition, top management must be committed to the adoption and implementation of BIM and provide the necessary resources and funding to support BIM projects (Al-Mohammad et al., 2022; Olanrewaju et al., 2022). This commitment must be reflected in the organization's culture and strategy to ensure the successful implementation of BIM.

11.4.2.2 Absence of Sufficient Technology and Infrastructure

The implementation of BIM in the CI requires adequate technology and infrastructure to support it. However, the absence of sufficient technology and infrastructure is often caused by the lack of demand from clients, which can be attributed to several factors (Liu, Xie, Tivendal, et al., 2019). One of the main factors that contribute to the lack of demand from clients for BIM is the lack of awareness and understanding of its benefits. According to research by Arayici et al. (2011), many clients are not familiar with BIM and are therefore reluctant to demand its use in their projects. This lack of awareness and understanding of BIM can be attributed to inadequate communication and education about its benefits and applications. Another factor that contributes to the lack of demand from clients for BIM is the perception that it is an expensive and time-consuming technology (Xu et al., 2018; Zhang et al., 2020). It was stated by researchers that clients may be hesitant to demand the use of BIM because they believe it will increase project costs and duration. This perception can be attributed to the lack of understanding of BIM's potential to improve project efficiency and reduce costs (Matarneh & Hamed, 2017).

11.4.2.3 Low Level of BIM Technical Awareness

The low level of BIM technical awareness can be attributed to the additional costs associated with training and software. The implementation of BIM requires specialized technical knowledge and training, which can be costly for firms (Akerele & Moses, 2016; Babatunde & Perera, 2020). Without sufficient technical knowledge, the benefits of BIM cannot be fully realized, and firms may be hesitant to invest in the technology. Research has shown that the cost of training is a significant barrier to the adoption and implementation of BIM in the CI (Babatunde et al., 2021; Ding et al., 2015). In addition to the initial costs of training, ongoing training, and development are necessary to keep up with new advancements in BIM technology (Davila Delgado et al., 2020; Liu, Xie, & Tivendal, 2019). These costs can be a significant burden for firms, especially small and medium-sized enterprises (SMEs), and may deter them from adopting BIM. The cost of software is another factor that contributes to the low level of BIM technical awareness. BIM software can be expensive, and firms may not have the resources to purchase and maintain the necessary software licenses. This can result in firms resorting to outdated software or not investing in BIM at all, which can limit their competitiveness and ability to take on complex projects. To address these barriers, governments and industry stakeholders can provide support to firms in the form of subsidies and funding for training and software (Ezeokoli et al., 2016). In addition, the development of open-source BIM software can provide more affordable options for firms, particularly SMEs (Ahuja et al., 2020; Coates et al., 2019). Collaboration between firms and universities can also provide opportunities for training and development, as well as research and development of new BIM technologies.

11.4.2.4   *Staff Resistance*

The low level of BIM technical awareness can be attributed to the additional costs associated with training and software. The implementation of BIM requires specialized technical knowledge and training, which can be costly for firms (Farouk et al., 2023; Zhang et al., 2014). Without sufficient technical knowledge, the benefits of BIM cannot be fully realized, and firms may be hesitant to invest in the technology. Research has shown that the cost of training is a significant barrier to the adoption and implementation of BIM in the CI. In addition to the initial costs of training, ongoing training and development are necessary to keep up with new advancements in BIM technology (Davila Delgado et al., 2020; Zhang et al., 2014). These costs can be a significant burden for firms, especially small and medium-sized enterprises (SMEs), and may deter them from adopting BIM. The cost of software is another factor that contributes to the low level of BIM technical awareness. BIM software can be expensive, and firms may not have the resources to purchase and maintain the necessary software licenses (Babatunde et al., 2021; Olanrewaju et al., 2022). This can result in firms resorting to outdated software or not investing in BIM at all, which can limit their competitiveness and ability to take on complex projects.

11.5   Discussion

The SLR presented in this chapter has shown that BIM adoption can provide significant benefits to QM in the CI. The review identified four major barriers of BIM implementation, including the high cost of software and training, low levels of technical awareness among industry professionals, and inadequate support from top management and policymakers. These challenges underscore the importance of addressing these barriers to ensure successful implementation of BIM in the CI. To address these challenges, it is crucial to provide sufficient resources and training for industry professionals to develop their technical skills and knowledge of BIM. In addition, government and policymakers must provide adequate support for the development and implementation of BIM guidelines and standards, which will facilitate a more consistent and standardized approach to BIM implementation across the CI.

Furthermore, the implementation of BIM should not be viewed as a one-time solution, but rather as a continuous process of improvement and adaptation to changing industry needs and advancements in technology. The development of effective BIM-based QM systems will require ongoing collaboration and communication among all stakeholders involved in the construction process. This systematic review has demonstrated that BIM adoption can provide significant benefits to QM in the CI. However, several challenges and barriers must be addressed to ensure successful implementation. The development of effective BIM-based QM systems will require ongoing collaboration and communication among all stakeholders involved in the construction process.

11.6   Conclusion and Way Forward

According to the current study, it is evident that QM has significant importance within the context of CI, while requiring a substantial investment of time. The implementation of a systematic strategy is essential in ensuring quality in construction and bridging the existing knowledge gap regarding the use of BIM in the construction industry. BIM has the potential to enhance the efficiency of construction endeavors, namely in the domains of building and maintenance. However, integration of BIM in construction projects has recently encountered several obstacles.

For instance, adoption of a BIM system in any organization demands expenses such as the *purchase of BIM software* as well as the *cost of training staff* to use the technology. One of the primary factors to the major cost hurdles of BIM adoption was the initial cost necessary to update software, replace equipment, and educate employees. The cost of the BIM was discovered to be the main sub-barrier to BIM adoption. *Policies and guidelines* were also identified as the second most significant sub-barrier to BIM implementation. Because there were no rules regulating the use of BIM at a particular project size, it was rarely used. The construction project contract lacks legal stability, considering the use of BIM irrelevant. The legal aspects included the establishment of BIM standards and project contracts. Countries without BIM use regulations were also found to have no national BIM standard in place. *Client demand for the use of BIM* became the subsequent sub-barrier to BIM implementation. CI was found to be lacking the necessary readiness to effectively apply BIM. Insufficient demand for the installation of BIM has arisen due to uncertainty about its immediate benefits especially during the planning phase.

Based on the findings of the review paper, several key measures were identified as significant in enhancing the adoption of BIM. These measures, listed in order of priority, include the implementation of education and training programs, integration of BIM into academic curricula, enactment of guidelines and regulations pertaining to BIM, provision of sufficient technology and infrastructure, mandatory inclusion of BIM in all procurement processes and contracts, and establishment of a BIM council.

References

Ahuja, R., Sawhney, A., Jain, M., Arif, M., & Rakshit, S. (2020). Factors influencing BIM adoption in emerging markets – The case of India. *International Journal of Management, 20*(1), 65–76. 10.1080/15623599.2018.1462445

Akerele, A. O., & Moses, E. (2016). Assessment of the level of awareness and limitations on the use of building information modeling in Lagos State. *International Journal of Scientific and Research Publications*, 6(2), 229. www.ijsrp.org

Al-Mohammad, M. S., Haron, A. T., Esa, M., Aloko, M. N., Alhammadi, Y., Anandh, K. S., & Rahman, R. A. (2022). Factors affecting BIM implementation: Evidence from countries with different income levels. *Construction Innovation*. 10.1108/CI-11-2021-0217

Arayici, Y., Coates, P., Koskela, L., Kagioglou, M., Usher, C., & O'Reilly, K. (2011). Technology adoption in the BIM implementation for lean architectural practice. *Automation in Construction*, 20(2), 189–195. 10.1016/j.autcon.2010.09.016

de Araújo, M. C. B., Alencar, L. H., & de Miranda Mota, C. M. (2017). Project procurement management: A structured literature review. *International Journal of Project Management*, 35(3), 353–377. 10.1016/j.ijproman.2017.01.008

Azhar, S. (2011). Building information modeling (BIM): Trends, benefits, risks, and challenges for the AEC industry. *Leadership and Management in Engineering*, 11(3), 241–252. 10.1061/(ASCE) LM.1943-5630.0000127

Babatunde, S. O., & Perera, S. (2020). An investigation into BIM uptake among contracting firms: An empirical study in Nigeria. *Journal of Financial Management of Property and Construction*, 26(1), 23–48. 10.1108/JFMPC-06-2019-0054

Babatunde, S. O., Udeaja, C., & Adekunle, A. O. (2021). Barriers to BIM implementation and ways forward to improve its adoption in the Nigerian AEC firms. *International Journal of Building Pathology and Adaptation*, 39(1), 48–71. 10.1108/IJBPA-05-2019-0047

Chan, D. W. M., Olawumi, T. O., & Ho, A. M. L. (2019). Critical success factors for building information modelling (BIM) implementation in Hong Kong. *Engineering, Construction and Architectural Management*, 26(9), 1838–1854. 10.1108/ECAM-05-2018-0204

Cheng, Y. M. (2018). Building information modeling for quality management. *ICEIS 2018 – Proceedings of the 20th International Conference on Enterprise Information Systems*, 2(Iceis 2018), 351–358. 10.5220/0006796703510358

Chien, K. F., Wu, Z. H., & Huang, S. C. (2014). Identifying and assessing critical risk factors for BIM projects: Empirical study. *Automation in Construction*, 45, 1–15. 10.1016/j.autcon. 2014.04.012

Coates, P., Arayici, Y., Koskela, L., Kagioglou, M., Usher, C., & O'Reilly, K. (2019). The key performance indicators of the BIM implementation process. *EG-ICE 2010 – 17th International Workshop on Intelligent Computing in Engineering*. United Kingdom.

Darwish, A. M., Tantawy, M. M., Elbeltagi, E., Wang, J., Sun, W., Shou, W., Wang, Oraee, M., Hosseini, M. R., Papadonikolaki, E., Palliyaguru, R., Arashpour, M., Wang, Q., Kim, M. K., Cheng, J. C. P., Sohn, H., Babatunde, S. O., Udeaja, C., Adekunle, A. O., … Love, P. E. D. (2017). Automated quality assessment of precast concrete elements with geometry irregularities using terrestrial laser scanning. *Automation in Construction*, 39(7), 180–191. 10.1007/s10846-014-0116-8

Darwish, A. M., Tantawy, M. M., & Elbeltagi, E. (2020). Critical success factors for BIM implementation in construction projects. *Saudi Journal of Civil Engineering*, 4(9), 180–191. 10.36348/sjce.2020.v04i09.006

Davila Delgado, J. M., Oyedele, L., Demian, P., & Beach, T. (2020). A research agenda for augmented and virtual reality in architecture, engineering and construction. *Advanced Engineering Informatics*, 45(December 2019), 101122. 10.1016/j.aei.2020.101122

Ding, L., Li, K., Zhou, Y., & Love, P. E. D. (2017). An IFC-inspection process model for infrastructure projects: Enabling real-time quality monitoring and control. *Automation in Construction*, 84(August), 96–110. 10.1016/j.autcon.2017.08.029

Ding, Z., Zuo, J., Wu, J., & Wang, J. Y. (2015). Key factors for the BIM adoption by architects: A China study. *Engineering, Construction and Architectural Management*, 22(6), 732–748. 10.1108/ECAM-04-2015-0053

Duttenhoeffer, B. R. (1992). *Cost and Quality Management'. Journal of Management in Engineering*, *8*(2), 167–175. 10.1061/(ASCE)9742-597X(1992)8:2(167)

Enegbuma, W. I., Dodo, Y. A., & Ali, K. N. (2014). Building information modelling penetration factors in Malaysia. *International Journal of Advances in Applied Sciences (IJAAS)*, *3*(1), 47–56.

Ezeokoli, F., Okoye, P., & Nkeleme, E. (2016). Factors affecting the adaptability of building information modelling (BIM) for construction projects in Anambra State Nigeria. *Journal of Scientific Research and Reports*, *11*(5), 1–10. 10.9734/jsrr/2016/26563

Farouk, A. M., Zulhisham, A. Z., Lee, Y. S., Rajabi, M. S., & Rahman, R. A. (2023). Factors, challenges and strategies of trust in BIM-based construction projects: A case study in Malaysia. *Infrastructures*, *8*(1), 13. 10.3390/infrastructures8010013

Hwang, B.-G., Zhao, X., & Yang, K. W. (2019). Effect of BIM on rework in construction projects in Singapore: Status quo, magnitude, impact, and strategies. *Journal of Construction Engineering and Management*, *145*(2), 1–16. 10.1061/(asce)co.1943-7862.0001600

Khan, K., Kunz, R., Kleijnen, J., & Antes, G. (2003). Five steps to conducting a systematic review. *Journal of the Royal Society of Medicine*, *96*(3), 118–121. 10.1177/014107680309600304

Lee, S., Asce, M., Peña-Mora, F., & Park, M. (2005). Quality and change management model for large scale concurrent design and construction projects. *Journal of Construction Engineering and Management*, *131*(8). 10.1061/ASCE0733-93642005131:8890

Liu, S., Xie, B., & Tivendal, L. (2019). The drivers, barriers, and enablers of building information modeling (BIM) innovation in developing countries: Insights from systematic lterature review and comparative analysis. *Buildings*, *7*(2), 288–302. 10.1016/j.autcon.2010.09.016

Liu, S., Xie, B., Tivendal, L., Liu, C., Darwish, A. M., Tantawy, M. M., Elbeltagi, E., Tokarsky, A., Topchiy, D., Coates, P., Arayici, Y., Koskela, L., Kagioglou, M., Usher, C., & O'Reilly, K. (2019). The key performance indicators of the BIM implementation process. *EG-ICE 2010 – 17th International Workshop on Intelligent Computing in Engineering*, *258*(6), 162. 10.1051/e3sconf/2 02125809028

Lukichev, S., & Romanovich, M. (2016). The quality management system as a key factor for sustainable development of the construction companies. *Procedia Engineering*, *165*, 1717–1721. 10.1016/j.proeng.2016.11.914

Matarneh, R., & Hamed, S. (2017). Barriers to the adoption of building information modeling in the Jordanian Building Industry. *Open Journal of Civil Engineering*, *07*(03), 325–335. 10.4236/ojce.2017.73022

Muhammad, H., Shehzad, F., Anwar, K., Khaidzir, M., Ibrahim, R. B., & Yusof, A. F. (2019). *The Adoption in the AEC Industry*. 20–25.

Olanrewaju, O. I., Kineber, A. F., Chileshe, N., & Edwards, D. J. (2022). Modelling the relationship between Building Information Modelling (BIM) implementation barriers, usage and awareness on building project lifecycle. *Building and Environment*, *207*, 1–23. 10.1016/j.buildenv. 2021.108556

Paneru, S., Ghimire, P., Kandel, A., Thapa, S., Koirala, N., & Karki, M. (2023). An exploratory investigation of implementation of building information modeling in Nepalese Architecture–Engineering–Construction Industry. *Buildings*, *13*(2). 10.3390/buildings13020552

Ringena, G., Aschehouga, S., Holtskogb, H., & Ingvaldsena, J. (2014). Integrating quality and lean into a holistic production system. *Procedia CIRP*, *17*, 242–247. 10.1016/j.procir. 2014.01.139

Seuring, S., & Gold, S. (2012). Conducting content-analysis based literature reviews in supply chain management. *Supply Chain Management*, *17*(5), 544–555. 10.1108/13598541211258609

Shah, A., & Varghese, S. (2008). Application of BIM as a model for quality management. *International Research Journal of Engineering and Technology*, June 1197. www.irjet.net

Tranfield, D., Denyer, D., & Smart, S. (2003). Towards a methodology for developing evidence-informed management knowledge by means of systematic review. *British Journal of Management*, *14*, 207–222. 10.1007/978-3-030-92836-0_28

Wang, Q., Kim, M. K., Cheng, J. C. P., & Sohn, H. (2016). Automated quality assessment of precast concrete elements with geometry irregularities using terrestrial laser scanning. *Automation in Construction, 68*, 170–182. 10.1016/j.autcon.2016.03.014

Wong, A., Wong, F., & Nadeem. (2010). Attributes of building information modelling implementations in various countries. *Architectural Engineering and Design Management, 6*(SPECIAL ISSUE), 288–302. 10.3763/aedm.2010.IDDS6

Xu, Z., Huang, T., Li, B., Li, H., & Li, Q. (2018). Developing an IFC-based database for construction quality evaluation. *Advances in Civil Engineering, 2018*. 10.1155/2018/3946051

Zhang, R., Tang, Y., & Wang, L. (2014). Barriers to the implementation of building information modeling among Jordanian AEC companies. *Automation in Construction, 46*(2), 102–114. 10.5220/0006796703510358

Zhang, R., Tang, Y., Wang, L., & Wang, Z. (2020). Factors influencing BIM adoption for construction enterprises in China. *Advances in Civil Engineering, 2020*. 10.1155/2020/8848965

12 Transforming Construction Site Safety with iSAFE

An Automated Safety Management Platform

Chansik Park, Mehrtash Soltani, Akeem Pedro,
Jaehun Yang, Doyeop Lee, and Rahat Hussain
School of Architecture and Building Science, Chung-Ang University,
Seoul, Republic of Korea

12.1 Introduction

Construction safety management can be divided into three stages: pre-work, during, and after work. However, current practices within these stages are associated with several issues. In the before-work stage, manual procedures and document-based data may lead to time-consuming processes, unreliable planning, and training which can be problematic. Additionally, the monitoring methodology during work is often inefficient, which can result in numerous injuries and fatalities at job sites. Furthermore, these problems are compounded by a lack of positive safety culture, making it a significant challenge for organizations to maintain a safe working environment and improve productivity throughout all stages of the work process (Martínez-Rojas et al., 2016). In the subsequent sections, various problems associated with current practices in construction safety management will be explored. Construction projects can be complex and risky, with many potential hazards that must be considered (Raftery, 1993). Without proper safety planning and training, workers may be at risk of injury or death (Seokho Chi et al., 2013). In addition, workers may not be aware of the specific risks associated with their tasks or may lack the knowledge and skills needed to perform their duties safely. Therefore, safety planning and training are critical components of any construction project, helping to identify potential hazards and develop strategies to mitigate them (Sooyoung Choe & Fernanda Leite, 2017).

Document-based risk assessments are frequently used in the construction industry to identify potential hazards (Andrew Rae et al., 2014). However, these assessments are not without their challenges. One of the primary difficulties is that they may not capture all the risks associated with a given project, leaving workers vulnerable to unanticipated hazards. Additionally, these assessments may not be tailored to the specific work plan and site situation, increasing the potential for accidents and injuries. Furthermore, even when safety plans are developed, they are often not utilized during construction, leading to confusion and increased risks. To create safer work environments and reduce the incidence of accidents and injuries, it is crucial to address these challenges by developing comprehensive and site-specific safety plans, providing updated planning to workers, and ensuring that safety plans are adhered to throughout the project (Numan Khan et al., 2022).

The construction industry encounters an additional predicament concerning standardized training programs that lack sufficient preparation for the hazards and risks associated with individual worksites (Oughton, 2013). Additionally, training programs

DOI: 10.1201/9781003408949-12

that are solely classroom-based or rely on short toolbox meetings may not provide workers with the hands-on experience and practical knowledge they need to perform tasks safely (Ricardo Eiris et al., 2020). These challenges can increase the risk of accidents and injuries on the job. To overcome these problems, it is important to prioritize site-specific, interactive, and practical training methods that equip workers with the knowledge and skills to perform their tasks safely and confidently.

The construction industry is not immune to safety risks even with proper training and planning, leading to accidents and injuries during the work process (Lei Hou et al., 2021). Neglecting safety risks can be minimized by establishing a proactive safety approach that prioritizes safety through regular inspections, ongoing training, and clear protocols. By doing so, the industry can reduce accidents caused by negligence and promote a safer working environment (Christopher Frazier et al., 2013). Safety inspection and monitoring are essential for identifying and addressing potential hazards in the construction industry (Kyungki Kim et al., 2021). Monitoring workers' unsafe behavior to minimize risks in construction sites is crucial. However, the manual monitoring process requires safety managers to be physically present at construction sites to identify potential hazards or non-compliance with safety rules, leading to delays in identifying and addressing potential hazards, which increases the risk of accidents and injuries. Implementing safety rules and monitoring Personal Protective Equipment (PPE) compliance in large and complex construction sites is expensive and impractical. Hence, there is a need to incorporate technological advancements to ensure the real-time safety of workers at hazardous elevations and reduce the workload on safety managers who monitor workers on-site (Muhammad Khan et al., 2022). One such advancement is the use of Tag and IoT-based safety hook monitoring for fall prevention from height, which can alert workers and safety managers in real-time to take appropriate action to prevent accidents and injuries (Muhammad Khan et al., 2022).

As discussed, the construction industry faces several challenges in after-work safety management, including the repetitive and error-prone nature of paper-based inspection records (Zhang, Chi et al., 2017) and the vulnerability of centralized data management systems to data falsification (Hunhevicz et al., 2020; Richard McCuen et al., 2011; Wang et al., 2017). Construction industry accident records indicate that safety accidents are often the result of safety rule violations due to mistakes or negligence (Lei Hou et al., 2021). Effective safety management is crucial for promoting safety measures and ensuring occupational safety, which can significantly impact unsafe behaviors. Scholars have found that safety policies, punishment, and a safe environment can positively influence workers' behaviors. However, conflicting goals between safety and production can lead to workers ignoring safety measures, rendering incentives ineffective (Jianbo ZhuCe et al., 2022).

This study proposes the iSAFE platform, a comprehensive construction safety management solution comprising three modules to address these challenges. The first module, iSAFE, assists in safety planning and training before the work phase. The second module, iSAFEGuard, monitors the construction site for compliance with safety rules during the work. Finally, the iSAFEIncentive module measures safety performance and provides incentives after work to encourage safe behavior. By implementing the iSAFE platform, construction companies can enhance safety management and reduce accidents caused by safety rule violations, ultimately promoting a safer working environment.

12.2   Research Core

Several academic articles have been disseminated to confront the difficulties above, which may culminate in a transformative advancement in construction safety management. Such progress may be realized through the technological enhancement of the iSAFE platform. These papers cover most technical parts and implementation of this platform, from planning, training, monitoring, and blockchain technology for storing safety performance data and incentive token-based applications. In terms of planning, one paper discusses the use of generative planning to identify optimal locations for construction safety surveillance cameras in a 4D BIM environment (Si Van-Tien Tran et al., 2018). Another paper proposes an approach for hazard identification that integrates 4D BIM and accident case analysis (Si Van-Tien Tran et al., 2022). Regarding training, some articles suggest new methods for safety training using immersive technologies such as cross-platform VR and extended reality. One recent paper presents an artificial intelligence (AI)-based safety helmet recognition system to enhance the safety monitoring process (Lan Bao et al., 2022). In terms of monitoring, several papers propose new approaches to iSAFE for fall prevention and worker safety using Internet of Things (IoT)-based monitoring, CV, and other technologies (Sharjeel Anjum et al., 2022; Muhammad Khan et al., 2022). Finally, some articles explore the potential of blockchain for construction safety management, including a proposed network concept model for reliable and accessible fine dust management at construction sites and the use of blockchain for scaffolding work management (Seungwon Cho et al., 2021; Jaehun Yang et al., 2022). Table 12.1 presents the different components within the iSAFE platform published in academic journals.

12.3   Introducing the iSAFE Platform

The iSAFE platform represents a novel approach to safety management for the construction industry, addressing challenges associated with traditional safety planning and monitoring methods. Safety management comprises various stages, including pre-work, during-work, and post-work, and the iSAFE platform provides a comprehensive solution for each stage. Traditional classroom training poses a significant challenge in safety planning, and the iSAFE platform proposes an innovative approach to overcome this issue. The platform integrates 4D Building Information Modeling (BIM)-based risk assessment and safety planning, virtual reality (VR), and 360-panorama-based training, providing a holistic solution for safety planning and training. Monitoring safety performance for a large workforce on complex job sites during the work phase is challenging. Manual inspection methods are time-consuming and inefficient, and the iSAFE platform proposes computer vision (CV) and sensor-based technologies for safety inspection and monitoring. The iSAFEGuard platform incorporates these technological advancements, resulting in a robust worker and job site monitoring process. Safety managers can proactively monitor safety compliance, reducing the need for hands-on experience on-site. The iSAFE platform integrates advanced technologies, providing a cutting-edge solution to safety management for the construction industry. Adopting the iSAFE platform would lead to a safer and more efficient work environment, enhancing the overall productivity and profitability of the construction industry.

The post-work phase of the safety management process is a crucial element involving document-based activities vital for fulfilling legal requirements and assessing safety performance. However, using hardcopy data sheets to collect this data can be problematic,

Table 12.1 Published papers related to the development of the iSAFE platform

No	Subject	Research Title	Summary	Main Technology	Journal Name	References
1	**Training**	Cross-platform virtual reality for real-time construction safety training using immersive web and industry foundation classes	The paper proposes a VR training program for construction safety that can be accessed from multiple platforms.	VR, Web technology, Industry classes	Automation in Construction	(Lan Bao et al., 2022)
2	**Planning**	A hazard identification approach of integrating 4D BIM and accident case analysis of spatial–temporal exposure	The paper proposes a novel hazard identification approach through spatial–temporal exposure analysis.	4D BIM	Sustainability	(Si Van-Tien Tran et al., 2018)
3	**Planning**	Generative planning for construction safety surveillance camera installation in 4D BIM environment	The paper proposes a generative planning method for installing surveillance cameras in construction sites using 4D BIM technology.	4D BIM	Automation in construction	(Si Van-Tien Tran et al., 2022)
4	**Planning**	Suggestions for improving South Korea's fall accidents prevention technology in the construction industry: focused on analyzing laws and programs of the United States	The paper proposes suggestions for improving fall accident prevention technology in the South Korean construction industry by analyzing the laws and programs of the United States.	Analysis of laws and programs	Sustainability	(Jeeyoung Lim et al, 2021)
5	**Monitoring**	Computer Vision Process Development regarding Worker's Safety Harness and Hook to Prevent Fall Accidents: Focused on System Scaffolds in South Korea	The paper proposes a computer vision system to prevent fall accidents by monitoring worker's safety harness and hook in construction scaffolds.	CV	Advances in Civil Engineering	(Jeeyoung Lim et al, 2022)

#	Category	Title	Description	Technology	Journal	Reference
6	Monitoring	Fall prevention from scaffolding using computer vision and IoT-based monitoring	The paper proposes an IoT-based monitoring system that uses computer vision to prevent fall accidents from scaffolding.	Internet of Things, CV	Journal of Construction Eng and Management	(Muhammad Khan et al., 2022)
7	Monitoring	Fall Prevention from Ladders Utilizing a Deep Learning-Based Height Assessment Method	The paper proposes a fall prevention system that uses deep learning for height assessment.	Deep learning	IEEE Access	(Sharjeel Anjum et al., 2022)
8	Monitoring	Tag and IoT-based safety hook monitoring for prevention of falls from height	IoT-based safety hook monitoring system to prevent falls from height.	Internet of Things	Automation in Const	(Muhammad Khan et al., 2022)
9	Monitoring	Utilizing safety rule correlation for mobile scaffolds monitoring leveraging deep convolution neural networks	Mobile scaffold monitoring system that uses deep convolution neural networks to correlate safety rules.	Deep convolution neural networks	Computers in Industry	(Numan Khan et al., 2021)
10	Performance assessment	Blockchain-based network concept model for reliable and accessible fine dust management system at construction sites	The paper proposes a blockchain-based network model for fine dust management in construction sites.	Blockchain	Applied Sciences	(Seungwon Cho et al., 2021)
11	Performance assessment	Leveraging blockchain for scaffolding work management in construction	Blockchain-based system for scaffolding work management in construction.	Blockchain	IEEE Access	(Jaehun Yang et al., 2022)

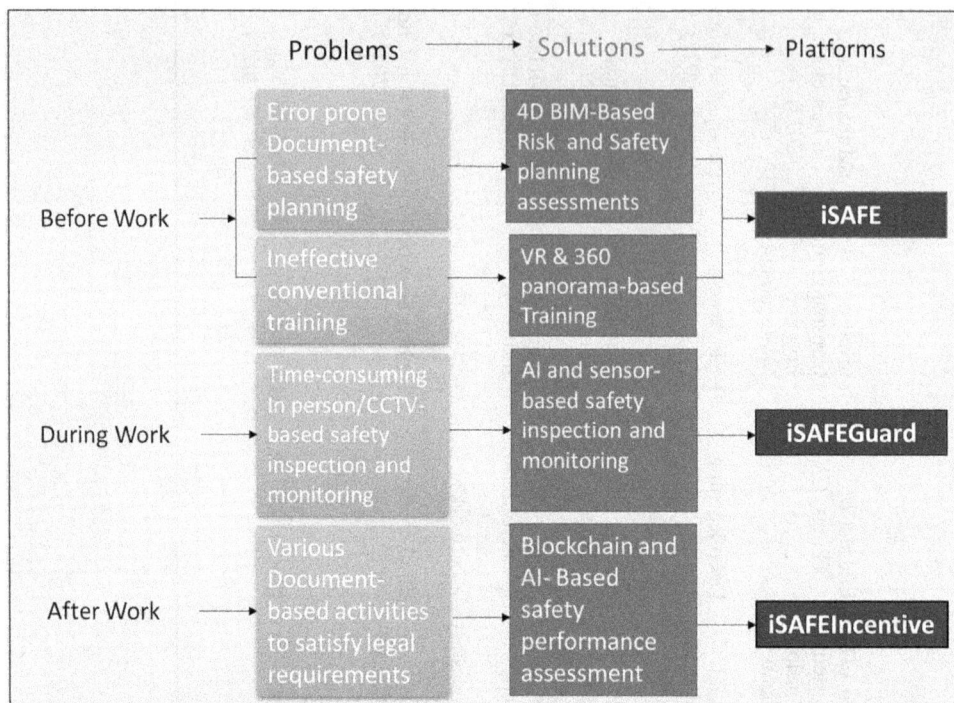

Figure 12.1 Problems and suggested solutions related to construction safety management.

particularly in job sites with a large workforce. This approach makes it challenging to evaluate individual workers' safety performance accurately. In response to this challenge, the iSAFE platform proposes the iSAFEIncentive system, which leverages blockchain and AI technologies. This system is a safety reward performance assessment mechanism that integrates advanced technologies, such as CV, to enhance worker and company safety performance evaluation. The system aims to establish a transparent and trustworthy platform for assessing safety performance that is impartial and unbiased. The iSAFEIncentive system incorporates token-based incentive technology, which rewards workers and companies for good safety performance. Safety managers award tokens to workers, while regulatory inspection governmental agencies award tokens to companies. This incentive-based approach motivates workers and companies to maintain excellent safety performance and cultivate a safety culture in the construction industry.

Figure 12.1 provides an overview of the current challenges in construction safety management and the proposed solutions offered by iSAFE.

Figure 12.2 depicts that the iSAFE platform utilizes advanced technologies such as 4D BIM, Digital Twin, augmented and VR, AI, and blockchain to automate the safety management cycle. The following sections provide additional details on each one.

12.3.1   iSAFE (Planning/Training)

Safety planning is a crucial part of the construction process on job sites. Several activities, if conducted in parallel, may trigger risks and hazards. To tackle this issue, iSAFE

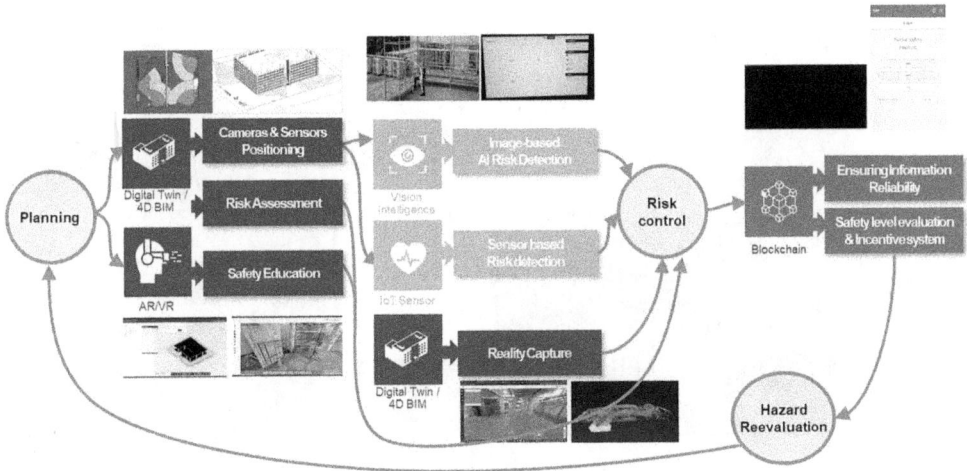

Figure 12.2 iSAFE platform framework.

provides an efficient work plan by combining the risk assessment database and specific construction job site conditions into the BIM model. It identifies the specific potential hazard situations and prepares prevention methods. Some of the functions of the systems include pre-risk assessment of the job site, safety information sharing amongst workers and training, and remote site inspection and instruction.

12.3.1.1 iSAFE Training

Traditional training methods may not keep pace with evolving technological and industrial practices, leading to worker frustration, disengagement, and ill-preparedness to tackle emerging challenges. Conversely, real-time learning offered in iSAFE utilizes a segmented approach, providing workers with tailored training and work instructions, resulting in efficient and customized education that enhances worker performance and engagement. Situated learning reflects new technological and industrial advancements, ensuring workers are well-equipped to handle emerging challenges. It utilizes technologies such as Digital Twin/4D BIM and AR/VR. Workers can access this education from iSAFE's high-fidelity virtual environment, which reflects daily job site conditions.

Additionally, iSAFE's planning toolkits combine risk assessment databases with construction job site characteristics to provide an efficient work plan and identify potential hazards. Options for surveillance camera locations are also generated to facilitate remote monitoring and AI-based risk detection during work. Figure 12.3 shows how educational content and instructions can be accessed at a job site.

12.3.1.2 iSAFE Planning

iSAFE planning is another advanced functionality of the iSAFE platform. Its algorithm is integrated with BIM and provides in-depth work report-based daily/weekly schedule information linking, work plan information extraction, and risk and change visualization rule development. After uploading the work schedule in the BIM application, the algorithm will evaluate all the parallel work to find any contradiction job that their same time

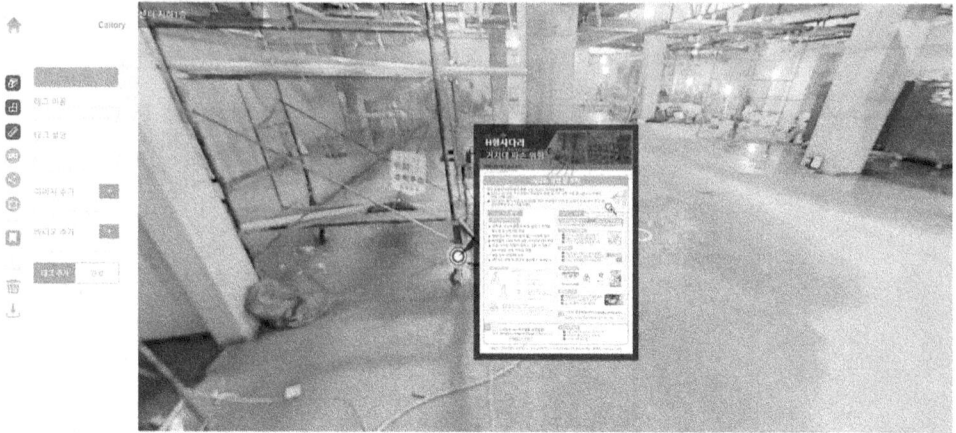

Figure 12.3 Reality capture view of iSAFE training.

is happening may result in danger. As appears in Figure. 12.4-A, activities such as painting and welding, any job defined near the opening or edge will be highlighted in the red zone to provide visualized risk zone to workers. All these adjustments will be made based on different rule compliance, and the algorithms are easily tailored to apply various rules and codes. iSAFE planning also offers a new framework for camera planning in construction, which includes functions such as selecting the appropriate camera type and positions based on the construction schedule by integrating into BIM 4D. A simulated viewpoint of the camera view in BIM is demonstrated in Figure. 12.4-B. Camera placement in construction is typically done manually, which can be time-consuming and costly (Si Van-Tien et al., 2022). This approach provides a cost-efficient, accurate solution with improved visualization by integrating BIM. The subsequent section elaborates on the monitoring capability of the iSAFE platform throughout the construction process stage.

12.3.2   *iSAFEGuard (Monitoring/Inspection)*

Despite efforts to plan for safety and provide training, unpredictable hazards can still occur during construction due to unsafe behaviors, unsafe working conditions, and violations of regulations. To address this persistent problem, iSAFEGuard has been developed as a technological solution utilizing video-based AI and sensor technologies to detect and prevent accidents and injuries on construction sites proactively. iSAFEGuard provides a comprehensive solution that detects and alerts safety managers and workers to take proactive measures to avoid accidents and injuries. The following sections explain the CV technology and sensor-based technologies of the iSAFEGuard. One of the fundamental features of iSAFEGuard is its video-based AI risk detection platform, which utilizes a deep neural network model for hazard detection. This system consists of 127 unsafe behavior detectors trained with synthetic and real data for various types of hazards, such as falling from a height, being hit by an object, collapsing, fire, and being struck by equipment. The detectors examine all tools and monitor individuals working with various equipment, including scaffolds, ladders, staircases, forklifts, etc. This model is designed to identify unsafe behaviors, working conditions, and violations of regulations in diverse scenarios. To develop this model,

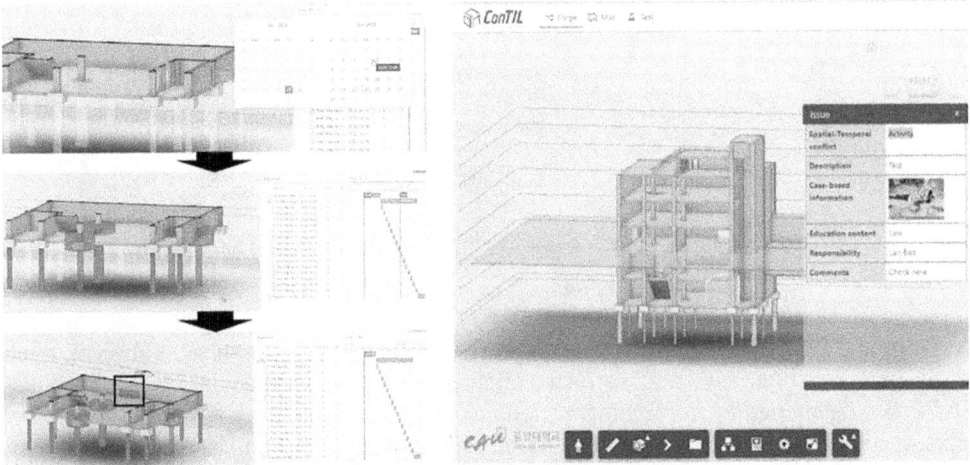

(A)

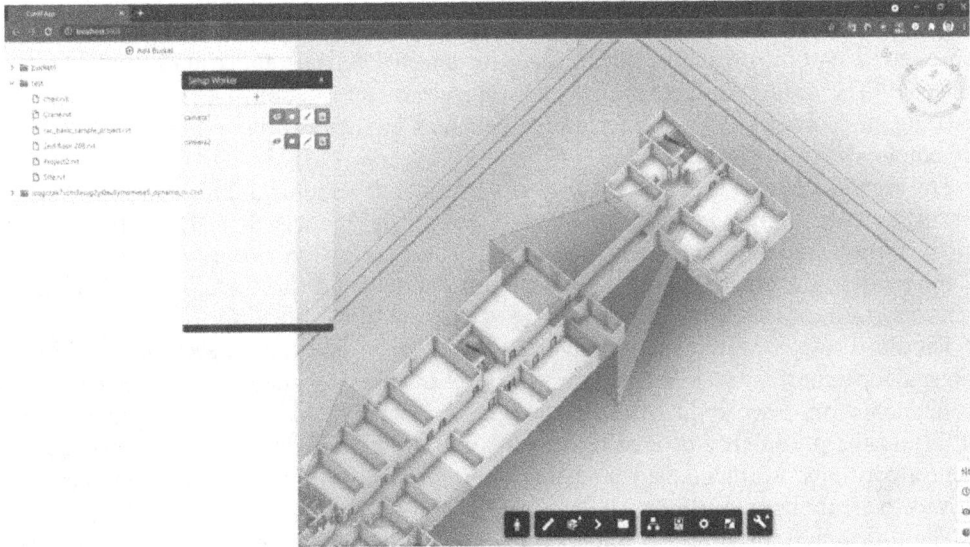

(B)

Figure 12.4 (A) Visualized risk zones in BIM. (B) Simulated version of the camera view in BIM.

large datasets are collected and processed to train and validate the system before deployment. The research project has been conducted to accumulate real site data, approximately 1 million images with location information (i.e., bounding boxes and polygons). In addition, a dataset augmentation technique using synthetic images was applied to overcome difficulties in collecting real-site photos such as difficulties in directing dangerous situations. The system utilizes a state-of-the-art model in the CV domain, considering detection speed and accuracy to enable real-time site monitoring and identifying any unsafe conditions or behaviors that may pose a risk to workers. The subsequent section will discuss the procedure for developing detectors and selecting paramount unsafe behaviors for their construction.

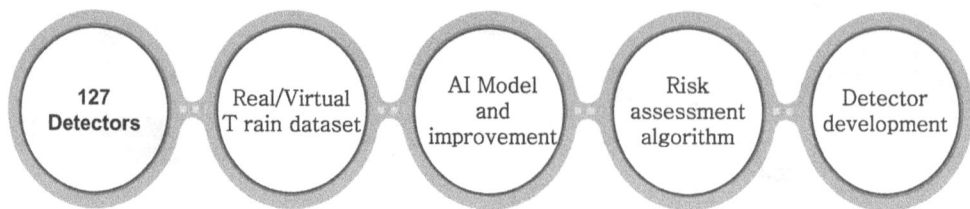

Figure 12.5 Development of an AI-based construction risk factor identification algorithm.

12.3.2.1 Detector Development

Developing a CV detector is a complex task that requires expertise in several areas, such as machine learning, CV, and software engineering. As presented in Figure 12.5, the iSAFE has developed by combining various technologies consisting of four main stages: dataset collection and preparation, AI model development and improvement, risk assessment algorithm development, and detector development.

The first stage of this methodology is data collection and preparation. To initiate this stage, the various unsafe behaviors are defined. Then, a dataset of images or videos representative of the objects or phenomena the detector will encounter in the real world is collected. The dataset is labeled with the appropriate annotations to enable supervised learning, and various tools and techniques, such as image resizing and dataset splitting, are employed to prepare the data.

The second stage of the methodology is AI model development and improvement. In this stage, deep neural networks are used to develop AI models. Transfer learning, data augmentation, and fine-tuning enhance accuracy and reduce overfitting. The performance of the models is evaluated using a separate test dataset, and the models are retrained as necessary to improve their performance.

The third stage of the methodology is risk assessment algorithm development. In this stage, algorithms are developed to assess the level of risk associated with a detected object or phenomenon. Factors like the unsafe behavior of workers, lack of compliance with PPE, and the probability of a collision are considered when developing the risk assessment algorithms. Mathematical modeling and machine learning are the techniques employed to create the algorithms.

The fourth and final stage of the methodology is detector development. In this stage, the AI models and risk assessment algorithms are integrated into a complete detector system, and software engineering techniques are employed to ensure that the system meets the specific requirements and constraints of the application. Object tracking and object classification are among the features included in the detector systems to enhance their performance.

12.3.2.2 Risk Situation Definition and Classification

By analyzing previous accident data, this study aimed to assess the various risks associated with the construction industry and classify them according to their potential hazards. The focus was on evaluating the probability of accidents to facilitate the categorization of these hazards. The outcomes of this analysis are presented in Table 12.2, wherein it is observed that the risk probability associated with accidents can be attributed to pre-work activities, such as training and planning, as well as during-work activities,

Table 12.2 Risk probability associated with accidents based on work steps

Work Steps	Importance
Before work	32.95%
Working (at all times)	41.71%
Working (Specific time)	8.05%
After work	13.09%
Management related	4.2%
Sum	100%

Table 12.3 Number of developed detectors based on hazard types

Hazard Type	Number of Classes
Fall from height	53
Collapse	17
Hit by objects	21
Struck by	17
Fire	15
Electric Shock	4
Sum	127

which include inspection and monitoring. These findings highlight the importance of undertaking relevant measures to manage and mitigate these risks during work.

During the working stage of construction job sites, the potential for unsafe behavior to occur presents a significant risk to the safety of workers. To address this issue, a comprehensive study was conducted to analyze previous accident reports, identifying 127 cases of such incidents. These unsafe conditions were further studied and analyzed to identify potential CV detectors. As a result, the iSAFEGuard system was developed to detect and mitigate these hazards. Table 12.3 presents the number of detector classes designed according to the type of hazard.

Workers typically operate various equipment at job sites, including external scaffolding, mobile scaffolding, ladders (Type A), aerial work platforms (Scissor Type), and others. Each equipment type has specific risks that must be carefully evaluated to ensure workplace safety. To develop the iSAFEGuard platform, the most hazardous activities, and behaviors linked to these equipment types were identified, and detectors were developed accordingly. These detectors were designed to monitor equipment use at job sites and to alert workers and supervisors of potential safety risks. The categories of risks identified in this study are outlined in Table 12.4.

12.3.2.3 AI Model Selection and Optimization

The iSAFEGuard platform employs a classification system consisting of three distinct categories of object properties: regular objects (e.g., persons, hardhats), irregular objects (e.g., ladders, scaffolding), and behavior analysis. For each classification category, different AI models are employed (refer to Figure 12.6). Regarding regular objects, the iSAFE platform utilizes bounding boxes with various algorithms such as YOLO, SSD, and Mask R-CNN. For irregular objects, instant segmentation is employed using

Table 12.4 Number of developed detectors based on types of equipment

Equipment	Number of Classes
External scaffolding	14
Mobile scaffolding	9
Ladder (Type A)	8
Aerial Work Platform (Scissor Type)	8
Aerial work platform (vehicle type)	7
Mobile Crane/Tower Crane	7
Horse scaffolding	4
Etc.	70
Sum	127

multiple algorithms such as YOLACT, SpineNet, Mask R-CNN, and SOLOv2. The behavior analysis classification uses the key point method, following Openpose and HR Net. Once the risks at job sites have been identified, the first step in developing detectors involves creating a dataset. The following sections will explore the procedure for developing the dataset.

12.3.2.4   Development of Real/Virtual Learning Data Construction (Dataset Build)

To ensure the quality of the dataset construction process, guidelines were established and followed for each phase of the process. As demonstrated in Figure 12.7, the first phase of dataset construction was a collection, which involved recording video footage of construction sites. To ensure the effectiveness of the dataset, it is important to recruit suitable sites and types of work for data collection. Additionally, shooting distance,

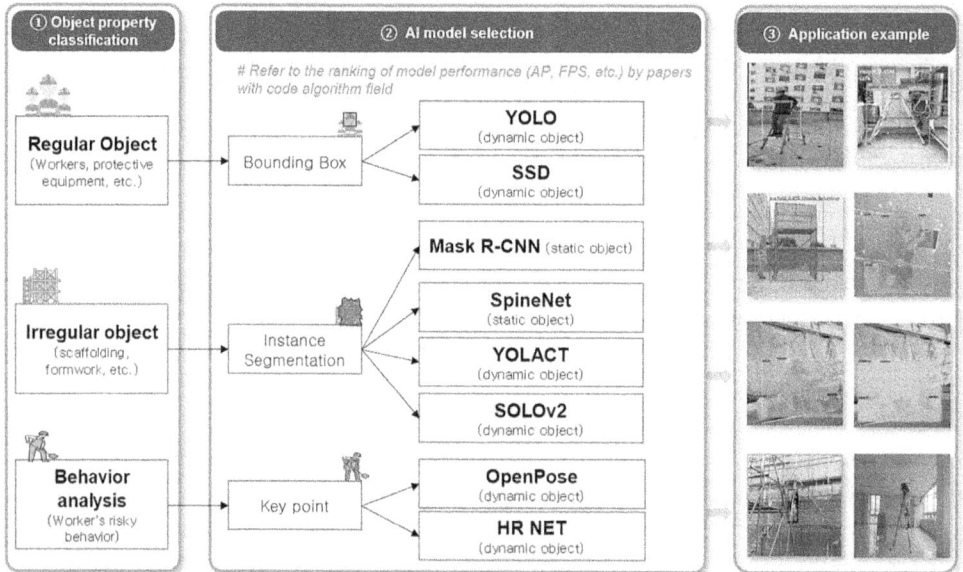

Figure 12.6 Classification of systems and AI model selection.

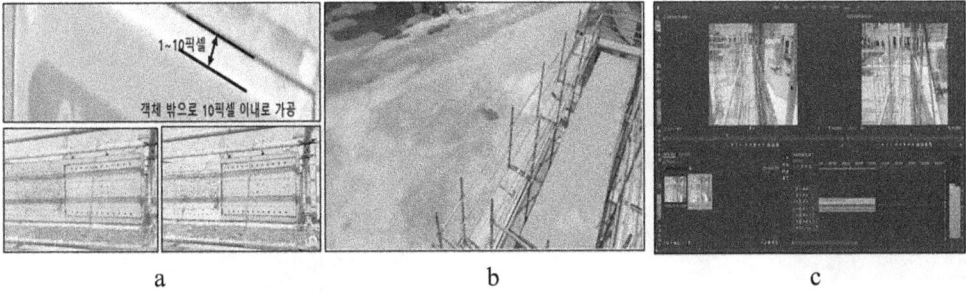

Figure 12.7 (a) Collection, construction site video recording; (b) refining, image extraction; (c) processing, creating learning data.

angle, and illumination standards were established for each risk situation to ensure consistent and accurate data collection. The second phase of dataset construction was refining, which involved extracting images from the video footage. A similarity review was conducted for image extraction to prevent overfitting and ensure the quality of the dataset. Meta information such as shooting location, time, time series, and environment information was also recorded during this phase. The third phase of dataset construction is processing, which involves creating learning data. To ensure the validity of the processing, criteria were established, such as an error range of within 10 pixels. Additionally, processing standards for virtual detection objects were selected for each risk situation.

12.3.2.5   Development of Virtual Data Generation Technology for Dataset Augmentation

During the modeling phase, 3D objects are created based on real object images, and real construction sites are modeled as 2D or 3D backgrounds. This phase also establishes shooting distance, angle, and illumination standards for each risk situation. The simulation phase is focused on directing dangerous situations by simulating the movement of workers and equipment. This phase involves randomizing environmental factors such as illuminance, angle, and color to simulate real-world scenarios. Finally, the extraction phase consists in creating virtual data using the automatic generation of learning data for each object, including bounding box and polygon data, and automated data format conversion for each AI model. By developing virtual data generation technology, more diverse data was generated and utilized to improve the accuracy of AI models used for construction site risk detection. Realistic virtual scenarios helped to supplement real-world data and ensure that the AI models are trained to detect a wider range of potential risks on construction sites. Figure 12.8 presents some samples of virtual objects, simulation, and virtual data creation. The subsequent sections elaborate on various techniques and algorithms for developing the detectors.

12.3.2.6   Techniques Development and Applications

12.3.2.6.1   TRACKING

The iSafeGuard platform has been developed to enhance safety measures in the construction sector using tracking technology. This platform leverages Mask R-CNN,

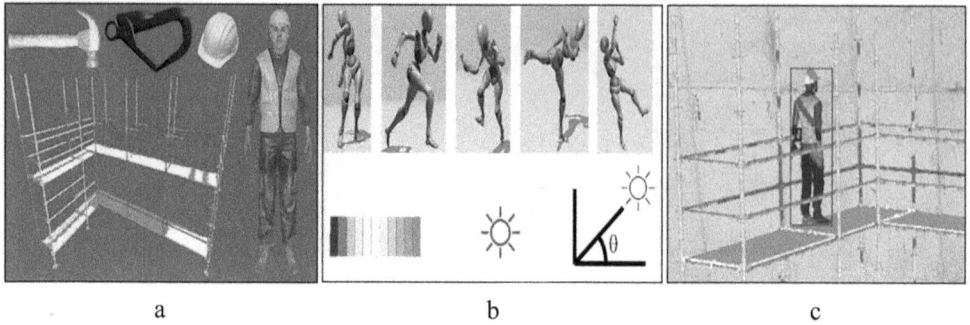

Figure 12.8 (a) Modeling, creating virtual objects; (b) simulation, directing the situation; (c) extracting, creating virtual data.

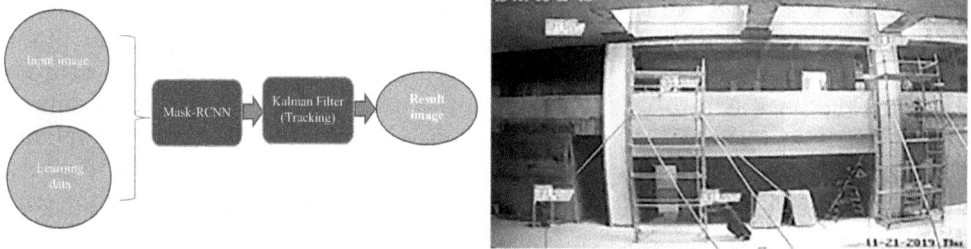

Figure 12.9 Utilization of displacement map after object detection and time-dependent object movement/position tracking and prediction.

an instance segmentation model that identifies objects within an image using a fully convolutional architecture. Additionally, a Kalman Filter is applied to track the movement of these objects. A visual representation of this process can be seen in Figure 12.9.

12.3.2.6.2 DEPTH ESTIMATION

Depth estimation is a well-established CV technique for measuring video and image-based object distance. By calculating the distance from the shooting point to the objects in a scene, the distance between objects can be accurately determined. This has been widely applied in various CV applications, where it plays a crucial role in obstacle detection and avoidance. Depth estimation has been achieved through multiple algorithms such as stereo vision, time-of-flight, and structured light. These techniques have been extensively studied and optimized for accurate and efficient depth estimation results. For the iSAFEGuard, the obtained depth information has been used for further processing or analysis such as object segmentation or 3D reconstruction. Figure 12.10 illustrates the flowchart and applicability of depth estimation.

12.3.2.6.3 COORDINATES

Coordinates is another application and technique employed in iSAFEGuard. It is a set of values that describe the position or location of an object in a particular system. In CV,

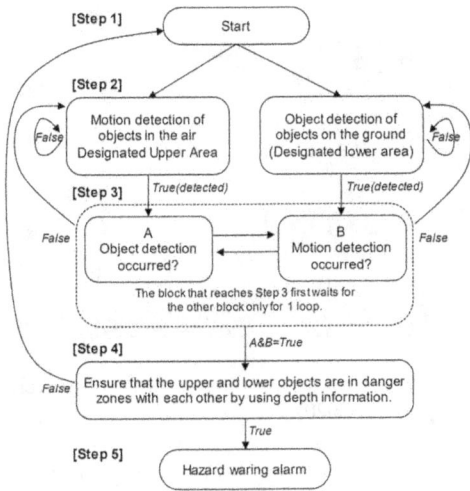

Figure 12.10 Stereo image-based object distance measurement (Doyeop et al., 2020).

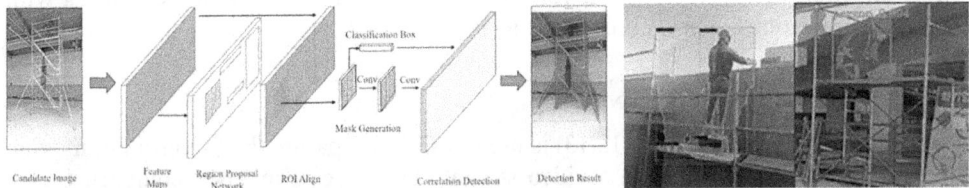

Figure 12.11 Applying the correlation logic between objects (Numan Khan et al., 2021).

coordinates are commonly used to represent the position of objects detected in an image or video. The coordinates are represented as a pair of numbers (x,y), which correspond to the horizontal and vertical positions of the object in the image. For the iSAFEGuard, detecting an object in an image or video involves analyzing the pixels in the image to determine the object's location. Once the object's location is selected, the location information is usually converted into coordinates for further processing. Figure 12.11 presents the flowchart and applicability of correlation logic between objects.

12.3.2.6.4   EDGE DETECTION

Edge detection is a technique that is also utilized in the development of iSAFEGuard. It is used to identify object boundaries and extract them from the background. Edge detection aims to find the areas in an image where the intensity or color changes rapidly, which typically corresponds to the objects or features' boundaries. Several methods are used to perform edge detection, including filters or mathematical algorithms that analyze the intensity or color values of adjacent pixels in the image. These methods identified the edges based on intensity, texture, or color changes. Once the edges of an object are detected and extracted, they are used for various applications, including image segmentation, object recognition, and tracking. For example, in object recognition, for the

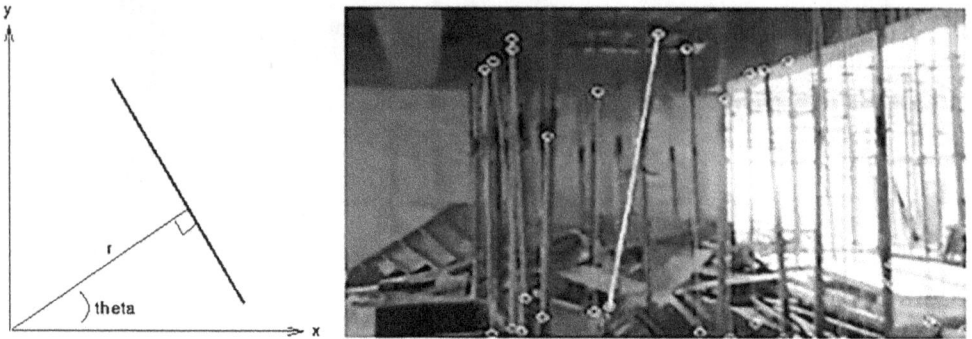

Figure 12.12 Identification of object boundaries and vertical lines using image processing technology (Guobo Xie & Wen Lu, 2013; Lin et al., 2020).

development of iSAFEGuard, the extracted edges were used to match the detected object to a database of known objects based on their shape or contour. In addition to identifying the boundaries of objects, edge detection is also used to determine the orientation of objects in an image. Through the analysis of vertical and horizontal edges in an image, the vertical direction of objects, such as formwork support, can be determined, along with their relative positions. This is illustrated in Figure 12.12.

12.3.2.6.5 BACKGROUND SUBTRACTION

Background subtraction (Figure 12.13) is used in CV to separate the foreground, moving objects, from the background in a video stream. The process involves comparing each video stream frame to a reference or background image and identifying the changed areas. The background image is generated by averaging a series of frames with little to no movement, resulting in an image representing the scene's static background. The moving objects or foreground is isolated by subtracting the background image from each video frame. One of the advantages of background subtraction of the iSAFEGuard is that it can detect even small or slender objects that may be difficult to detect using other methods. However, it can also be susceptible to false positives, where small changes in lighting or shadows can be mistaken for moving objects. Consequently, background subtraction is used in applications such as video surveillance, where it is used to detect and track objects in real time. The next section discusses how iSAFEGuard utilizes various image acquisition devices such as sever, edge, and mobile-based applications.

Figure 12.13 Background subtraction by distinguishing between moving or changing objects and non-moving objects (Lendave, 2021).

12.3.2.7   Operational Applications Using Various Image Acquisition Devices

12.3.2.7.1   SERVER BASE HIGH-PERFORMANCE DETECTOR

Developing risk detectors for various applications involves creating systems that accurately identify potential risks and dangerous situations in real time. To achieve this, iSAFEGuard utilizes a server-based high-performance detector with a high-performance AI model to detect potential hazards. In this platform, the video stream from the CCTV cameras is transmitted to a gateway and then to the server where the detectors are located. The high-performance AI model analyzes the video stream in real-time to detect potential risks at job sites. As demonstrated in Figure 12.14, the data from the detectors is then post-processed based on the specific type of dangerous situation. For example, if the system detects unsafe behavior or hazards, it alerts workers and safety managers. One of the advantages of this system is that the platform is used in conjunction with on-site CCTV systems, allowing for real-time monitoring of potential risks.

12.3.2.7.2   EDGE BASE LIGHTWEIGHT DETECTOR

The edge-based lightweight detector oversees the areas where CCTV blind spots occur, or temporary installation is necessary at hazardous work sections. As depicted in Figure 12.15, the edge device is furnished with detectors that undergo post-processing after object detection via a lightweight AI object detection model. Subsequently, all identified objects and unsafe behaviors are scrutinized, and the collected data is conveyed to a gateway before transmission to the field office or control room.

12.3.2.7.3   MOBILE BASE LIGHTWEIGHT DETECTOR

iSAFEGuard offers greater scalability through its mobile-based monitoring system when the server or edge-based monitoring is unavailable, particularly for small and medium-sized projects. The application is equipped with a lightweight object detection system, and

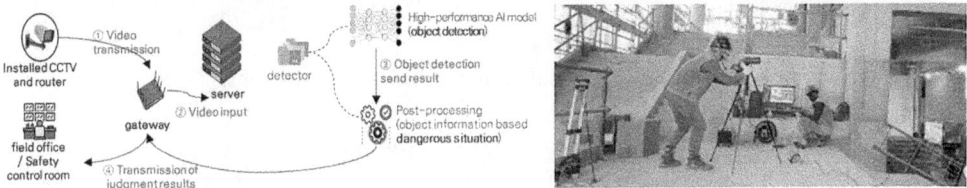

Figure 12.14 Server base high-performance detector.

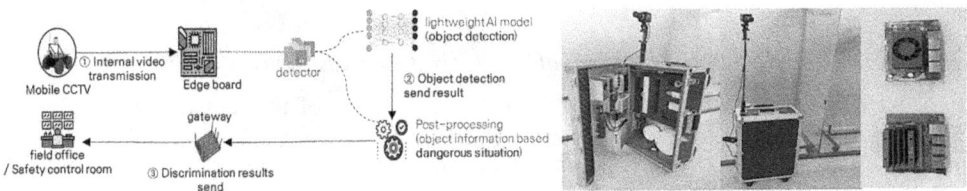

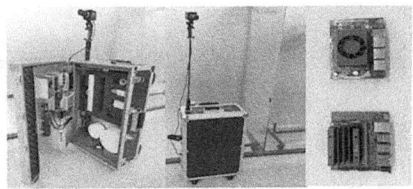

Figure 12.15 Edge base lightweight detector.

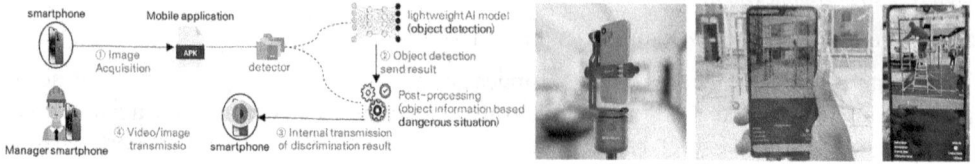

Figure 12.16 Mobile base lightweight detector.

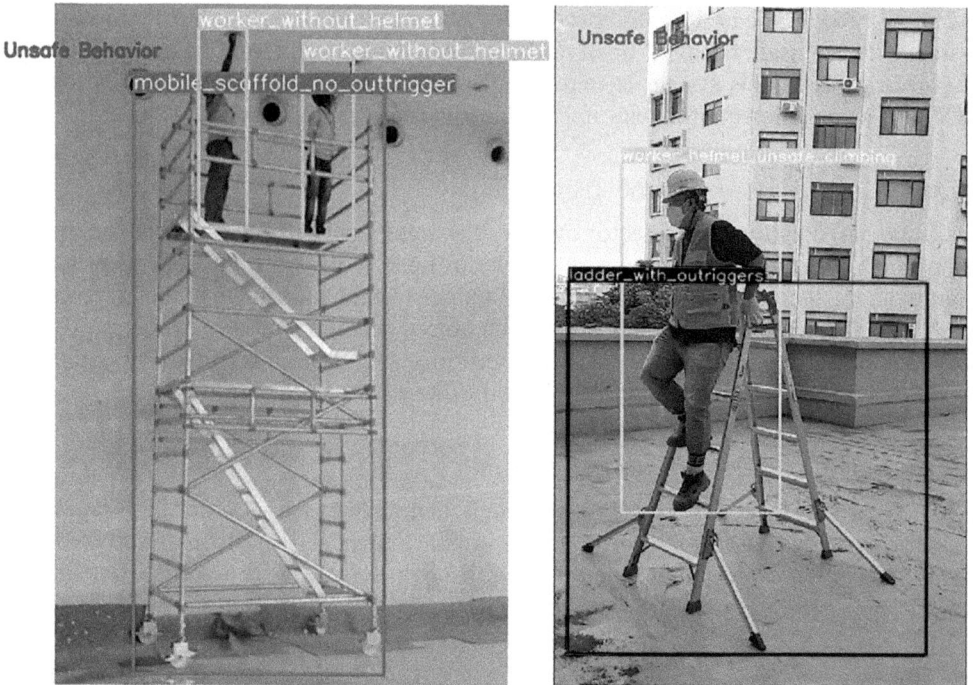

Figure 12.17 Unsafe detection scenarios related to scaffold and ladder jobs.

the data is sent for post-processing to detect any unsafe behaviors. As shown in Figure 12.16, the output is demonstrated on the mobile phone screen through internal transmission, providing a convenient method for safety managers to detect any defects at the job site or identify all unsafe behaviors.

Figure 12.17 depicts two examples of iSAFEGuard unsafe activity detection scenarios associated with scaffold and ladder jobs.

12.3.3   *iSAFEIncentive (Data Management/Token-Based Rewards System)*

The iSAFEIncentive aims to transform the safety culture of the construction industry by incentivizing and rewarding safety performance, shifting the focus from blame and punishment to a collaborative worker-driven approach. By leveraging cutting-edge technologies like CV, AI, blockchain, and IoT, iSAFEIncentive streamlines safety inspection and reporting circumventing conventional methods' subjective and

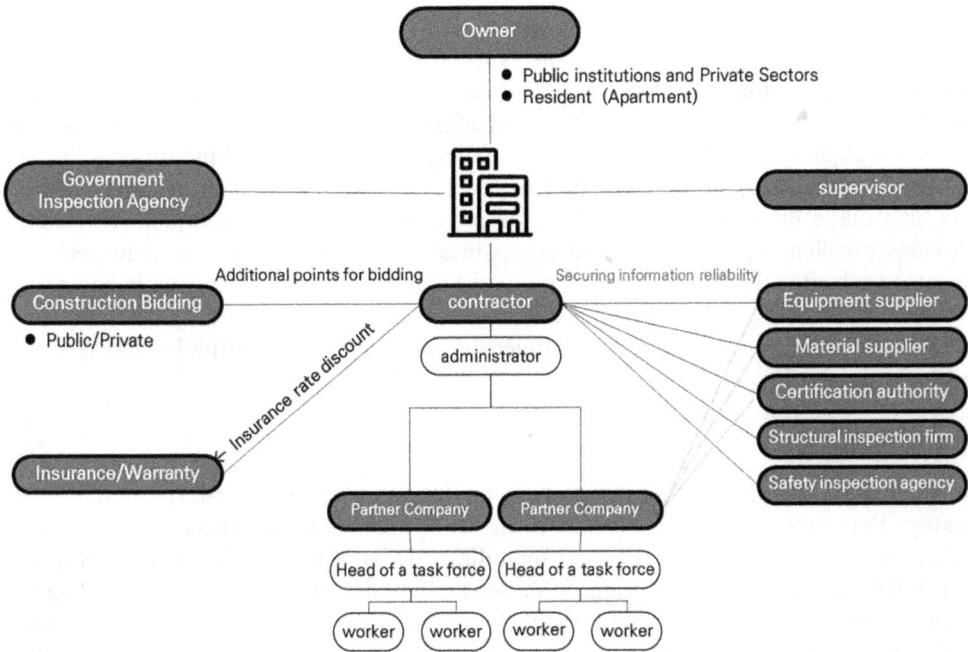

Figure 12.18 The iSAFEIncentive token economy system.

time-intensive nature. This includes using decentralized identifier certificates, CVs, and smart contracts to verify and reward safety performance. Tokens are issued to workers, safety managers, and contractors based on their safety ratings. As illustrated in Figure 12.18, the system interconnects regulatory entities with contractors, partner companies, managers, and workers. The safety proficiency of everyone is assessed based on the nature of their services, and their performance is evaluated through the employment of Blockchain technology. To ensure sustainability and efficacy, governmental agencies employ the token-burning mechanism. The iSAFEIncentive platform encourages worker-led safety management, ensuring the dependability of safety information, facilitating the evaluation of safety standards and remuneration, and ultimately enhancing safety performance and culture in the construction domain.

12.4   Discussion

The iSAFE platform has emerged as a promising safety solution for construction job sites. The platform offers three critical safety barriers that aim to enhance the safety culture within the industry and prevent accidents. The first barrier, the iSAFEIncentive program, is designed to promote safety culture among workers, thereby improving the industry's overall safety. The program rewards workers with positive safety behaviors and a high safety rating. Moreover, the entire company's performance is evaluated based on the safety ratings of individual workers, and this information is reported to safety agencies and regulatory bodies. Companies with higher safety risks are incentivized, just as workers are rewarded for safe behavior. The second barrier involves providing proper education and training sessions, which companies and contractors typically mandate. Effective safety planning is also crucial in understanding the risks and hazardous activities involved in the work. iSAFE provides robust and comprehensive training materials based on the type of work and activities involved and can visualize dangers and hazards in BIM format. The third barrier involves real-time monitoring of workers and job sites. Experienced and inexperienced workers may become negligent towards safety rules, particularly when tired, working on complicated tasks, or in a rush. Additionally, inadequate human resources, such as safety inspectors, can make monitoring all workers on a large job site challenging. iSAFEGuard provides excellent opportunities for companies to monitor workers and inspect any defects at job sites that may put workers at risk. Early detection of unsafe behaviors or workers significantly reduces the chances of risky situations. The system can alert workers and safety managers about dangerous activities, ensuring the highest levels of safety in the workplace.

12.5   Conclusion

In summary, the iSAFE platform presents a comprehensive safety management strategy that encompasses all aspects of the safety management process. The platform utilizes state-of-the-art technologies, including 4D BIM-based risk assessment, VR and 360-panorama-based training, CV, sensor-based technologies, and blockchain, to proactively solve safety-related problems in the pre-work, during-work, and post-work stages. Moreover, the token-based incentive system motivates workers and companies to prioritize safety performance, resulting in a safer and more productive work environment.

Acknowledgment

This work was supported by the National Research Foundation of Korea (NRF) grant funded by the Korean government (MSIT) (No. NRF-2022R1A2B5B02002553) and Basic Science Research Program through the National Research Foundation of Korea (NRF) funded by the Ministry of Education (No. NRF-2020R1I1A1A01073167). We also need to acknowledge the support of the "National R&D Project for Smart Construction Technology (No.23SMIP-A158708-04)" funded by the Korea Agency for Infrastructure Technology Advancement under the Ministry of Land, Infrastructure and Transport and managed by the Korea Expressway Corporation.

References

Sharjeel Anjum, Numan Khan, Rabia Khalid, Muhammad Khan, Dongmin Lee, Chansik Park. 2022. "Fall prevention from ladders utilizing a deep learning-based height assessment method." *IEEE Access* 10: 36725–36742.

Lan Bao, Si Van-Tien Tran, Truong Linh Nguyen, Hai Chien Pham, Dongmin Lee, Chansik Park. 2022. "Cross-platform virtual reality for real-time construction safety training using immersive web and industry foundation classes." *Automation in Construction* 143: 1–19.

Seokho Chi, Sangwon Han. 2013. "Analyses of systems theory for construction accident prevention with specific reference to OSHA accident reports." *International Journal of Project Management* 1027–1041.

Seungwon Cho, Muhammad Khan, Jaeho Pyeon, Chansik Park. 2021. "Blockchain-based network concept model for reliable and accessible fine dust management system at construction sites." *Applied Sciences* 11 (18): 1–17.

Sooyoung Choe, Fernanda Leite. 2017. "Construction safety planning: Site-specific temporal and spatial information integration." *Automation in Construction* 335–344.

Ricardo Eiris, Anujeet Jain, Masoud Gheisari, Andrew Wehle. 2020. "Safety immersive storytelling using narrated 360-degree panoramas: A fall hazard training within the electrical trade context." *Safety Science* 1–14.

Christopher Frazier, Timothy D. Ludwig, Brian Whitaker, Steve Roberts. 2013. "A hierarchical factor analysis of a safety culture survey." *Journal of Safety Research* 15–18.

Lei Hou, Shaoze Wu, Guomin (Kevin) Zhang, Yongtao Tan, Xiangyu Wang. 2021. "Literature review of digital twins applications in construction workforce safety." *Applied Sciences* 1–21.

Jens J. Hunhevicz, Daniel M. Hall. 2020. "Do you need a blockchain in construction? Use case categories and decision framework for DLT design options." *Advanced Engineering Informatics* 45: 1–14.

Muhammad Khan, Rabia Khalid, Sharjeel Anjum, Numan Khan, Seungwon Cho, Chansik Park. 2022. "Tag and IoT based safety hook monitoring for prevention of falls from height." *Automation in Construction* 136: 1–13.

Numan Khan, Muhammad Rakeh Saleem, Doyeop Lee, Man-Woo Park, Chansik Park. 2021. "Utilizing safety rule correlation for mobile scaffolds monitoring leveraging deep convolution neural networks." *Computers in Industry* 129: 14.

Numan Khan, Ahmed Khairadeen Ali, Si Van-Tien Tran, Doyeop Lee, Chansik Park. 2022. "Visual language-aided construction fire safety planning approach in building information modeling." *Applied Sciences* 1–19.

Kyungki Kim, Sungjin Kim, Daniel Shchur. 2021. "A UAS-based work zone safety monitoring system by integrating internal traffic control plan (ITCP) and automated object detection in game engine environment." *Automation in Construction* 128 (128): 1–12.

Doyeop Lee, Numan Khan, Chansik Park. 2020. "Stereo Vision based Hazardous Area Detection for Construction Worker's Safety." *The International Symposium on Automation and Robotics in Construction (ISARC 2020)*.

Vijaysinh Lendave. 2021. *Using Background Subtraction Methods in Image Processing*. 8 25. https://analyticsindiamag.com/using-background-subtraction-methods-in-image-processing/.

Jeeyoung Lim, Kiyoung Son, Chansik Park, Daeyoung Kim. 2021. "Suggestions for improving South Korea's fall accidents prevention technology in the construction industry: focused on analyzing laws and programs of the United States." *Sustainability* 13 (8): 1–13.

Jeeyoung Lim, Dae Gyo Jung, Chansik Park, Dae Young Kim. 2022. "Computer vision process development regarding worker's safety harness and hook to prevent fall accidents: focused on system scaffolds in South Korea." *Advances in Civil Engineering* 1–12.

Y. Lin, S. Pintea, J.C. van Gemert. 2020. "Deep Hough-Transform Line Priors." *Computer Vision – ECCV 2020 - 16th European Conference, 2020, Proceedings*.

María Martínez-Rojas, Nicolás Marín, and M. Amparo Vila. 2016. "The role of information technologies to address data handling in construction project management." *Journal of Computing in Civil Engineering* 1–20.

Richard H. McCuen, Edna Z. Ezzell, Melanie K. Wong. 2011. *Fundamentals of Civil Engineering*. Boca Raton: Taylor & Francis Group.

Nicholas Oughton. 2013. "Managing occupational risk in creative practice: a new perspective for occupational health and safety." *Archives of Environmental & Occupational Health* 47–54.

Andrew Rae, Rob Alexander, John McDermid. 2014. "Fixing the cracks in the crystal ball: A maturity model for quantitative risk assessment." *Reliability Engineering & System Safety* 67–81.

J. Raftery. 1993. *Risk Analysis in Project Management*. London: Routledge.

Si Van-Tien Tran, Numan Khan, Doyeop Lee, Chansik Park. 2018. "A hazard identification approach of integrating 4D BIM and accident case analysis of spatial–temporal exposure." *Sustainability* 13 (4): 1–19.

Si Van-Tien Tran, Truong Linh Nguyen, Hung-Lin Chi, Doyeop Lee, Chansik Park. 2022. "Generative planning for construction safety surveillance camera installation in 4D BIM environment." *Automation in Construction* 134: 1–14.

J. Wang, P. Wu, X. Wang, W. Shou 2017. "The outlook of blockchain technology for construction engineering management." *Frontiers of Engineering Management* 4 (1): 67–75.

Guobo Xie, Wen Lu. 2013. "Image edge detection based on Opencv." *Computer Science, International Journal of Electronics and Electrical Engineering* 1 (2).

Jaehun Yang, Dongmin Lee, Chanwoo Baek, Chansik Park, Bao Quy Lan, Doyeop Lee. 2022. "Leveraging blockchain for scaffolding work management in construction." *IEEE Access* 10: 39220–39238.

Hao Zhang, Seokho Chi, Jay Yang, Madhav Nepal. 2017. "Development of a safety inspection framework on construction sites using mobile computing." *Journal of Management in Engineering* 33 (3): 1–10.

Jianbo Zhu, Ce Zhang, Shuyi Wang, Jingfeng Yuan, Qiming Li. 2022. "Evolutionary game analysis of construction workers' unsafe behaviors based on incentive and punishment mechanisms." *Frontiers in Psychology* 13.

13 BIM with Blockchain for Decentralised Circular Construction Supply Chain

Faris Elghaish[1], M. Reza Hosseini[2],
Tuba Kocaturk[3], Mehrdad Arashpour[4], and
Masoomeh Bararzadeh Ledari[5]

[1]*School of Natural and Built Environment, Queen's University Belfast, Belfast, Northern Ireland, United Kingdom;* [2]*School of Architecture and Built Environment, Mediated Intelligence in Design lab, Deakin University, Geelong, Australia;* [3]*Department of Civil Engineering, Monash University, Australia;* [4]*School of Architecture and Built Environment, Deakin University, Australia;* [5]*Energy Department, Sharif University of Technology, Tehran, Iran*

13.1 Introduction

The concept of circular economy is gaining momentum in the construction industry; it is touted as an effective measure towards employing resource-efficient tools, methods, and procedures that reduce the consumption of resources and eliminate waste (Ghisellini et al., 2016; Shooshtarian et al., 2022). The problem of waste for the construction industry is particularly acute, compared to other industries (Baek et al., 2013; Nassri et al., 2021). That is because, the construction industry consumes 30% of earth's raw materials, 12% of its land, 25% of its water resources and 40% of its energy, generating up to 3 billion tonnes of construction and demolition (C&D) waste each year (Adabre et al., 2022). The waste generated in the construction industry causes environmental degradation, due to its – high volume and – low recycling rates (Brandão et al., 2021).

The concept of circular economy in the construction industry strongly relies on reverse logistics of construction products and materials (Adabre et al., 2022). Reverse logistics in the construction context is defined as a coalesce of processes and tools for reusing, re-cycling, and remanufacturing building materials at their end of life, mostly the outcome of demolition and deconstruction processes (Ghobakhloo et al., 2013; Wibowo et al., 2022). Pushpamali et al. (2019) assert that reverse supply chain for buildings should be considered for both upstream (design and construction) and end-of-the-life to maximise the salvage value of building materials and components. As such, particular emphasis should be put on procedures and practices that enable the adoption of reverse construction supply chains to facilitate the uptake of circular economy, and as a result, reducing waste and consumption of virgin products and materials (Wijewickrama et al., 2021a). In this regard, the construction industry is currently exploring the emerging potential of Industry 4.0 technologies. This is to enable new processes to achieve reverse logistics towards the goal of transitioning to circular construction practices (Elghaish et al., 2022; Yu et al., 2022).

The relationship between Industry 4.0 technologies and reverse logistic is presented to enable sharing materials/items in a real-time mode, as well as, enabling the automated estimation of return for reverse supply chains (Dev et al., 2020). A wide range of Industry 4.0 technologies have been employed. Examples are the Internet of Things (IoT) to enable

DOI: 10.1201/9781003408949-13

collecting real-time information (Damianou et al., 2019), Artificial Intelligence (AI) to appraise the value and functionality of building components (Chidepatil et al., 2020) and blockchain technology to automate payments and checking information reliability (Shojaei et al., 2021; Upadhyay et al., 2021).

Recently, the blockchain technology has received growing attention. In a nutshell, the blockchain technology refers to a digitally distributed, decentralised and highly secured way of recording transactions and tracking assets in a business network (Li et al., 2019). Various areas of application in the construction sector include improving contractual aspects and the operation of buildings (Elghaish et al., 2020); supporting sharing risk/ reward among Integrated Project Delivery (IPD) project parties (Hamledari and Fischer, 2021); automating construction payments in integration with robotic reality capture technology; and improving the traceability of prefabricated components across the building's supply chain (Wang et al., 2020b). Likewise, researchers have begun to consider blockchain and smart contracts to minimise the inherent fragmentation in construction reverse supply chain processes in support of the transition to a circular economy (Elghaish et al., 2022; Yu et al., 2022). For example, Upadhyay et al. (2021) evaluated the role of blockchain for circular economy and indicated that blockchain offers a promising avenue for improving performance and communication through enhancing trust and transparency among suppliers. Shojaei et al. (2021) developed a blockchain-based solution to enable tracking building components and making proactive planning. Despite the added value, these studies fall short in providing an information channel between designers and asset operators. Besides, they remain silent on tracking the treatment of residual waste that is deemed hazardous or unsalvageable.

In essence, the blockchain and smart contract solutions are proven to offer effective measures to speed up the diffusion of circular economy (Yu et al., 2022). Nevertheless, integrated solutions that cover all the aspects of circular supply chain including network structure, business processes, and the products are still missing (Shooshtarian et al., 2022). This represents a gap in the now available literature on the topic (Montag et al.; Wijewickrama et al., 2021b). A review run on the related literature reveals that few studies have provided potentially applicable solutions to promote circular supply chain using blockchain networks, therefore, there is a conspicuous need for comprehensive and integrated solutions (Adabre et al., 2022; Elghaish et al., 2022).

To address this gap, this chapter aims to provide a workable solution to address the disconnected nature of blockchain adoption across construction operations. Objectives are formulated to develop a solution that enables construction practitioners to share their buildings components information on a safe and immutable medium of information; enables governmental authorities (i.e., city council) to track the treatment and delivery of hazardous unsalvageable items/materials; supports the upstream circular economy design through enabling collaboration among 'designers to designers' and 'designers to asset owners' in a blockchain network; and support developing 'a bank of reusable BIM families' to encourage designers to consider these items in their designs. The proposed solution integrates Blockchain, smart contract and BIM in an integrated workflow to facilitate the transition to circular economy for the construction industry. Moreover, this solution is a practical step towards employing industry 4.0 technologies (i.e., blockchain and smart contract) towards digitalising the circular economy process and this attempt opens new horizons to employ further technologies such as IoT, digital twin, AI, etc.

The rest of this chapter is structured as follows. Section 13.2 represents the conceptual background of circular supply chain and various applications of blockchain for circular economy. This is followed by Section 13.3, which presents the research gap. The research methods and logic are presented in Section 13.4, followed by a description of the solution development and evaluation of the study, under Section 13.5. Section 13.6 offers a brief discussion on findings, followed by a summary of practical implications in Section 13.7. This chapter concludes with an overall assessment of findings from a broad perspective, concluding remarks and acknowledging the limitations, in Section 13.8.

13.2 Contextual Background

13.2.1 Circular Economy and Industry 4.0

The terms of circular economy and Industry 4.0 have appeared together in many recent studies (Benachio et al., 2020; Martínez-Rocamora et al., 2021; Nascimento et al., 2019; Piscitelli et al., 2020; Rajput and Singh, 2019; Sanchez and Haas, 2018). Apart from proving the value of Industry 4.0 for supporting the adoption of circular economy, past research has pointed to the organic interrelationship that can develop between the processes and tasks of these two emerging concepts (Piscitelli et al., 2020). In this regard, Martínez-Rocamora et al. (2021) introduced a methodological-technological framework based on Industry 4.0 to enable circular economy; the authors called for new business and development models to be developed for facilitating the adoption of Industry 4.0 solutions to enable circular economy in the construction industry. Different business and technical models were developed by Rahimian et al. (2021) to implement Industry 4.0 technologies to adopt circular supply chain, however, more validations using large-scale case studies are needed to check the scalability of their proposed solutions. Besides, lean design and production concepts and methodologies should be employed to bolster the power of Industry 4.0 for supporting circular economy, as argued by Ciliberto et al. (2021).

With the above in mind, some studies have focused on certain technologies – components of Industry 4.0 – for supporting the adoption of circular economy. Craveiroa et al. (2019) proposed additive manufacturing to enhance resource efficiency in construction, however, additive manufacturing requires integration with other Industry 4.0 technologies like the IoT for large-scale applications (Craveiroa et al., 2019; Sauerwein et al., 2019). That is, Ashima et al. (2021) asserted that employing IoT with additive manufacturing can significantly reduce waste, as well as, making the process and application of additive manufacturing customer-friendly.

Digital twin as a replica of the physical building is also proposed to improve the remanufacturing process through facilitating the tracking, recycling and managing construction waste in remanufacturing processes (Chen and Huang, 2020). However, digital twin applications for construction circular economy rely on the development of workable solutions that track the performance of a wide range of materials and services from assets and link their data with information models (Boje et al., 2020; Lu et al., 2020). Thence, Tagliabue et al. (2021) applied digital twin in conjunction with IoT to assess the sustainability factors for an educational building through the whole life cycle of the building; findings demonstrated the superior performance of the digital twin and IoT compared to the use of traditional checklists. Hämäläinen (2020) extended previous efforts and recommended the employment of a dynamic digital twin (DDT) to track smart city buildings and services performance through automatically detecting and managing the reuse of elements at the end-of-life point. It is argued that the use of digital twins and IoT can minimise the supply chain lead time toward developing a lean, flexible, and smart supply chain (Abideen et al., 2021). There are, however, few attempts to develop integrated platforms for circular supply chains, as discussed next.

13.2.2   *Circular Supply Chain*

Kovacic et al. (2020) proposed a digital platform for construction circular economy that enables all stakeholders to manage building construction and operation resources from a circular economy perspective – from cradle to grave. And Leising et al. (2018) developed a framework to explore how new approaches of supply chain collaboration can support a transition to a circular building sector. Bressanelli et al. (2021) stated that policymakers should provide mandatory regulations that foster circular product design and highlight the rights and obligations of stakeholders. There are even some studies that propose selecting suppliers according to their level of circular economy adoption. For example, Haleem et al. (2021) identified some criteria for supplier selection based on how they adopt circular economy in their supply chains. Automated tools were developed to assess suppliers based on circular economy competencies level too. This included the study by Alavi et al. (2021), who developed a dynamic decision support system based on a fuzzy best-worst method (BWM) to select optimal suppliers according to suppliers' weight of economic, social, and circular criteria; Kusi-Sarpong et al. (2021) developed a multi-criteria decision-making support tool using Industry 4.0 technologies to select sustainable suppliers. There are some attempts to adopt circular supply chain with BIM (Nikmehr et al., 2021). This includes Akinade and Oyedele (2019) who developed a tool to analyse building waste analytics and categorise items based on BIM computational capabilities. Moreover, many studies recommended a transition to deconstruction by BIM, to avoid demolition (Akbarnezhad et al., 2014; Rios et al., 2015; Tatiya et al., 2018).

Within the last few years, the synergies between circular economy and blockchain have received growing attention from researchers, as discussed in the following section.

13.2.3   *Circular Economy and Blockchain*

Various studies have explored the application of blockchain in a transition to circular economy (Böckel et al., 2021; Figueiredo et al., 2022; Shojaei et al., 2021). Upadhyay et al. (2021) studied the implications of employing blockchain to achieve sustainability goals in terms of minimising the transaction cost, enhancing trust among industry parties and providing a secure communication environment for interconnected supply chain process. These benefits of blockchain to foster the adoption of circular economy concept have been similarly acknowledged by researchers like Böckel et al. (2021), who asserted that trust and verification features in blockchain can be the main drivers for developing sustainable supply chain processes. There are also practical applications of implementing blockchain and multi-sensor-driven AI for circular economy. For example, Chidepatil et al. (2020) developed a system to trace plastic waste, as well as, managing purchasing tasks and other tender procedures using blockchain – smart contracts – to speed up the processing of plastic waste. Integrating products and services in an interconnected process based on blockchain can enable the life cycle assessment of products, to justify the need for adopting circular economy (Kouhizadeh et al., 2019; Vogel et al., 2019). As discussed, reserve logistics is a vital dimension of circular economy. As such, Bekrar et al. (2021) proposed a nexus of transportation, reserve logistic, and blockchain to enable tracking of all equipment components automatically – using immutable ledge blockchain feature – to acquire an accurate reserve supply chain system.

The concept of directly implementing blockchain to adopt circular economy for construction industry is further emphasised by Shojaei et al. (2021). Their findings show that blockchain can enable construction practitioners to trace and predict material values and energy consumption throughout the project and asset lifecycle. This gives the designer the ability to optimise design and include materials – in design – that give a high salvage value by the end of their lifetime (Shojaei et al., 2021). Blockchain can be also used in large scale to track the values of assets throughout the operation phase by giving an ID for each facility and enabling stakeholders to invoke transactions to change values for their assets regularly, therefore, authorities can get a precise value of all existing assets (Maciel, 2020). Moreover, blockchain is a prominent tool to support a transition to green energy through using automated and digital measurements, reporting, and verification systems that track energy consumption levels and enable the building sector to contribute to the carbon credit market (Woo et al., 2021).

So too, the blockchain has been used to develop collaborative construction design platforms to track changes automatically. These platforms accordingly enable a wide range of stakeholders to provide views towards sustainable design (Nawari and Ravindran, 2019; Singh and Ashuri, 2019). Blockchain is also currently employed to enhance urban design by developing a 'synergetic landscapes' system; this system allows considering living, non-living, physical, analogue, digital and virtual aspects of design by representing design factors as tokens, which enable all stakeholders to share their contribution through a blockchain network (Davidova and McMeel, 2020).

Li and Kassem (2021) explored integrating blockchain, BIM and IoT to allow reciprocal communications from the built asset to BIM models. Shojaei et al. (2021) investigated the comprehensive advantages of circular economy-based blockchain for the built environment sector; their findings show that blockchain can trace construction materials from source to the end of the life, to maximise the opportunity of early planning and enhance the chance of reuse to eliminate waste.

In summary, almost all existing construction circular economy solutions based on blockchain fall short of providing workable and practical solutions that are applicable to large-scale projects (Adabre et al., 2022). Besides, there is a need for a digital ecosystem to integrate all these technologies and built upon their synergy (Adabre et al., 2022; Damianou et al., 2019).

13.3   Research Methods

The aim and objectives of this chapter point towards the necessity for the improvement of a circular supply chain – in construction – through the adoption of a technological innovation, namely the blockchain system. This is in close alignment with design science research (DSR) activities that aim to design and investigate the application of an artefact – as a treatment for the problem – in a context; design scientists suggest and develop an artefact to interact with a problem context, in order to improve something in that context (Wieringa, 2014c). As a bridge between theory and practice, DSR is recommended in construction research, to create practical solutions that improve performance, along with contributing to theoretical knowledge. The use of DSR allows researcher to become problem solvers, as opposed to applying methods that focus on observing and evaluating problem-solving activities in the world of practices (Agrawal et al., 2022; AlSehaimi et al., 2013).

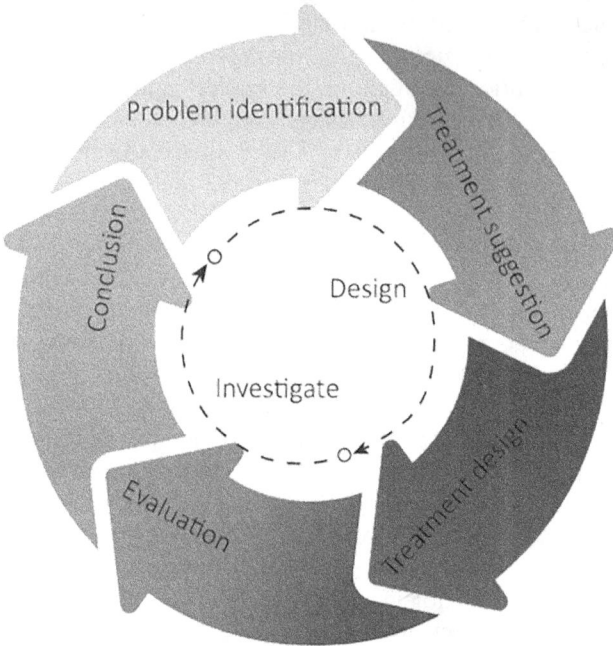

Figure 13.1 Developing the treatment based on a design research cycle (adapted from Wieringa (2014a) and Kuechler and Vaishnavi (2008)).

This chapter hence relies on the overall approach of DSR, borrowed from information systems, following the cycle of activities, as recommended by Wieringa (2014a) and Kuechler and Vaishnavi (2008), as illustrated in Figure 13.1. This follows the five-staged procedure: (1) problem identification, (2) treatment suggestion, (3) treatment design, (4) evaluation and (5) conclusion (Kuechler and Vaishnavi, 2008; Wieringa, 2014a).

A project informed by the DSR iterates over the activities of design and investigation (see Figure 13.1). The design task itself is decomposed into three tasks, namely, problem identification, treatment suggestion and treatment design. These are followed by evaluating the treatment in a real-life context, to investigate the interaction between the treatment and the context, with the aim of improving the design of the treatment (Wieringa, 2014a; 2014c). The cycle starts with identifying a problem in the real-life context of the domain at hand and establishing the need for and the value of a treatment, to be designed and provided. This is discussed next.

13.4 Problem Identification

A DSR project proposes a treatment that interacts with a normative context of regulations, constraints, values, desires and goals. Chief among all the elements of the context is those stakeholders who have goals that the research project must contribute to (Wieringa, 2014b). Hence, this part provides an overview of various stakeholders associated with the treatment, as proposed, and how the treatment can address their aims and objectives.

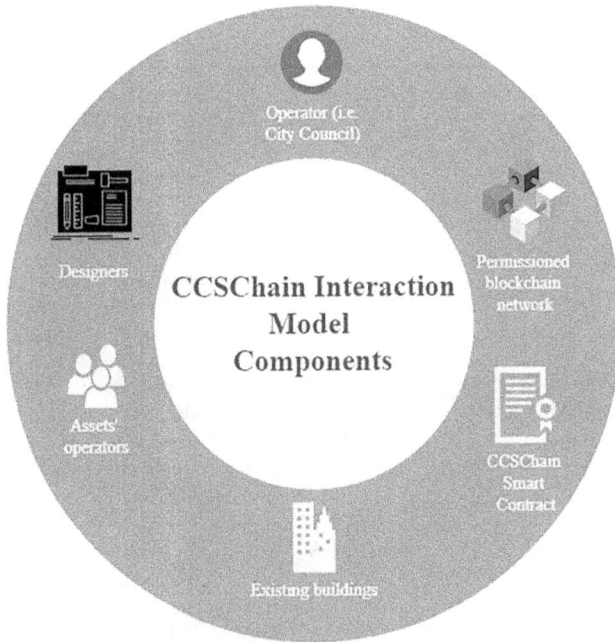

Figure 13.2 CCSChain interaction model.

These speak to the nature of the problems as observed and lived experiences within the industry, to identify the problem(s) that this chapter aims to address and define the first stage of the DSR approach adopted in this chapter.

Figure 13.2 shows that there are six main components in the proposed treatment: the main operator party – city council – which will be mainly responsible to launch the process given all the information of existing assets in the city are available – for the city council. Technological solutions that rely on blockchain, including developing a blockchain network and smart contracts to enable the main operator, asset owners and designers to interact and record on the ledger. Once the main operator has developed the blockchain network and smart contract functions, asset operators (owners) should register their assets and start sending/receiving functions. The main purpose of circular supply chain is to maximise the value of resources and reuse them, therefore, designers will have access to the blockchain network to share information regarding new design families, as well as, developing BIM families for existing items.

As tabulated in Table 13.1, there are few studies on investigating the role of blockchain to support circular supply chain, despite its promising capabilities to manage the main circular supply chain process – recycle, redistribute and remanufacture. Moreover, Table 13.1 shows that there is only one study in the built environment domain; that is, the paper by Shojaei et al. (2021), which does not cover all circular supply chain processes and functions. As such, there is a need for a solution able to cover (1) smart contract functions to recycle, redistribute and remanufacture, and (2) enable safe interconnected collaboration among 'owner to owner' and 'designer to owner' and 'designer to designer' arrangements.

Table 13.1 Blockchain for circular economy (noteworthy studies)

Item	Author	Focus of Study	Methods	Limitations
1	Shojaei et al. (2021)	To enable adopting circular economy in built environment using blockchain network	'Proof of Concept' development	The provided smart contract does not include functions to classify the types of salvageable and unsalvageable items. Moreover, it does not enable government authorities to track the treatment of hazardous waste.
2	Davidova and McMeel (2020)	To employ blockchain for developing synergetic design that considers circular economy factors	Conceptual solution	The proposed solution does not provide smart contract functions and how the solution is technically validated.
3	Chidepatil et al. (2020)	Using blockchain and smart contract to track information of recycled materials such as plastic.	Conceptual framework	The proposed solution is not tested in a real-life setting.
4	Wang et al. (2020a)	To develop a conceptual solution for utilizing blockchain to support circular supply chain through overcoming critical challenge of tracing the reuse of materials over multiple life cycles.	System architecture of blockchain-enabled CSCM	The proposed system architecture was not implemented in a real-life case, and the study was limited to the fashion industry.
5	Centobelli et al. (2021)	Designing a circular blockchain platform to manage the three circular supply chain processes including recycle, redistribute and remanufacture.	Integrated Triple Retry framework	Findings should be investigated regarding the impact of blockchain on different processes and actors in the supply chain.

13.5   Treatment Suggestion

A treatment is suggested in the form of developing a Construction Circular Supply Chain-based blockchain (CCSChain-blockchain). Therefore, a framework should be developed to provide a conceptual model to connect the technological solution (blockchain), design process-based BIM and proposed users.

Figure 13.3 shows the structure of CCSChain smart contract functions including assets, items that will be managed, as well as transactions. There are five assets to be developed in the smart contract including 'reusable items' to enable asset operators to invoke transactions to record all 'reusable items' and all relevant information such as specifications, quantities and corresponding offered prices. Meanwhile, salvaged asset items will be also uploaded to the blockchain network to enable other asset operators and designers to consider them in new design, as well as refurbishment of other assets. To enhance the process of the CCSChain, two main functions should be developed like plans

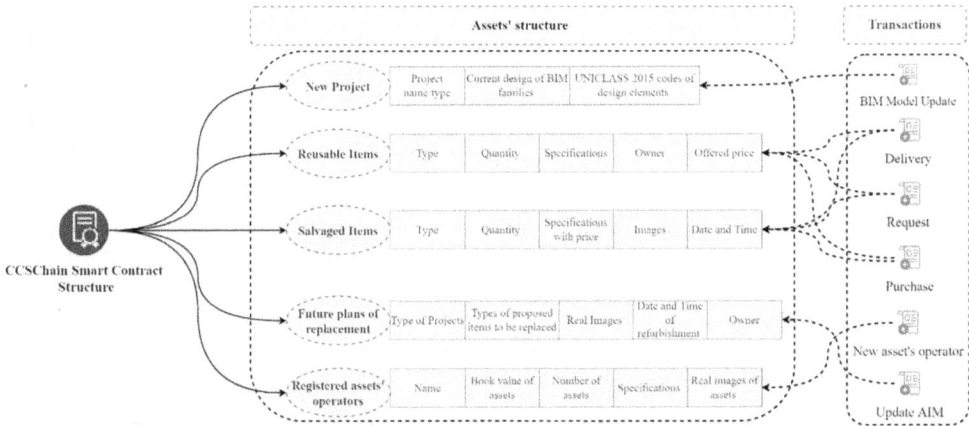

Figure 13.3 Assets and transactions structure of the CCSChain smart contract.

of replacement to enable designers to adapt new project designs according to items that will be available on the mentioned dates in the transactions. Given that most of designs are currently developed using BIM, an asset should be developed to enable asset operators to get BIM families for all required elements for a new project. To interact and manage the processes between all the smart contract assets, a set of transaction is developed to enable parties to add information to these assets and inquire the recorded information in the ledger for all the assets.

Figure 13.3 shows that there are six transactions to run the blockchain network efficiently and effectively. These include the 'bim model update', which will be used to update all new BIM families over the design process, as well as a transaction to update the 'Asset Information Model' (AIM) in case the existing building was designed and implemented using BIM. This therefore can significantly enhance collaboration between asset operators and designers. Moreover, there is a set of transactions to allow network parties to complete purchasing and selling processes including request, purchase and delivery tractions. The network should be scalable; therefore, new asset operators can register their asset characteristics and book value anytime using 'new asset's operator' transaction. The interrelationship between transactions and assets is presented in Figure 13.3, therefore, a user can be directed based on the purpose of each transaction. As such, the proposed smart contract includes all key functions that allow practitioners to interact and leverage the reuse and recycle of building products and materials.

Figure 13.4 shows the development process of CCSChain-based blockchain system and the interaction processes with BIM. End users such as city councils can follow this process to develop similar solution in their local areas.

The main operator such as council of a city will be responsible to develop a permissioned blockchain network and to enable all aforementioned smart contract functions (see Figure 13.3). To enable asset operators, regardless of their IT skills competency, an API should be developed to make posting or getting information from the blockchain network ledger possible. Two API(s) should be developed; the first one is to enable authorised but not registered parties to invoke transactions, and the second API is a private one to enable only registered blockchain parties (assets' operators and designers) to

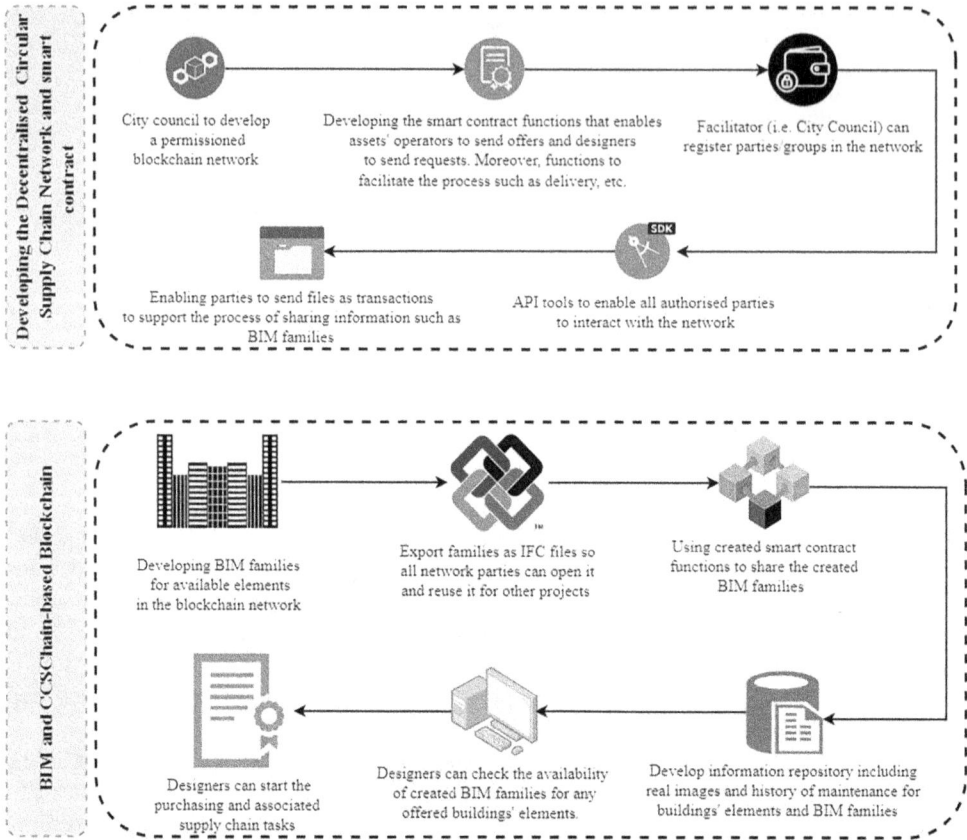

Figure 13.4 CCSChain development process and integration tasks with BIM.

record new blocks. To make the CCSChain system a workable solution in integration with BIM, there is a need to enable recording all formats of information like pdf, JPG, DWG, etc.

To enable the sustainability of data, the proposed CCSChain solution supports developing BIM families that are similar to existing real building elements; these families should be exported as IFC files, hence all asset operators can open them using any BIM platform. Once the network is active, designers for different buildings would record many BIM families in the blockchain network, with no need to 'design as exist' from scratch for new projects.

The proposed CCSChain solution can be used during the maintenance of an existing building and during the preparation of demolition too. During the maintenance, the operator of the asset should classify elements that will be replaced or upgraded to three categories: reusable items, salvaged items and virgin items. Afterwards, the operator should use the smart contract to (1) check all requests of items from designers, (2) invoke transactions to record a list of new items with detailed specifications and real images (see Figure 13.5). Moreover, this solution considers the option of 'Scan to BIM' families, which can save designers' time to develop BIM families for reusable buildings items or facilities from scratch.

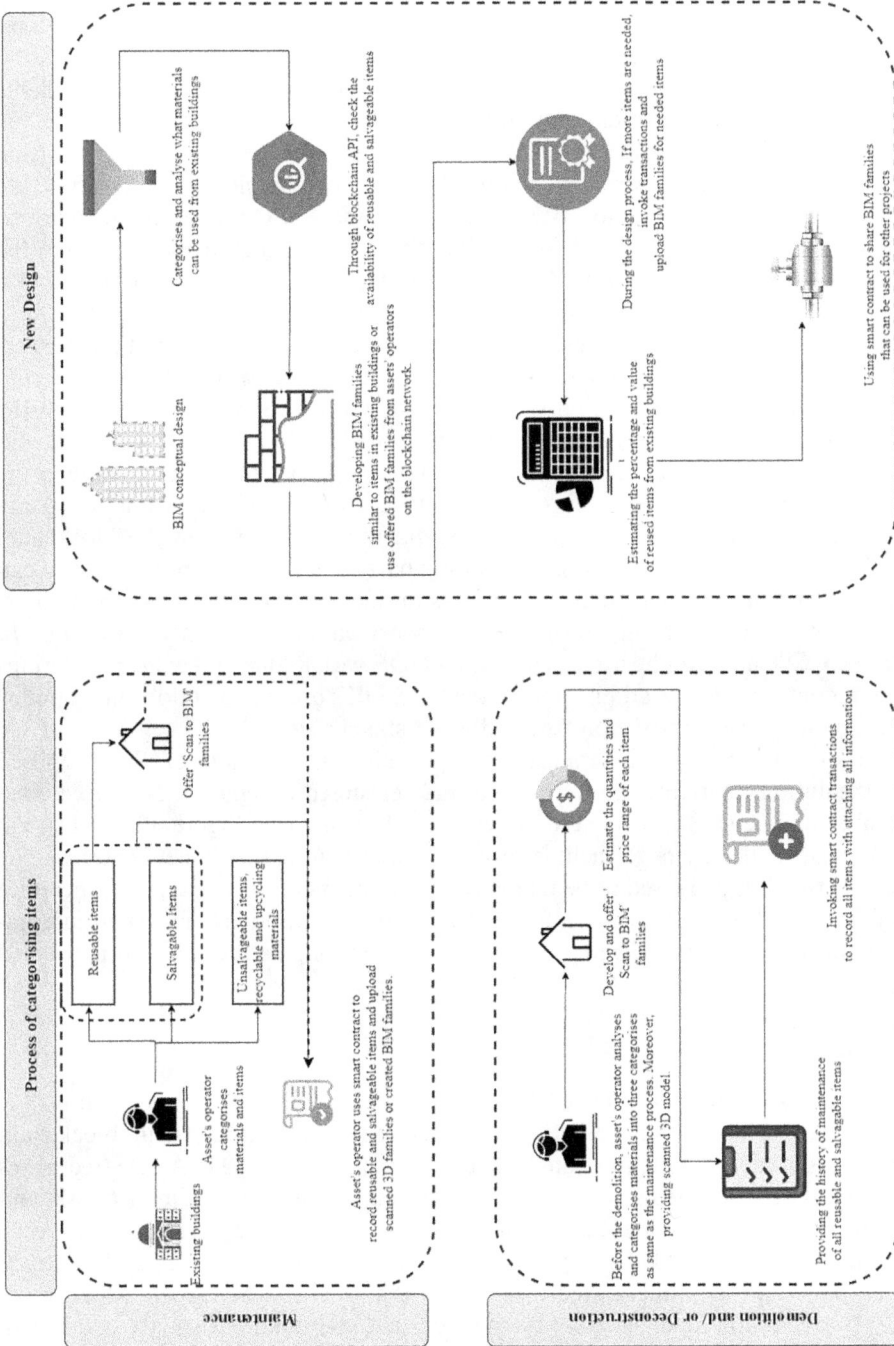

Figure 13.5 The flow of information for CCSChain-based blockchain technology.

In order to minimise the cost of demolition and inventory of items that have a high salvage value, the CCSChain-based blockchain solution enables building owners to scan buildings before demolition and categorises all items and materials into three categories, namely, 'reusable items', 'salvaged items', 'unsalvageable items/materials'. The next stage can be estimating items and desired prices to sell these items to be used in new buildings or recycling process for salvaged or un-salvaged items. After that, building owners should register the asset on the blockchain network for offering the types of items, quantities, price lists, specifications and real images.

On the other side – see Figure 13.5 – designers of new buildings can check any available items in the network through using the authorised API to receive information from the network. They can design BIM families like existing items and use the purchasing function to buy these items. However, the needed items on the blockchain network are not available; designers can develop BIM families and share IFC files on the network so that building owners can offer more items when they observe high demand. This is the main purpose of developing the CCSChain-based blockchain solution: to leverage a circular supply chain using blockchain and smart contract technology.

Given that ISO 19650 is being used to govern the delivery of construction projects using BIM, therefore, this CCSChain solution can be aligned with ISO 19650 design stages. For example, owners (appointing party) can specify the need to develop BIM families in the Employer Information Requirements (EIR) during the 'assessment and need' stage and the lead appointed party will have to consider this while justifying the resources as a part of the tender response stage. Moreover, the Common Data Environment (CDE) can store information containers that come from the blockchain network during design stage, therefore, appointing party and or lead appointed party can check the performance of their items or facilities during the operation stage, as well as, using this information in future projects to support circular economy transition. In addition, the CCSChain can be used as a secured CDE with adding a few more functions to the smart contract such as adding and retrieving CDE's areas and endorsement policy to govern the access of information from different stakeholders.

Information sustainability is important to sustain the function of the proposed CCSChain solution, therefore, all authorised and registered designers can get recorded BIM families and reuse them in new projects, as illustrated in Figure 13.5. This will encourage most of designers to include more reusable items in their new designs. The blockchain network is proposed to be managed by city councils, therefore, the experience can be globalised through sharing data locally to enable adopting the system internationally. Items can even be delivered internationally, as long as the solution is deemed economic.

13.6 Treatment Design

Figure 13.6 shows the graph of the developed smart contract as proposed in Figure 13.3, illustrating relationships between assets and transactions. The 'Simba Chain' blockchain platform is used to develop the presented smart contract in Figure 13.6. Assets (red boxes in Figure 13.6) refer to items that will be managed in the blockchain network and transactions refer to tasks that are associated with assets. For example, for new projects, BIM families should be added using the 'BIM_mode_update' transaction. Besides, if an existing building was implemented using BIM, the AIM can be used to share information; so too it can be updated once the replacement and maintenance are completed.

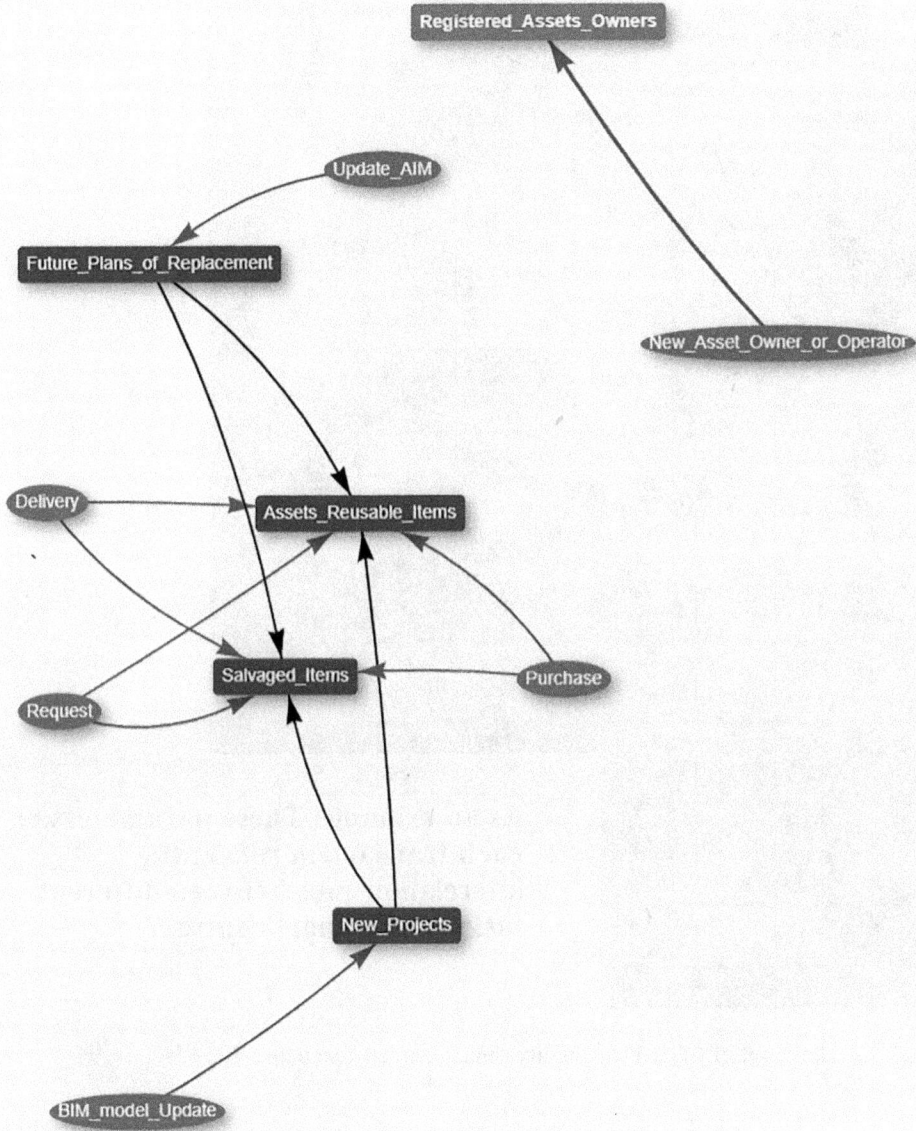

Figure 13.6 The graph of the Smart contract including assets and transactions.

Moreover, the COBie file can be recovered in the blockchain network. This enables designers to receive information about various items.

The smart contract graph in Figure 13.6 was developed using Solidity version 0.5.10; all assets and function variables are provided in Figure 13.7, where a wide range of variables are coded for each function. Variables include 'strings' and 'Unit 64' and attaching documents which will enable recording all types of information on the

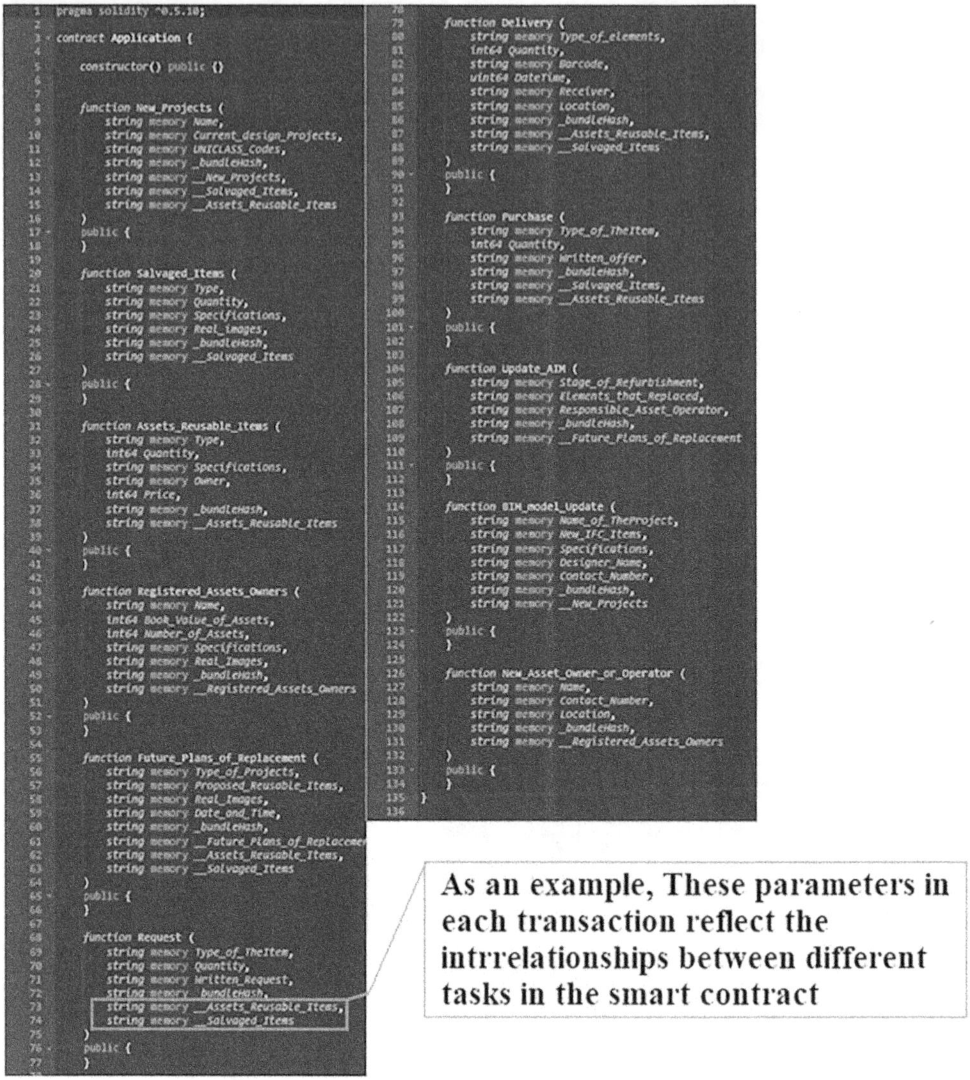

```solidity
1   pragma solidity ^0.5.10;
2
3   contract Application {
4
5       constructor() public ()
6
7
8       function New_Projects (
9           string memory Name,
10          string memory Current_design_Projects,
11          string memory UNICLASS_Codes,
12          string memory _bundleHash,
13          string memory _New_Projects,
14          string memory _Salvaged_Items,
15          string memory _Assets_Reusable_Items
16      )
17      public {
18      }
19
20      function Salvaged_Items (
21          string memory Type,
22          string memory Quantity,
23          string memory Specifications,
24          string memory Real_Images,
25          string memory _bundleHash,
26          string memory _Salvaged_Items
27      )
28      public {
29      }
30
31      function Assets_Reusable_Items (
32          string memory Type,
33          int64 Quantity,
34          string memory Specifications,
35          string memory Owner,
36          int64 Price,
37          string memory _bundleHash,
38          string memory _Assets_Reusable_Items
39      )
40      public {
41      }
42
43      function Registered_Assets_Owners (
44          string memory Name,
45          int64 Book_Value_of_Assets,
46          int64 Number_of_Assets,
47          string memory Specifications,
48          string memory Real_Images,
49          string memory _bundleHash,
50          string memory _Registered_Assets_Owners
51      )
52      public {
53      }
54
55      function Future_Plans_of_Replacement (
56          string memory Type_of_Projects,
57          string memory Proposed_Reusable_Items,
58          string memory Real_Images,
59          string memory Date_and_Time,
60          string memory _bundleHash,
61          string memory _Future_Plans_of_Replacement,
62          string memory _Assets_Reusable_Items,
63          string memory _Salvaged_Items
64      )
65      public {
66      }
67
68      function Request (
69          string memory Type_of_TheItem,
70          string memory Quantity,
71          string memory Written_Request,
72          string memory _bundleHash,
73          string memory _Assets_Reusable_Items,
74          string memory _Salvaged_Items
75      )
76      public {
77      }
78
79      function Delivery (
80          string memory Type_of_elements,
81          int64 Quantity,
82          string memory Barcode,
83          uint64 DateTime,
84          string memory Receiver,
85          string memory Location,
86          string memory _bundleHash,
87          string memory _Assets_Reusable_Items,
88          string memory _Salvaged_Items
89      )
90      public {
91      }
92
93      function Purchase (
94          string memory Type_of_TheItem,
95          int64 Quantity,
96          string memory Written_offer,
97          string memory _bundleHash,
98          string memory _Salvaged_Items,
99          string memory _Assets_Reusable_Items
100     )
101     public {
102     }
103
104     function Update_AIM (
105         string memory Stage_of_Refurbishment,
106         string memory Elements_that_Replaced,
107         string memory Responsible_Asset_Operator,
108         string memory _bundleHash,
109         string memory _Future_Plans_of_Replacement
110     )
111     public {
112     }
113
114     function BIM_model_Update (
115         string memory Name_of_TheProject,
116         string memory New_IFC_Items,
117         string memory Specifications,
118         string memory Designer_Name,
119         string memory Contact_Number,
120         string memory _bundleHash,
121         string memory _New_Projects
122     )
123     public {
124     }
125
126     function New_Asset_Owner_or_Operator (
127         string memory Name,
128         string memory Contact_Number,
129         string memory Location,
130         string memory _bundleHash,
131         string memory _Registered_Assets_Owners
132     )
133     public {
134     }
135 }
136
```

As an example, These parameters in each transaction reflect the intrrelationships between different tasks in the smart contract

Figure 13.7 The script of CCSChain solidity smart contract functions.

blockchain network. Designers will be able to share specifications and BIM families of elements that are requested; these can be used from existing elements. Similarly, building owners can share real images of reusable and salvaged items so that designers can make informed decisions regarding the suitability of these items, for making a transaction to purchase.

One of the main purposes of CCSChain system is to enable proactive decisions. Figures 13.6 and 13.7 depict that most transactions are correlated, therefore, once the purchasing order has been confirmed and endorsed, the delivery function can enable tracking the status of items. This can minimise the fragmentation of a typical construction supply chain. Likewise, functions such as 'future plan of replacement', 'reusable items' and 'salvaged items/materials' are correlated as designers need to check future plan records to contact registered owners to make a deal and invoke purchase transactions.

With any change in the business, new smart contract functions can be added and the same blockchain network can be used. For example, items can be further divided to sub-categorises like adding new asset (function) about 'reusable items from historical buildings', 'hazardous unsalvageable materials/items', etc. Suppliers of different types of materials can participate in the network and purchase these materials for selling them to designers or helping building owners to manage hazardous materials waste according to regulations and legislations. As such, the CCSChain-based blockchain solution provides a comprehensive solution to maximise the value of various types of materials; it also facilitates managing hazardous waste materials with minimal impact on the environment.

To assist all parties to record blocks (information) on the blockchain network, a private API can be developed as illustrated in Figure 13.8. This will enable all authorised

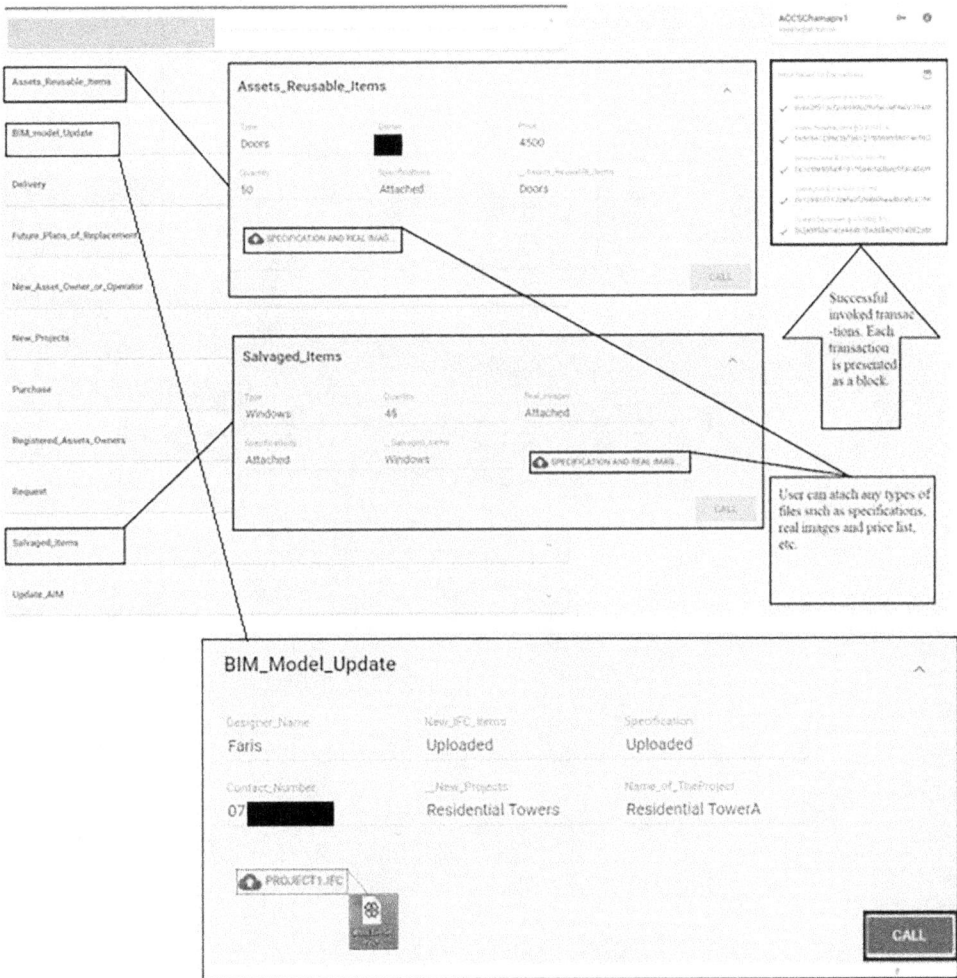

Figure 13.8 Private API to enable network participants to call any transaction.

parties to invoke transactions. It offers a user-friendly interface where all participants, regardless of their IT skills competency would be able to call transactions and interact with the blockchain network. They would be allowed to upload and record documents like IFC files, real images and specifications (see Figure 13.8). Accordingly, designers can reuse these BIM families in new projects, which will leverage the concept and the adoption of a construction circular supply chain. Moreover, this will encourage designers to develop a huge BIM objects library for existing buildings' items over the time. This can foster the adoption of circular economy since all these objects will be accessible from any designers for new projects and then using the created blockchain network and smart contract to look for these items.

Where an asset operator or designer is not registered in the blockchain network, the city council can authorise parties using the API key, as presented in Figure 13.9.

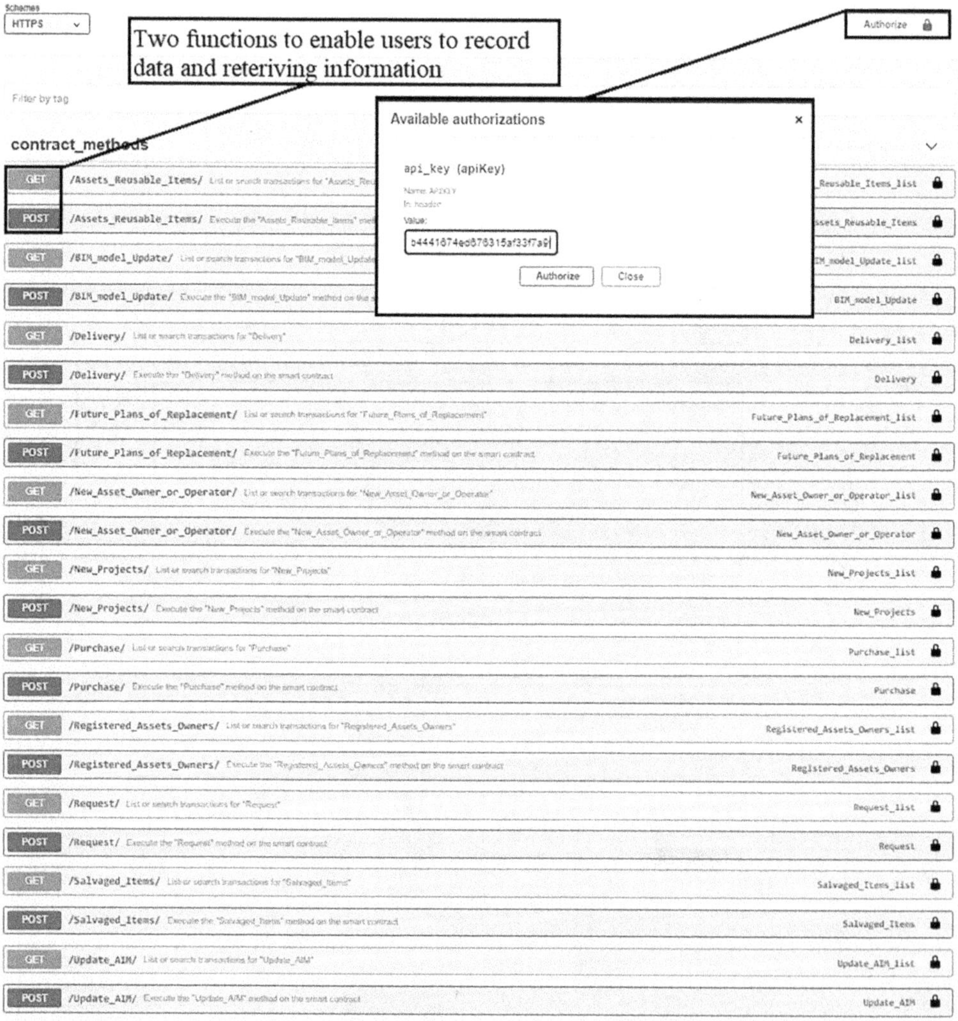

Figure 13.9 Public API for the developed CCSChain smart contract.

Notification > Manage

ACCSChain_API
✉ Email:████████@hotmail.com

ACCSChain_API
💬 SMS: +44████████

ACCSChain_API
✉ Email: ████████@icloud.com

ACCSChain_API
✉ Email: ████████@gmail.com

Figure 13.10 Linking smart contract and blockchain network update with parties' communication means.

This will enable them to execute transactions; parties would still be able to send all types of information and upload documents to be recorded as blocks in the blockchain network.

A platform called 'Simba Chain' is used to develop this PoC. It enables linking the blockchain network with different servers through email or SMS. All parties, therefore, would be able to get notifications once new materials have been offered as shown in Figure 13.10. This capability is especially valuable where the sold material is categorised as a 'hazardous material'. The city council hence would be able to monitor the progress of managing the waste of materials and recycling, where some of sold items can enter the recycling process – according to the agreed upon site waste management plan (SWMP). The moderator (city council) can record all contacts for all registered asset operators, designers and suppliers.

13.7 Evaluation

Four transactions were called to test the usability of the proposed solution. As illustrated in Figure 13.11 the developed PoC was tested through invoking a set of transactions that included different types of information like 'update AIM', 'adding salvaged items', 'adding r-usable items' and' update BIM'. All mentioned, transactions require sending

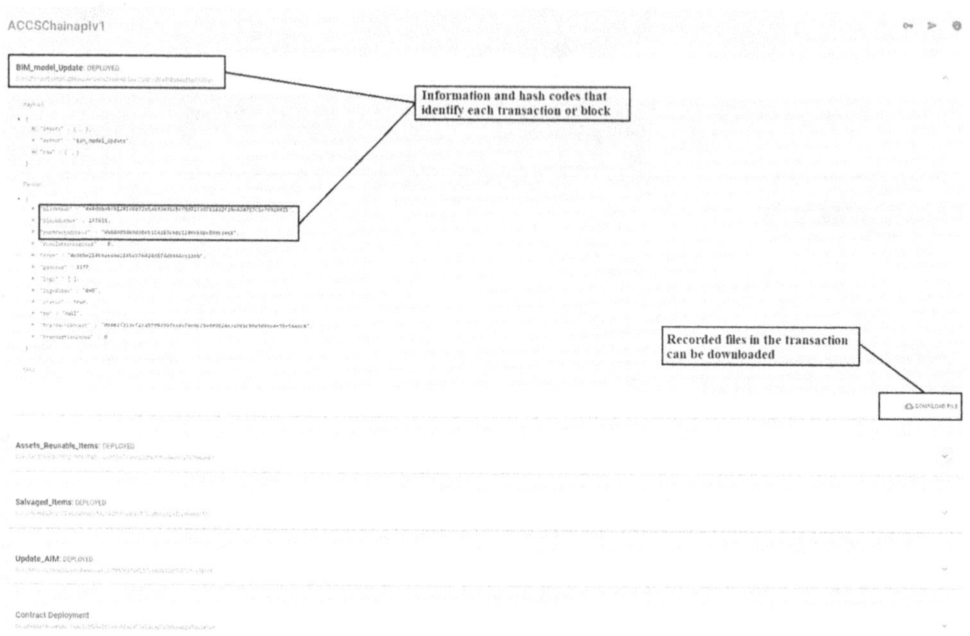

Figure 13.11 A set of recorded blocks in the blockchain network.

documents such as IFC files, DWG, pdf, JPG, etc. As in Figure 13.11, all these transactions were successfully recorded, including all types of information and blocks that
were recorded in the ledger; each block is represented by a specific number and hash code,
which cannot be repeated when new blocks are added to the ledger.

13.8   Discussion of Findings

In view of a lack of practical solutions to assist a transition to circular supply chains in
the construction context, this chapter provides a potentially workable solution for
facilitating the adoption of circular construction supply chain. The solution is based on
blockchain and smart contract technology; a system called CCSChain is created to
enable construction practitioners to collaboratively share information about their asset's maintenance and deconstruction replacement items that can be reused for new
buildings. The developed CCSChain system builds upon informational and technological solutions such as the blockchain and smart contracts that enable sharing of all
types of information in an immutable, safe and interconnected medium. The CCSChain
solution supports the integrating of BIM through enabling owners to share 'Scanned to
BIM' families for reusable items to be employed in new designs, as well as enabling
designers to create and share BIM families in case that owners do not have BIM objects
for their reusable items. CCSChain provides an integrated solution since it integrates
BIM, blockchain and smart contract in a single information workflow to support the
transition to circular economy.

The development of CCSChain contributes to the field in several ways, as discussed
next. This is through extending the body of knowledge to address some of the

major barriers to the adoption of construction circular supply chain as identified in past research;

- **Supporting circular supply chain through a digital ecosystem:** Mastos et al. (2021) assert that the adoption of circular supply chain requires a digital ecosystem to enable the transformation of products in the end of their functional life to new products. Technology is a main component in the digital ecosystem (Gamidullaeva et al., 2021), this chapter hence provides an interactive and interconnected tool to combine all stakeholders in a user-friendly and safer platform using blockchain technology to share all types of items that come out from their assets maintenance and demolition.

- **Providing drivers and facilitating adopting reverse logistics in the construction industry:** The adoption of circular supply chain relies on enhancing the impacts of some drivers associated with raising awareness and facilitating the adoption process for construction practitioners (Chen et al., 2021; Shooshtarian et al., 2022). This chapter proposes a workable solution, as a new component for shifting towards a digital ecosystem to support the adoption of circular economy. The proposed solution would enable practitioners to sell their reusable and recyclable items without a need to engage a third party (i.e., suppliers). Besides, the proposed solution makes possible the co-planning of maintenance and deconstruction among designers and asset owners to maximise the added value of reusable items and salvaged materials/items in buildings.

- **Enabling designing with reused building components:** The reuse of building components is crucial to minimise generating the waste of building materials during refurbishment and demolition (Charef et al., 2021). However, there is a need for a platform to enable designers to check the availability of building components and evaluate their usability for new projects. Besides, any demand for certain items should be shared among asset owners in a safe environment. CCSChain-based blockchain solution as proposed in this chapter offers a medium to enable sharing building components including types, quantities, prices, real images, etc. So too, the solution makes it possible to share new design BIM families as blocks in the blockchain network, where only authorised parties can see information via a designated API(s).

- **Lack of demand for second-hand, reused building components:** The reuse of building items is not mainstream practice yet. Studies that develop solution to raise awareness of the economic and environmental feasibility of re-using building components are much needed, a point argued by Shooshtarian et al. (2022). So too, providing an interconnected and decentralised medium to manage a large-scale circular supply chain process has remained wanting (Rakhshan et al., 2020). The CCSChain system is developed as a response to these demands in the domain. The proposed system hence contributes to the field by providing a solution that enables practitioners to interact in a decentralised platform and encourages designers to share BIM families that are designed like existing components. It stands out as with its application more designers can reuse these BIM families in new projects, facilitating as transition to making the reuse of components a mainstream practice within the construction contexts.

- **Making adopting deconstruction cost-effective and justified:** As a major barrier to the adoption of circular supply chains, the cost of deconstruction is higher than that of conventional demolition (Coelho and de Brito, 2011). An effective solution is the reuse of building components and materials to make deconstruction more cost-effective (Tatiya et al., 2018). The CCSChain solution enables building owners to predict the value of money for their reusable and salvaged items. This can justify adopting deconstruction and makes practitioners capable of making informed decisions about deconstruction earlier on the project.

- **Providing grounds to assess the feasibility of deconstruction:** Current practices of deconstruction rely on highly skilled labours; it takes longer than demolition. Accordingly, there is a need for planning, ahead of maintenance and deconstruction to allow building owners to determine the feasibility of deconstruction (Rios et al., 2015). CCSChain enables owners-designers to cooperate during the pre-deconstruction stage so that designers can check items that can be purchased, and owners can determine the outcome of deconstruction and make data-informed decisions.

13.9   Practical Implications

Apart from theoretical contributions, this chapter offers several practical implications as discussed below.

- Enabling governmental authorities like city council to check the compliance of circular supply chain processes with regulations: recycle, reuse, remanufacture.
- Leveraging the consideration of circular economy in early design stages through enabling 'designer to designer' and 'designer to asset owners' collaboration'. This will facilitate the development of BIM families for existing elements; these families can be shared with local/international designers to make them capable of repeating the experience and extending the adoption of circular supply chain.
- The proposed solution creates structural flexibility in supply chains as it allows asset owners, regardless of the type and size of their assets, to get involved in the process and sell salvageable items.
- Given that all transactions are conducted in an immutable Hyperledger, this solution enables government authorities to automatically determine any applicable tax for the sold items. Besides, incentives can be provided to encourage building owners to go for deconstruction rather than the demolition of their buildings.

13.10   Conclusion

The proposed CCSChain-based blockchain technology solution provides an integrated solution for some major barriers and challenges of construction circular supply chain. This enables considering existing elements in new project designs through allowing designers to develop BIM families like existing buildings and share the developed families with other designers to leverage the concept of circular design. So too, the proposed CCSChain-based blockchain solution allows local agencies like city councils to track the details of materials, their quantities, delivery destinations and treatment process in a safe and interconnected platform.

The CCSChain-based blockchain for the first time integrates BIM into blockchain to correlate the design process with a circular supply chain and enables developing of a 'bank of BIM families for reusable items' for local zones. Moreover, if an asset is delivered through the use of BIM, the CCSChain-based blockchain enables sharing the reusable offered items as BIM families from the AIM model and COBie sheet. These can be reused in new designs. This creates a desired closed circular supply chain loop.

CCSChain stands out among similar solutions as it addresses the problem of early decision-making of maintenance and deconstruction, which has remained a barrier towards a transition to a circular supply chain. The CCSChain enables designers and asset owners to prepare and agree to purchase proposed salvaged and reusable items before the maintenance and deconstruction start. This can encourage owners to deconstruct rather the demolish.

There are some limitations to be acknowledged. The proposed solution is tested using an illustrative case study; the validity of the solution components including smart contract functions has been assessed through submitting various types of information (i.e., strings, pdf, RVT) and checking the security capabilities of the solution. The proposed solution shows acceptable levels of security, and it is deemed user-friendly. However, this solution should be further validated using large-scale cases through adopting it in various contexts and settings.

As with other limitations, the full automation of the solution remains to be achieved in future research studies. This can be through integrating the IoT into CCSChain solution to support using IoT sensors. Future researchers hence can check the validity of building components to be used in new designs and automatically warn asset owners to invoke transactions in a blockchain network to offer these buildings components.

References

Abideen, A. Z., Sundram, V. P. K., Pyeman, J., Othman, A. K., & Sorooshian, S. (2021). Digital twin integrated reinforced learning in supply chain and logistics. *Logistics*, *5*(4), 84. 10.3390/logistics5040084

Adabre, M. A., Chan, A. P. C., Darko, A., & Hosseini, M. R. (2022). Facilitating a transition to a circular economy in construction projects: intermediate theoretical models based on the theory of planned behaviour. *Building Research & Information*, 1–20. 10.1080/09613218.2022.2067111

Agrawal, A., Fischer, M., & Singh, V. (2022). Digital twin: from concept to practice. *Journal of Management in Engineering*, *38*(3), 06022001. 10.1061/(ASCE)ME.1943-5479.0001034

Akbarnezhad, A., Ong, K. C. G., & Chandra, L. R. (2014). Economic and environmental assessment of deconstruction strategies using building information modeling. *Automation in Construction*, *37*, 131–144. 10.1016/j.autcon.2013.10.017

Akinade, O. O., & Oyedele, L. O. (2019). Integrating construction supply chains within a circular economy: An ANFIS-based waste analytics system (A-WAS). *Journal of Cleaner Production*, *229*, 863–873. 10.1016/j.jclepro.2019.04.232

AlSehaimi, A., Koskela, L., & Tzortzopoulos, P. (2013). Need for alternative research approaches in construction management: case of delay studies. *Journal of Management in Engineering*, *29*(4), 407–413. 10.1061/(ASCE)ME.1943-5479.0000148

Alavi, B., Tavana, M., & Mina, H. (2021). A dynamic decision support system for sustainable supplier selection in circular economy. *Sustainable Production and Consumption*, *27*, 905–920. 10.1016/j.spc.2021.02.015

Ashima, R., Haleem, A., Bahl, S., Javaid, M., Mahla, S. K., & Singh, S. (2021). Automation and manufacturing of smart materials in additive manufacturing technologies using Internet of Things towards the adoption of Industry 4.0. *Materials Today: Proceedings*, *45*, 5081–5088. 10.1016/j.matpr.2021.01.583

Baek, C., Park, S.-H., Suzuki, M., & Lee, S.-H. (2013). Life cycle carbon dioxide assessment tool for buildings in the schematic design phase. *Energy and Buildings*, *61*, 275–287. 10.1016/j.enbuild.2013.01.025

Bekrar, A., El Cadi, A. A., Todosijevic, R., & Sarkis, J. (2021). Digitalizing the closing-of-the-loop for supply chains: a transportation and blockchain perspective. *Sustainability*, *13*(5), 2895. 10.3390/su13052895

Benachio, G. L. F., Freitas, M. d. C. D., & Tavares, S. F. (2020). Circular economy in the construction industry: a systematic literature review. *Journal of Cleaner Production*, *260*, 121046. 10.1016/j.jclepro.2020.121046

Boje, C., Guerriero, A., Kubicki, S., & Rezgui, Y. (2020). Towards a semantic construction digital twin: directions for future research. *Automation in Construction, 114*, 103179. 10.1016/j.autcon.2020.103179

Brandão, R., Edwards, D. J., Hosseini, M. R., Silva Melo, A. C., & Macêdo, A. N. (2021). Reverse supply chain conceptual model for construction and demolition waste. *Waste Management & Research, 39*(11), 1341–1355. 10.1177/0734242X21998730

Bressanelli, G., Pigosso, D. C. A., Saccani, N., & Perona, M. (2021). Enablers, levers and benefits of circular economy in the electrical and electronic equipment supply chain: a literature review. *Journal of Cleaner Production, 298.* 10.1016/j.jclepro.2021.126819

Böckel, A., Nuzum, A.-K., & Weissbrod, I. (2021). Blockchain for the circular economy: analysis of the research-practice gap. *Sustainable Production and Consumption, 25*, 525–539. 10.1016/j.spc.2020.12.006

Centobelli, P., Cerchione, R., Del Vecchio, P., Oropallo, E., & Secundo, G. (2021). Blockchain technology for bridging trust, traceability and transparency in circular supply chain. *Information & Management*, 103508. 10.1016/j.im.2021.103508

Charef, R., Morel, J.-C., & Rakhshan, K. (2021). Barriers to implementing the circular economy in the construction industry: A critical review. *Sustainability, 13*(23), 12989. 10.3390/su132312989

Chen, Q., Feng, H., & de Soto, B. G. (2021). Revamping construction supply chain processes with circular economy strategies: A systematic literature review. *Journal of Cleaner Production*, 130240. 10.1016/j.jclepro.2021.130240

Chen, Z., & Huang, L. (2020). Digital twin in circular economy: remanufacturing in construction. *IOP Conference Series: Earth and Environmental Science, 588*(3), 032014. 10.1088/1755-1315/588/3/032014

Chidepatil, A., Bindra, P., Kulkarni, D., Qazi, M., Kshirsagar, M., & Sankaran, K. (2020). From trash to cash: how blockchain and multi-sensor-driven artificial intelligence can transform circular economy of plastic waste? *Administrative Sciences, 10*(2), 23. https://doi.org/10.3390/admsci10020023

Ciliberto, C., Szopik-Depczyńska, K., Tarczyńska-Łuniewska, M., Ruggieri, A., & Ioppolo, G. (2021). Enabling the circular economy transition: a sustainable lean manufacturing recipe for Industry 4.0. *Business Strategy and the Environment.* 10.1002/bse.2801

Coelho, A., & de Brito, J. (2011). Economic analysis of conventional versus selective demolition—a case study. *Resources, Conservation and Recycling, 55*(3), 382–392. 10.1016/j.resconrec.2010.11.003

Craveiroa, F., Duartec, J. P., Bartoloa, H., & Bartolod, P. J. (2019). Additive manufacturing as an enabling technology for digital construction: a perspective on Construction 4.0. *Sustainable Development, 4*, 6. 10.1016/j.autcon.2019.03.011

Damianou, A., Angelopoulos, C. M., & Katos, V. (2019). An architecture for blockchain over edge-enabled IoT for smart circular cities. 2019 15th International Conference on Distributed Computing in Sensor Systems (DCOSS). IEEE, 465–472. 10.1109/DCOSS.2019.00092

Davidova, M., & McMeel, D. (2020). *Codesigning with blockchain for synergetic landscapes: the cocreation of blockchain circular economy through systemic design.* Paper presented at the AADRIA 2020: e:Anthropocene – Design in the Age of Humans, Bangkok, Thailand. https://orca.cardiff.ac.uk/id/eprint/134033/

Dev, N. K., Shankar, R., & Qaiser, F. H. (2020). Industry 4.0 and circular economy: Operational excellence for sustainable reverse supply chain performance. *Resources, Conservation and Recycling, 153*, 104583. 10.1016/j.resconrec.2019.104583

Elghaish, F., Abrishami, S., & Hosseini, M. R. (2020). Integrated project delivery with blockchain: An automated financial system. *Automation in Construction, 114*, 103182. 10.1016/j.autcon.2020.103182

Elghaish, F., Matarneh, S. T., Edwards, D. J., Pour Rahimian, F., El-Gohary, H., & Ejohwomu, O. (2022). Applications of Industry 4.0 digital technologies towards a construction circular economy: Gap analysis and conceptual framework. *Construction Innovation, 22*(3), 647–670. 10.1108/CI-03-2022-0062

Figueiredo, K., Hammad, A. W., Haddad, A., & Tam, V. W. (2022). Assessing the usability of blockchain for sustainability: Extending key themes to the construction industry. *Journal of Cleaner Production*, 131047. 10.1016/j.jclepro.2022.131047

Gamidullaeva, L., Tolstykh, T., Bystrov, A., Radaykin, A., & Shmeleva, N. (2021). Cross-sectoral digital platform as a tool for innovation ecosystem development. *Sustainability*, *13*(21), 11686. https://doi.org/10.3390/su132111686

Ghisellini, P., Cialani, C., & Ulgiati, S. (2016). A review on circular economy: The expected transition to a balanced interplay of environmental and economic systems. *Journal of Cleaner Production*, *114*, 11–32. 10.1016/j.jclepro.2015.09.007

Ghobakhloo, M., Tang, S. H., Zulkifli, N., & Ariffin, M. (2013). An integrated framework of green supply chain management implementation. *International Journal of Innovation, Management and Technology*, *4*(1), 86. 10.7763/IJIMT.2013.V4.364

Haleem, A., Khan, S., Luthra, S., Varshney, H., Alam, M., & Khan, M. I. (2021). Supplier evaluation in the context of circular economy: A forward step for resilient business and environment concern. *Business Strategy and the Environment*, *30*(4), 2119–2146. 10.1002/bse.2736

Hamledari, H., & Fischer, M. (2021). Construction payment automation using blockchain-enabled smart contracts and robotic reality capture technologies. *Automation in Construction*, *132*, 103926. 10.1016/j.autcon.2021.103926

Hämäläinen, M. (2020). Smart city development with digital twin technology. Proceedings of the 33rd Bled eConference-Enabling Technology for a Sustainable Society, Bled, Slovenia, 28–29. 10.18690/978-961-286-362-3.20

Kouhizadeh, M., Sarkis, J., & Zhu, Q. (2019). At the nexus of blockchain technology, the circular economy, and product deletion. *Applied Sciences*, *9*(8), 1712. 10.3390/app9081712

Kovacic, I., Honic, M., & Sreckovic, M. (2020). Digital platform for circular economy in AEC industry. *Engineering Project Organization Journal*, *9*. 10.25219/epoj.2020.00107

Kuechler, B., & Vaishnavi, V. (2008). On theory development in design science research: Anatomy of a research project. *European Journal of Information Systems*, *17*(5), 489–504. 10.1057/ejis.2008.40

Kusi-Sarpong, S., Gupta, H., Khan, S. A., Chiappetta Jabbour, C. J., Rehman, S. T., & Kusi-Sarpong, H. (2021). Sustainable supplier selection based on industry 4.0 initiatives within the context of circular economy implementation in supply chain operations. *Production Planning & Control*, 1–21. 10.1080/09537287.2021.1980906

Leising, E., Quist, J., & Bocken, N. (2018). Circular Economy in the building sector: Three cases and a collaboration tool. *Journal of Cleaner Production*, *176*, 976–989. 10.1016/j.jclepro.2017.12.010

Li, J., & Kassem, M. (2021). Applications of distributed ledger technology (DLT) and Blockchain-enabled smart contracts in construction. *Automation in Construction*, *132*, 103955. 10.1016/j.autcon.2021.103955

Li, J., Greenwood, D., & Kassem, M. (2019). Blockchain in the built environment and construction industry: A systematic review, conceptual models and practical use cases. *Automation in Construction*, *102*, 288–307. 10.1016/j.autcon.2019.02.005

Lu, Q., Xie, X., Parlikad, A. K., & Schooling, J. M. (2020). Digital twin-enabled anomaly detection for built asset monitoring in operation and maintenance. *Automation in Construction*, *118*, 103277. 10.1016/j.autcon.2020.103277

Maciel, A. (2020). *Use of blockchain for enabling Construction 4.0*. In *Construction 4.0* (pp. 395–418): Routledge.

Martínez-Rocamora, A., Rivera-Gómez, C., Galán-Marín, C., & Marrero, M. (2021). Environmental benchmarking of building typologies through BIM-based combinatorial case studies. *Automation in Construction*, *132*, 103980. 10.1016/j.autcon.2021.103980

Mastos, T. D., Nizamis, A., Terzi, S., Gkortzis, D., Papadopoulos, A., Tsagkalidis, N., Ioannidis, D., Votis, K., & Tzovaras, D. (2021). Introducing an application of an industry 4.0 solution for circular supply chain management. *Journal of Cleaner Production*, *300*, 126886. 10.1016/j.jclepro.2021.126886

Montag, L., Klünder, T., & Steven, M. Paving the way for circular supply chains: Conceptualization of a circular supply chain maturity framework. *Frontiers in Sustainability, 101.* 10.3389/frsus.2021.781978

Nascimento, D. L. M., Alencastro, V., Quelhas, O. L. G., Caiado, R. G. G., Garza-Reyes, J. A., Rocha-Lona, L., & Tortorella, G. (2019). Exploring Industry 4.0 technologies to enable circular economy practices in a manufacturing context: A business model proposal. *Journal of Manufacturing Technology Management.* 10.1108/JMTM-03-2018-0071

Nassri, S., Talebi, S., Elghaish, F., Koohestani, K., McIlwaine, S., Hosseini, M. R., Poshdar, M., & Kagioglou, M. (2021). Labor waste in housing construction projects: An empirical study. *Smart and Sustainable Built Environment.* 10.1108/SASBE-07-2021-0108

Nawari, N. O., & Ravindran, S. (2019). Blockchain and the built environment: Potentials and limitations. *Journal of Building Engineering, 25,* 100832. 10.1016/j.jobe.2019.100832

Nikmehr, B., Hosseini, M. R., Wang, J., Chileshe, N., & Rameezdeen, R. (2021). BIM-based tools for managing construction and demolition waste (CDW): A scoping review. *Sustainability, 13*(15), 8427. 10.3390/su13158427

Piscitelli, G., Ferazzoli, A., Petrillo, A., Cioffi, R., Parmentola, A., & Travaglioni, M. (2020). Circular economy models in the industry 4.0 era: A review of the last decade. *Procedia Manufacturing, 42,* 227–234. 10.1016/j.promfg.2020.02.074

Pushpamali, N., Agdas, D., & Rose, T. M. (2019). A review of reverse logistics: An upstream construction supply chain perspective. *Sustainability, 11*(15), 4143. 10.3390/su11154143

Rahimian, F. P., Goulding, J. S., Abrishami, S., Seyedzadeh, S., & Elghaish, F. (2021). *Industry 4.0 solutions for building design and construction: A paradigm of new opportunities* (Vol. 1): Routledge. ISBN: 1000506363

Rajput, S., & Singh, S. P. (2019). Connecting circular economy and industry 4.0. *International Journal of Information Management, 49,* 98–113. 10.1016/j.ijinfomgt.2019.03.002

Rakhshan, K., Morel, J.-C., Alaka, H., & Charef, R. (2020). Components reuse in the building sector–A systematic review. *Waste Management & Research, 38*(4), 347–370. 10.1177/0734242X20910463

Rios, F. C., Chong, W. K., & Grau, D. (2015). Design for disassembly and deconstruction – Challenges and opportunities. *Procedia Engineering, 118,* 1296–1304. 10.1016/j.proeng.2015.08.485

Sanchez, B., & Haas, C. (2018). Capital project planning for a circular economy. *Construction Management and Economics, 36*(6), 303–312. 10.1080/01446193.2018.1435895

Sauerwein, M., Doubrovski, E., Balkenende, R., & Bakker, C. (2019). Exploring the potential of additive manufacturing for product design in a circular economy. *Journal of Cleaner Production, 226,* 1138–1149. 10.1016/j.jclepro.2019.04.108

Shojaei, A., Ketabi, R., Razkenari, M., Hakim, H., & Wang, J. (2021). Enabling a circular economy in the built environment sector through blockchain technology. *Journal of Cleaner Production, 294,* 126352. 10.1016/j.jclepro.2021.126352

Shooshtarian, S., Hosseini, M. R., Kocaturk, T., Arnel, T., & Garofano, N. T. (2022). Circular economy in the Australian AEC industry: Investigation of barriers and enablers. *Building Research & Information,* 1–13. 10.1080/09613218.2022.2099788

Singh, S., & Ashuri, B. (2019). Leveraging blockchain technology in AEC industry during design development phase. *Computing in Civil Engineering 2019: Visualization, Information Modeling, and Simulation,* 393–401. 10.1061/9780784482421.050

Tagliabue, L. C., Cecconi, F. R., Maltese, S., Rinaldi, S., Ciribini, A. L. C., & Flammini, A. (2021). Leveraging digital twin for sustainability assessment of an educational building. *Sustainability, 13*(2), 480. 10.3390/su13020480

Tatiya, A., Zhao, D., Syal, M., Berghorn, G. H., & LaMore, R. (2018). Cost prediction model for building deconstruction in urban areas. *Journal of Cleaner Production, 195,* 1572–1580. 10.1016/j.jclepro.2017.08.084

Upadhyay, A., Mukhuty, S., Kumar, V., & Kazancoglu, Y. (2021). Blockchain technology and the circular economy: Implications for sustainability and social responsibility. *Journal of Cleaner Production*, 126130. 10.1016/j.jclepro.2021.126130

Vogel, J., Hagen, S., & Thomas, O. (2019). *Discovering blockchain for sustainable product-service systems to enhance the circular economy*. Paper presented at the 14th International Conference on Wirtschaftsinformatik, Siegen, Germany. https://aisel.aisnet.org/wi2019/track12/papers/10/, Last Access:

Wang, B., Luo, W., Zhang, A., Tian, Z., & Li, Z. (2020a). Blockchain-enabled circular supply chain management: A system architecture for fast fashion. *Computers in Industry*, *123*, 103324. 10.1016/j.compind.2020.103324

Wang, Z., Wang, T., Hu, H., Gong, J., Ren, X., & Xiao, Q. (2020b). Blockchain-based framework for improving supply chain traceability and information sharing in precast construction. *Automation in Construction*, *111*, 103063. 10.1016/j.autcon.2019.103063

Wibowo, M. A., Handayani, N. U., Mustikasari, A., Wardani, S. A., & Tjahjono, B. (2022). Reverse logistics performance indicators for the construction sector: A building project case. *Sustainability*, *14*(2), 963. 10.3390/su14020963

Wieringa, R. J. (2014a). The design cycle. In *Design science methodology for information systems and software engineering* (pp. 27–34). Berlin, Heidelberg: Springer Berlin Heidelberg. 10.1007/978-3-662-43839-8_3

Wieringa, R. J. (2014b). Stakeholder and goal analysis. In *Design science methodology for information systems and software engineering* (pp. 35–40). Berlin, Heidelberg: Springer Berlin Heidelberg. 10.1007/978-3-662-43839-8_4

Wieringa, R. J. (2014c). What is design science? In *Design science methodology for information systems and software engineering* (pp. 3–11). Berlin, Heidelberg: Springer Berlin Heidelberg. 10.1007/978-3-662-43839-8_1

Wijewickrama, M., Chileshe, N., Rameezdeen, R., & Ochoa, J. J. (2021a). Information sharing in reverse logistics supply chain of demolition waste: A systematic literature review. *Journal of Cleaner Production*, *280*, 124359. 10.1016/j.jclepro.2020.124359

Wijewickrama, M. K. C. S., Rameezdeen, R., & Chileshe, N. (2021b). Information brokerage for circular economy in the construction industry: A systematic literature review. *Journal of Cleaner Production*, *313*, 127938. 10.1016/j.jclepro.2021.127938

Woo, J., Fatima, R., Kibert, C. J., Newman, R. E., Tian, Y., & Srinivasan, R. S. (2021). Applying blockchain technology for building energy performance measurement, reporting, and verification (MRV) and the carbon credit market: A review of the literature. *Building and Environment*, 108199. 10.1016/j.buildenv.2021.108199

Yu, Y., Yazan, D. M., Junjan, V., & Iacob, M.-E. (2022). Circular economy in the construction industry: A review of decision support tools based on Information & Communication Technologies. *Journal of Cleaner Production*, *349*, 131335. 10.1016/j.jclepro.2022.131335

14 Developing a Digitized Maintenance Supply Chain System for Sensitive Assets Using 'Blockchain of Things'

Faris Elghaish[1], Farzad Pour Rahimian[2], and Nashwan Dawood[2]

[1]*School of Natural and Built Environment, Queen's University Belfast, Belfast, Northern Ireland, United Kingdom;* [2]*School of Computing, Engineering & Digital Technologies, Teesside University, Middlesbrough, UK*

14.1 Introduction

The construction industry has been slow to incorporate blockchain and smart contracts in its practices. However, multiple attempts have been made to use them by constructing business models at various points of the project's lifecycle (Elghaish et al., 2021a). The engineering and construction stages of blockchain research are the most active. Several studies, for example, looked at the security of payments during the engineering and construction stages and found that blockchain-based smart contracts are an effective way to resolve interim payment difficulties in building projects (Ahmadisheykhsarmast and Sonmez, 2020; Chong and Diamantopoulos, 2020; Das et al., 2020). Other research focused on construction quality information management and indicated that blockchain implementation can improve construction quality information management by allowing information traceability (Sheng et al., 2020; Zhang et al., 2020; Zhong et al., 2020). Some studies have focused on multi-stage applications and investigated blockchain applications during the entire project lifecycle. For example, Pattini et al. (2020) looked into the capabilities of blockchain in information flow management in construction projects and found that it might be useful in the design, bidding, construction, and maintenance stages. The use of blockchain and smart contracts in the construction contracting process was also highlighted. Researchers found that they have the potential to reduce transaction costs (Dakhli et al., 2019), reduce paperwork (Mason and Escott, 2018), increase trust (Jun WANG, 2017), facilitate immediate payment (Jun WANG, 2017), secure payments (Salar, 2018), reduce construction disputes (Saygili et al., 2022), and improve procurement practice (Maciel, 2020).

The integration of Internet of Things (IoT) and blockchain has been discussed in previous studies regarding the technological potential and challenges (Alladi et al., 2019; Banafa, 2017; Panarello et al., 2018; Reyna et al., 2018). Studies have also discussed different integration use cases for different industries like food and automobiles, among others, as Ourad et al. (2018) proposed a blockchain-based solution using Ethereum smart contracts that provides authentication and secure communication to IoT devices. However, no available study has yet considered the potential use of integrating these two within the construction domain (Ghosh et al., 2020). Researchers like Li et al. (2019) and Perera et al. (2020) provide a holistic view of the potential uses of blockchain in the construction industry and conclude that blockchain has significant potential in construction. This is mainly due to the transformation in the industry regarding procurement

DOI: 10.1201/9781003408949-14

combining the change of onsite to offsite construction. As for IoT, extant literature, for the most part, comprises conceptual frameworks or is focused on creating point solutions for technical issues (Ding et al., 2018; Li et al., 2018; Pour Rahimian et al., 2020). Given that there is ample opportunity for the integration of IoT and blockchain (Alladi et al., 2019; Wang et al., 2020b) such as automated tracking of project resources, managing supply chain processes, solving the disconnectivity issues in complex projects, managing equipment remotely and supporting the transformation to smart cities. Combined with a conspicuous gap in the literature on this point, gaining a full understanding of various aspects of integrating these two technologies within the construction industry is much needed.

Miraz (2020) identified the Blockchain of Things (BCoT) as integration of blockchain technology and IoT to leverage the advantages of each individual technology, for example, blockchain can improve the security for IoT system. The BCoT is introduced to the construction industry by Elghaish et al. (2021a) to assess the potential of its applications to improve the delivery of construction projects. Blockchain is a proven technology to improve the transparency and trust among project parties through using its 'Peer to Peer' (P2P) technology to automate many processes (Alladi et al., 2019; Elghaish et al., 2020). Moreover, IoT can be used in the construction industry to automatically collect real-time information either from sites or facilities (Elghaish et al., 2021a).

Given that IEAA (2022), Martin and Abbt (2020) mentioned that there are several challenges in managing supply chain for highly sensitive assets such as nuclear power plants for energy generation such as the lack of finding qualified local suppliers, lack of collaboration among international suppliers, the lack of proactive material management. Given that the BCoT technology comprises blockchain and IoT advantages in terms of detecting real-time performance of elements using IoT sensors (Isyanto et al., 2020; Reyna et al., 2018) and automating supply chain processes including ordering, purchasing, carrying, and inventory automatically (Abrishami and Elghaish, 2019; Elghaish et al., 2021a; Elghaish et al., 2022; Gharaibeh et al., 2022; Wang et al., 2020a).

Tao et al. (2022) developed a framework for blockchain and BIM to boost design collaboration process. To ensure transparency, traceability, and immutability throughout its fragmented supply chain management, Li et al. (2022) developed a service-oriented system architecture of blockchain, IoT, and BIM for the data-information-knowledge-driven supply chain management in modular construction. Finally, to improve information visibility, traceability, and boost collaboration in working environment, Wu et al. (2022) developed a blockchain-enabled IoT-BIM platform for off-site production management in modular construction.

With the aforementioned considerations in mind, a proactive and automated supply chain system based on BCoT has been developed as a solution to address the challenges identified in the supply chain for highly sensitive assets such as nuclear power plants for energy production. A smart contract has been created to facilitate functions such as requesting offers from suppliers, receiving offers, managing deliveries, and updating inventory levels. This enables the asset operator to automate supply chain tasks and monitor national inventory levels of critical elements through the blockchain and smart contract. Additionally, a high-level architecture for an IoT system has been developed to detect and provide real-time information to the asset operator for proactive planning of required items.

The outcome of this research is expected to benefit asset owners/operators of highly sensitive assets by enabling proactive planning of their supply chain needs and minimizing

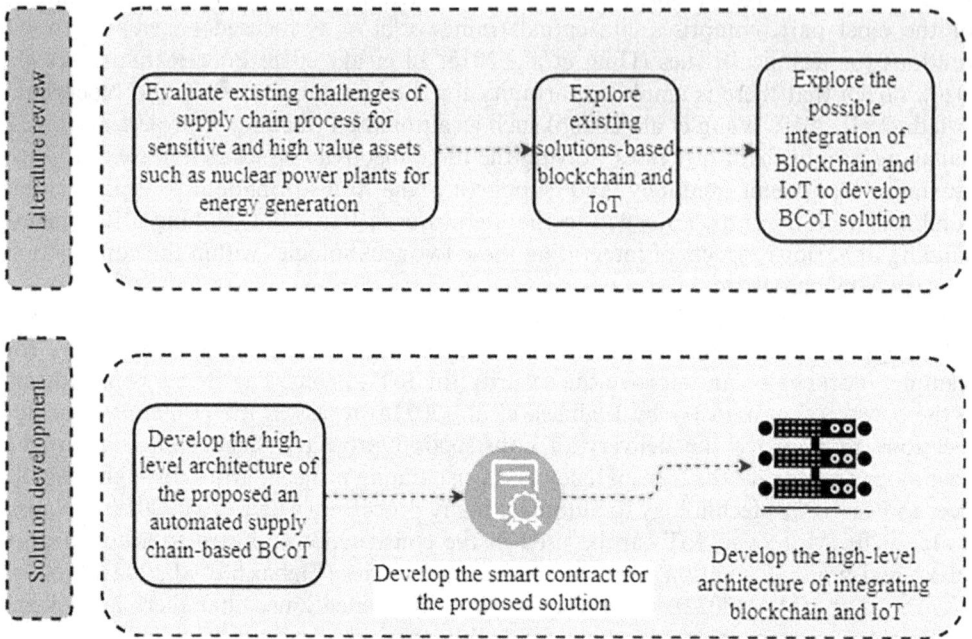

Figure 14.1 Research logic and steps of development.

interruptions in asset operations. It also has the potential to optimize the Total Material Cost (TMC) over the long term.

14.2   Methodology

The literature review was used to highlight the existing challenges of supply chain process for the high-sensitive assets such as nuclear power plants for energy production. The literature review is a robust method to highlight the existing knowledge gap for specific topic or practice (Elghaish et al., 2022; Elghaish et al., 2021b). The development of a framework can help researchers to investigate relevant variables of a phenomenon and develop a solution in a structured way (Regoniel, 2015). As such, a framework will be developed to provide a solution for revealed key issues in the literature review. Figure 14.1 shows the logic and steps of conducting the research and fulfilling the aim of this research.

14.3   Conceptual Solution Development

14.3.1   *Decentralized Supply Chain System-Based Hyperledger Fabric*

Given that there should be a proactive supply chain system for the highly sensitive assets such as nuclear power plants, the road to this proactive supply chain is to automate its processes and tasks. Blockchain and smart contracts are proven technology to develop decentralized system that depends on pre-agreed consensus mechanism to transfer and endorse information. This study proposes using Hyperledger fabric and chaincode to develop a proactive decentralized supply chain system for highly sensitive assets such as

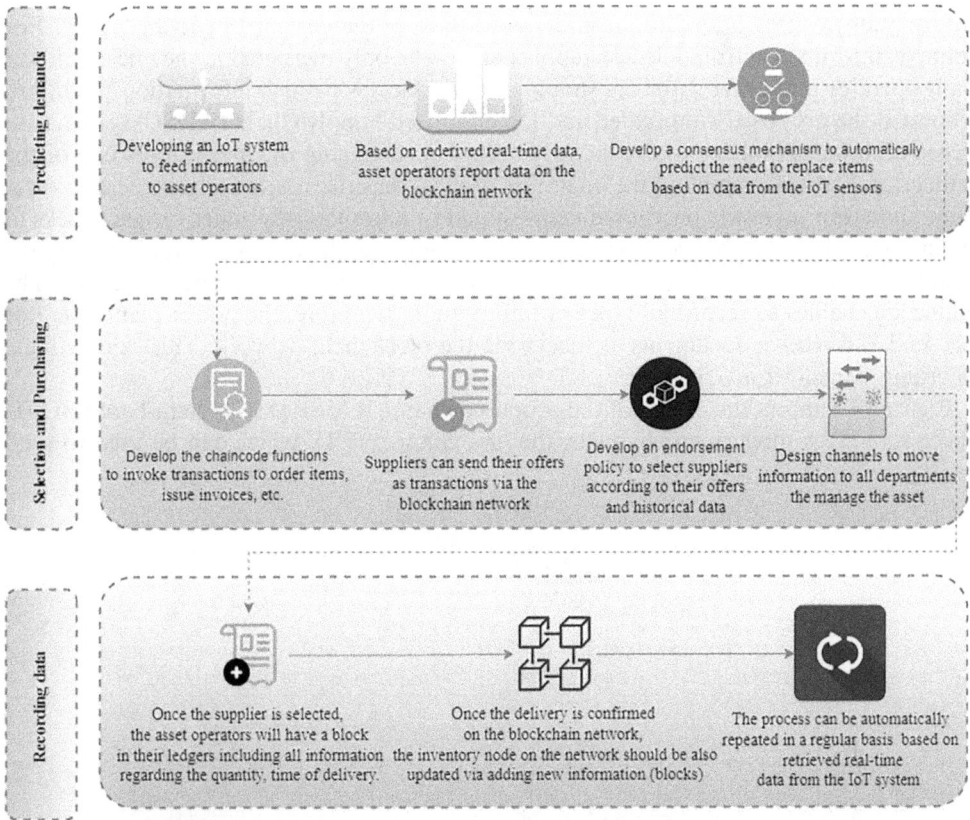

Figure 14.2 High-level architecture of the BCoT supply chain system.

nuclear power plants to combat issues that are published by the International Atomic Energy Agency (IAEA).

Figure 14.2 shows the high-level architecture of the proposed supply chain process-based Hyperledger and chaincode. It can be seen that the decentralized supply chain-based BCoT process is divided into three main stages, the first stage is the prediction of the need to replace items, this process will be conducted using IoT system. A set of IoT sensors should be installed to detect the performance of key facilities for the nuclear power plant, and rules are developed to determine based on received real-time data if the item needs to be replaced or not. This can be undertaken through developing an endorsement policy in the Hyperledger fabric to determine if the item needs to be replaced or not according to given ranges in the endorsement policy. The second process is to automatically select the suppliers using the pre-agreed endorsement policy, for examples, rules can be agreed to accept offers between specific cost ranges, and if offers from more than one supplier meet this range, then the nuclear power plant operator can check received blocks (offers) in the asset operator blockchain node and make the decision. A set of functions in the chaincode should be developed to enable all stakeholders (asset operators and suppliers) to undertake all tasks, namely, items to be replaced, offers from suppliers, payment, items in inventory. Once the supplier is selected either using the automated endorsement policy or by the asset operator if more than one offers are valid and meet criteria, the order will be placed, and a block (information) will be added to all nodes according to the designed ordering policy. For example, information from selected supplier should be only registered in the nuclear power plant operator and other suppliers should not have access to such information. The third process in the proposed Automated and Decentralized Supply Chain (ADSC) system is to record all information for future usage. Given that managing the inventory is one of the main critical tasks to maintain the operation of most sensitive assets such as nuclear power plant, therefore, all newly purchased items should be automatically added as new blocks to the inventory node in the blockchain network. Therefore, the asset operator can keep the inventory level at the optimal value to avoid any interruption of the operation. The chaincode enables to record all types of information, therefore, the power plant operator can send and receive documents as blocks via the blockchain network. This can facilitate recording signed-off invoices, etc.

Figure 14.3 shows the design of the proposed channels for ADSC system-based BCoT. There should be three main channels, the first channel (C1), which can be used to send

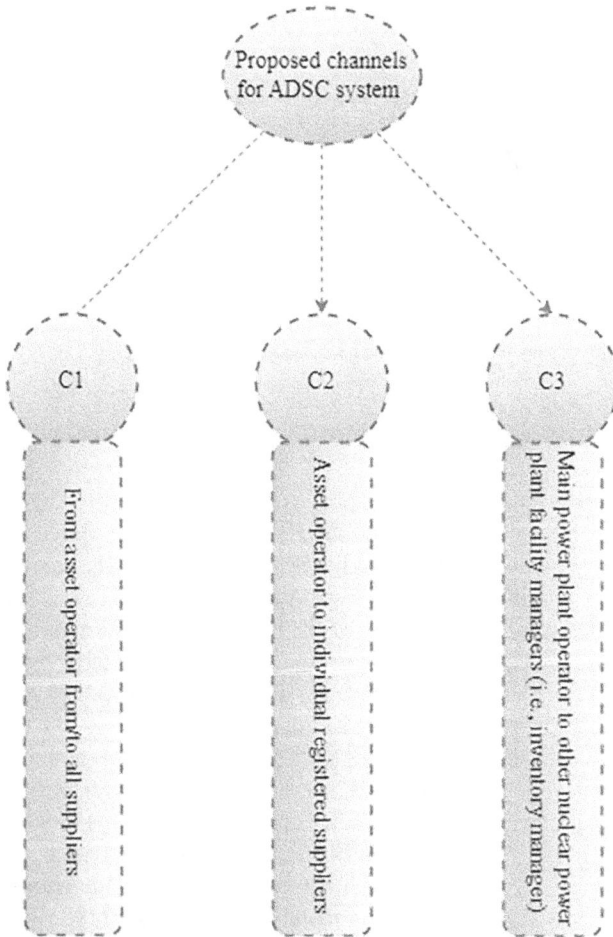

Figure 14.3 Blockchain channels design for ADSC system.

information to all registered suppliers such as items that needs to be replaced—call for offers. The second channel (C2), which can be used to send offer acceptance letter and invoices between operator and selected supplier. The third channel (C3), this is for internal information sharing, for example, operator can use it to share information with inventory manger to add the new items to the inventory node in the blockchain network.

14.3.2 Chaincode Functions Development

Figure 14.4 shows the structure and functions of the smart contract to manage all supply chain tasks on the blockchain network. As it can be seen from Figure 14.4 that there are main four functions in the smart contract to manage supply chain for a nuclear reactor power plant, these functions are 'items to be replaced', 'offers by suppliers', 'payments and invoices', and 'update the inventory statuses'. All these functions are linked with the asset (nuclear power plant) in the smart contract, therefore, if there are more than one plant, asset operator can create another asset in the smart contract and link same functions to the new asset (i.e., nuclear power plant (20), etc.). Therefore, a national blockchain and smart contract can be developed to serve and cover all power plants in the country. This can help to check the national inventory level for the most critical items to avoid any interruption in the operation and the supply of energy from nuclear power plants. Asset operators and suppliers will be able to share documents in the blockchain network, for example, specifications of required items can be sent and recorded in the right blockchain nodes.

```
contract Application {

    constructor() public {}          Asset name

    function NuclearPowerPlant (
        string memory __NuclearPowerPlant
    )
    public {
    }

    function ItemsTobeReplaced (
        string memory Name,
        string memory Quantity,
        string memory Date_of_Replacement,
        string memory Specification,
        string memory _bundleHash,
        string memory __NuclearPowerPlant
    )
    public {
    }

    function OffersBySupplier (
        string memory Type_of_items,
        string memory Specification,
        string memory Price_Per_Item,
        string memory Total_Price,
        string memory Date_of_Arrival,
        string memory _bundleHash,
        string memory __NuclearPowerPlant
    )
    public {            1
    }
```

```
    function Payments_Invoices (
        string memory Invoice_Number,
        string memory Invoice_Value,
        string memory Payment_Date,
        string memory Document4Record,
        string memory _bundleHash,
        string memory __NuclearPowerPlant
    )
    public {
    }

    function Update_Inventory (
        string memory Type_Of_Item,
        string memory Existing_Quantity,
        string memory New_Items,
        string memory Total_Quantity,
        string memory _bundleHash,
        string memory __NuclearPowerPlant
    )
    public {            2
    }
}
```

Figure 14.4 The smart contract for ADSC system.

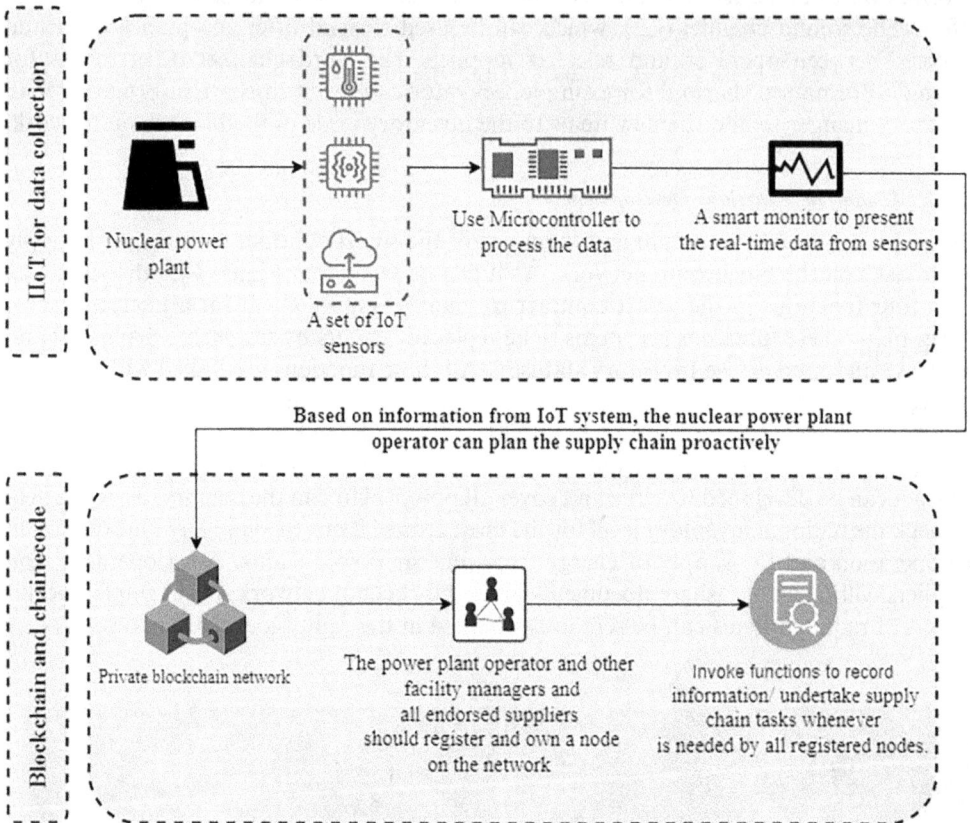

Figure 14.5 High-level architecture of integration IoT and blockchain.

14.3.3 *Blockchain and IoT Integration*

As aforementioned that BCoT relies on integrating IoT and blockchain to detect and collect real-time information, as well as processing and recording this information using blockchain technology. This research proposes developing an IoT system to detect and provide real-time information to the nuclear power operator and the blockchain and smart contract will be used to automatically manage the supply chain process. Figure 14.5 shows the integration of the IoT system and blockchain to develop the BCoT for ADSC system. According to the survey that the nuclear power plant facility managers do, a set of sensors should be selected for all items and facilities that needs to be monitored regularly, then a set of microcontroller or raspberry pi needs to be connected with IoT sensor in order to process the data and transfer it directly to the main server and shows all the data in a smart monitor. For example, if the degree of temperature of item A is higher than degree (n), therefore, a warning will be shown in the monitor to the power plant operator to take an action for replacement.

14.4 Discussion

According to Martin and Abbt (2020), there are challenges that face the current practices of supply chain for nuclear power plants for energy generation in Europe such as difficulties in finding suppliers, unuse of non-nuclear components that meet the criteria of safety, especially components that have lower level of nuclear safety. Therefore, there is a need for cross-border collaboration among European nuclear agencies. Moreover, IEAA (2022) asserted that a proactive supply chain system is highly needed for nuclear power plants to combat failures of structures, systems, or components to perform their planned function, or deal with the poor performance, which can negatively affect the global economy. As such, the proposed solution proposes the usage of BCoT to develop a proactive and automated cross-border supply chain system for nuclear power plants for energy production. The proposed solution provides the following features to manage revealed supply chain challenges in the literature review:

Integrating IoT into blockchain to automate the process of identifying the need to replace components and start the supply chain processes including ordering, purchasing, carrying, and inventory. All mentioned processes are integrated into a proactive approach.

Supporting the national and international collaboration among nuclear suppliers via registering all valid supplier who provides components that meet the national and international quality and safety standards in a blockchain network and enable them to see any request for offers to supply elements, as well as, submitting their offers using the proposed smart contract.

The proposed solution will enable power plant operator (i.e., facility manager) to automatically check the national inventory for all components since there is an inventory function in the smart contract to update the inventory level of specific items once new lot is received from suppliers. As such, this can minimize the existing fragmentation in supply chain process.

Adopting the concept of circular supply chain in nuclear power plants, especially for elements that have no negative impact on health and safety can raise the salvage value and maximize resources. Therefore, employing the proposed solution can enable a wide range of suppliers to collaboratively buy the items that come out from maintenance process to enter recycling or re-using them if possible.

The outcome of this research can be used as a point of departure to develop more functions for the created smart contract to expand the function of the proposed ADSC system.

Given that the proposed solution is conceptually presented, and a smart contract is developed, however, a full simulation is required to test the entire applicability, validity, and reliability of the solution and participants such as asset operators and suppliers should be involved in the evaluation process.

14.5 Conclusion

This research proposes a proactive solution to address the supply chain challenges for sensitive assets, such as nuclear power plants, which require automated monitoring of critical components and zero delays in the supply chain. A proactive supply chain system based on BCoT has been developed to connect the process of evaluating the need to order items with an IoT system and other tasks in the supply chain such as purchasing and inventory management. Considering the endemic issue of managing inventory for large-scale assets using traditional independent material management approaches, this research proposes an automated and integrated approach using blockchain and smart contracts to automatically reflect newly purchased items in the inventory levels. Furthermore, all similar facilities within a country can be managed through a single blockchain network to optimize the TMC.

The outcome of this research has the potential to be extended to future studies for implementation in real-life case studies to assess its applicability, feasibility, and scalability in different settings. Involving targeted stakeholders, such as facility managers and suppliers, to evaluate the usability of the solution would be beneficial.

Acknowledgment

We acknowledge that certain portions of this work were previously published in the proceedings of the Creative Construction Conference 2022 and EG-ICE 2022.

References

Abrishami, S., & Elghaish, F. (2019). *Revolutionising AEC financial system within project delivery stages: A permissioned blockchain digitalised framework*. Paper presented at the 36th CIB W78, Newcastle, UK. https://researchportal.port.ac.uk/en/publications/revolutionising-aec-financial-system-within-project-delivery-stag, Last Access: 15 January 2022.

Ahmadisheykhsarmast, S., & Sonmez, R. (2020). A smart contract system for security of payment of construction contracts. *Automation in Construction, 120*, 103401. 10.1016/j.autcon.2020.103401

Alladi, T., Chamola, V., Parizi, R. M., & Choo, K.-K. R. (2019). Blockchain applications for industry 4.0 and industrial IoT: A review. *IEEE Access, 7*, 176935–176951.

Banafa, A. (2017). IoT and blockchain convergence: Benefits and challenges. *IEEE Internet of Things, 9*(2017).

Chong, H.-Y., & Diamantopoulos, A. (2020). Integrating advanced technologies to uphold security of payment: Data flow diagram. *Automation in Construction, 114*, 103158. 10.1016/j.autcon.2020.103158

Dakhli, Z., Lafhaj, Z., & Mossman, A. (2019). The potential of blockchain in building construction. *Buildings, 9*(4), 77.

Das, M., Luo, H., & Cheng, J. C. P. (2020). Securing interim payments in construction projects through a blockchain-based framework. *Automation in Construction, 118*, 103284. 10.1016/j.autcon.2020.103284

Ding, K., Shi, H., Hui, J., Liu, Y., Zhu, B., Zhang, F., & Cao, W. (2018). Smart steel bridge construction enabled by BIM and Internet of Things in industry 4.0: A framework. 2018 IEEE 15th International Conference on Networking, Sensing and Control (ICNSC). IEEE, 1–5.

Elghaish, F., Abrishami, S., & Hosseini, M. R. (2020). Integrated project delivery with blockchain: An automated financial system. *Automation in Construction, 114*, 103182. 10.1016/j.autcon.2020.103182

Elghaish, F., Hosseini, M. R., Matarneh, S., Talebi, S., Wu, S., Martek, I., Poshdar, M., & Ghodrati, N. (2021a). Blockchain and the 'Internet of Things' for the construction industry: research trends and opportunities. *Automation in Construction, 132*, 103942. 10.1016/j.autcon.2021.103942

Elghaish, F., Matarneh, S. T., Talebi, S., Abu-Samra, S., Salimi, G., & Rausch, C. (2021b). Deep learning for detecting distresses in buildings and pavements: A critical gap analysis. *Construction Innovation*.

Elghaish, F., Matarneh, S. T., Edwards, D. J., Rahimian, F. P., El-Gohary, H., & Ejohwomu, O. (2022). Applications of Industry 4.0 digital technologies towards a construction circular economy: gap analysis and conceptual framework. *Construction Innovation* (ahead-of-print).

Gharaibeh, L., Eriksson, K. M., Lantz, B., Matarneh, S., & Elghaish, F. (2022). Toward digital construction supply chain-based Industry 4.0 solutions: Scientometric-thematic analysis. *Smart and Sustainable Built Environment*. (early cite.)

Ghosh, A., Edwards, D. J., & Hosseini, M. R. (2020). Patterns and trends in Internet of Things (IoT) research: Future applications in the construction industry. *Engineering, Construction and Architectural Management, 28*(2), 457–481.

IEAA. (2022). Management of the nuclear supply chain. Retrieved from https://www.iaea.org/topics/management-systems/management-of-the-nuclear-supply-chain.

Isyanto, H., Arifin, A. S., & Suryanegara, M. (2020). Design and implementation of IoT-based smart home voice commands for disabled people using Google Assistant. 2020 International Conference on Smart Technology and Applications (ICoSTA). IEEE, 1–6.

Li, C. Z., Xue, F., Li, X., Hong, J., & Shen, G. Q. (2018). An Internet of Things-enabled BIM platform for on-site assembly services in prefabricated construction. *Automation in Construction, 89*, 146–161.

Li, J., Greenwood, D., & Kassem, M. (2019). Blockchain in the built environment and construction industry: A systematic review, conceptual models and practical use cases. *Automation in Construction, 102*, 288–307.

Li, X., Lu, W., Xue, F., Wu, L., Zhao, R., Lou, J., & Xu, J. (2022). Blockchain-enabled IoT-BIM platform for supply chain management in modular construction. *Journal of Construction Engineering and Management, 148*(2). 10.1061/(ASCE)CO.1943-7862.0002229

Maciel, A. (2020). Use of blockchain for enabling Construction 4.0. pp. 395–418. 10.1201/9780429398100-20

Martin, O., & Abbt, M. (2020). *Current challenges of the European nuclear supply chain.* EUR 30309 EN, Publications Office of the European Union, Luxembourg, 2020, ISBN 978-92-76-20872-3. 10.2760/23903, JRC121103

Mason, J., & Escott, H. *Smart contracts in construction: Views and perceptions of stakeholders.* 2018.

Miraz, M. H. (2020). Blockchain of things (BCoT): The fusion of blockchain and IoT technologies. In *Advanced applications of blockchain technology* (pp. 141–159). Springer.

Ourad, A. Z., Belgacem, B., & Salah, K. (2018). Using blockchain for IOT access control and authentication management. International Conference on Internet of Things (pp. 150–164). Springer.

Panarello, A., Tapas, N., Merlino, G., Longo, F., & Puliafito, A. (2018). Blockchain and IoT integration: A systematic survey. *Sensors, 18*(8), 2575.

Pattini, G., Di Giuda, G. M., & Tagliabue, L. C. (2020). Blockchain application for contract schemes in the construction industry. Proceedings of International Structural Engineering and Construction. 10.14455/ISEC.res.2020.7(1).AAE-21

Perera, S., Nanayakkara, S., Rodrigo, M., Senaratne, S., & Weinand, R. (2020). Blockchain technology: Is it hype or real in the construction industry? *Journal of Industrial Information Integration, 17*, 100125.

Pour Rahimian, F., Seyedzadeh, S., Oliver, S., Rodriguez, S., & Dawood, N. (2020). On-demand monitoring of construction projects through a game-like hybrid application of BIM and machine learning. *Automation in Construction, 110*, 103012. 10.1016/j.autcon.2019.103012

Regoniel, P. A. (2015). Conceptual framework: A step by step guide on how to make one. *Simplyeducate.me.*

Reyna, A., Martín, C., Chen, J., Soler, E., & Díaz, M. (2018). On blockchain and its integration with IoT. Challenges and opportunities. *Future Generation Computer Systems, 88*, 173–190.

Salar, A., & Sönmez, R. (2018). *Smart Contracts in Construction Industry.* Paper presented at the 5th International Project and Construction Management Conference (IPCMC2018) Cyprus International University, Faculty of Engineering, Civil Engineering Department, North Cyprus. http://pcmc2018.ciu.edu.tr/index.php/ipcmc-2018-proceedings/

Saygili, M., Mert, I. E., & Tokdemir, O. B. (2022). A decentralized structure to reduce and resolve construction disputes in a hybrid blockchain network. *Automation in Construction, 134*. 10.1016/j.autcon.2021.104056

Sheng, D., Ding, L., Zhong, B., Love, P. E. D., Luo, H., & Chen, J. (2020). Construction quality information management with blockchains. *Automation in Construction, 120*, 103373. 10.1016/j.autcon.2020.103373

Tao, X., Liu, Y., Wong, P. K. Y., Chen, K., Das, M., & Cheng, J. C. P. (2022). Confidentiality-minded framework for blockchain-based BIM design collaboration. *Automation in Construction*, *136*. 10.1016/j.autcon.2022.104172

Wang, B., Luo, W., Zhang, A., Tian, Z., & Li, Z. (2020a). Blockchain-enabled circular supply chain management: A system architecture for fast fashion. *Computers in Industry*, *123*, 103324.

Wang, J., Wu, P., Wang, X., & Shou, W. (2017). The outlook of blockchain technology for construction engineering management. *Frontiers in Engineering*, *4*(1), 67–75. 10.15302/j-fem-2017006

Wang, Q., Zhu, X., Ni, Y., Gu, L., & Zhu, H. (2020b). Blockchain for the IoT and industrial IoT: A review. *Internet of Things*, *10*, 100081.

Wu, L., Lu, W., Xue, F., Li, X., Zhao, R., & Tang, M. (2022). Linking permissioned blockchain to Internet of Things (IoT)-BIM platform for off-site production management in modular construction. *Computers in Industry*, *135*. 10.1016/j.compind.2021.103573

Zhang, Z., Yuan, Z., Ni, G., Lin, H., & Lu, Y. (2020). The quality traceability system for prefabricated buildings using blockchain: An integrated framework. *Frontiers of Engineering Management*, *7*(4), 528–546. 10.1007/s42524-020-0127-z

Zhong, B., Wu, H., Ding, L., Luo, H., Luo, Y., & Pan, X. (2020). Hyperledger fabric-based consortium blockchain for construction quality information management. *Frontiers of Engineering Management*, *7*(4), 512–527. 10.1007/s42524-020-0128-y

15 Digital Twinning in the Malaysian Construction Industry

Shan Jiancheng[1], Chai Chang Saar[2], and Eeydzah Aminudin[2]

[1]*Faculty of Engineering, Computing and Science, Swinburne University of Technology Sarawak Campus, Kuching, Sarawak, Malaysia;* [2]*School of Civil Engineering, Faculty of Engineering, Universiti Teknologi Malaysia, Skudai Johor, Malaysia*

15.1 Introduction

The Fourth Industrial Revolution (IR 4.0) was first proposed in 2011 at the Hannover Messe Fair in Germany, and the term has been gradually becoming famous around the world. Three crucial technological factors are driving this revolution: connectivity, intelligence, and flexible automation (Ammar et al., 2021). IR 4.0 incorporates a set of technologies (Petrillo et al., 2018) that transform whole systems of production, management, and governance (Olsson et al., 2021). This revolution is characterized by the seamless connection of the real and virtual worlds, thereby bringing the transformation of digitization and automation to industries (Dallasega et al., 2020). Besides, in order to enhance the competitiveness of German manufacturing, Industry 4.0 becomes a global term from a vision (Ghobakhloo, 2020). More than a decade after IR 4.0 was proposed, most industries have completed their transformation of digitalization and automation. A series of strategic plans for Industry 4.0 has been released within this decade around the world, such as Germany Industrie 4.0, Made in China 2025, Advanced Manufacturing Partnership 2.0 (AMP 2.0) in the United States and I-KOREA 4.0. It has been observed that there is a growth in the productivity and effectiveness of all complex projects in almost every field through the technologies introduced by Industry 4.0, and the construction industry is one of the most benefited industries. With the slow and powerful changes in the processes and working flow in the construction industry by Industry 4.0, Construction 4.0 appears with those leading technologies introduced (Forcael et al., 2020). This term was first mentioned in 2016 and mainly revolved around the awareness of digitization in the construction industry for all stakeholders (Berger, 2016). Hence, some countries, especially Malaysia, have proposed specific strategic plans for digitalization transformation in the construction industry.

Malaysia is one of the fast-emerging developing countries, and the construction industry is one of the essential sectors of this country's economy (Alawag et al., 2022). It is because the construction industry plays an essential role in the country's economic growth and supports socioeconomic development (Mohsen et al., 2021). Hence, the Ministry of Works Malaysia and Construction Industry Development Board Malaysia (CIDB) have proposed the Construction 4.0 Strategic Plan (2021–2025) to allow all stakeholders within the construction industry to gear up and respond to the rapid changes of IR 4.0. According to the Construction Industry Development Board Malaysia (CIDB) (2020), a series of cutting-edge technologies have been introduced to enhance project productivity and effectiveness and to provide more sustainable approaches for the

DOI: 10.1201/9781003408949-15

construction industry by strengthening existing and emerging technologies. In other words, this strategic plan has been proposed to encourage the transformation of the construction industry towards digitalization. Twelve emerging technologies have been proposed within the Construction 4.0 Strategic Plan (2021–2025) for the betterment of productivity and competitivity in the Malaysian construction industry, which are Prefabrication and Modular Construction, Building Information Modeling (BIM), Autonomous Construction, Augmented Reality and Virtualization, Cloud and Realtime Collaboration, 3D Scanning and Photogrammetry, Big Data and Predictive Analytics, Internet of Things (IoT), 3D Printing and Additive Manufacturing, Advanced Building Materials, Blockchain, and Artificial Intelligence.

However, the productivity of the Malaysian construction industry has stagnated or even decreased over the past decade, despite significant changes in other industries, since it has not entirely adopted and utilized the latest technologies due to unwillingness and unawareness (Besklubova et al., 2021; Francis & Thomas, 2020). Besides, Iktisas Environment Sdn Bhd (2019) has listed the challenges and issues that may be faced while implementing Construction 4.0 in the Malaysian construction industry; the low productivity and lower working effectiveness are mainly due to the lack of understanding and utilization of existing and emerging technologies. In addition, previous studies have mainly focused on the development and implementation of a single or few technologies in the construction industry, such as BIM (Charehzehi, Chai, Md Yusof, Chong, & Loo, 2017), cloud-based BIM (Shan, Chai, Gui, & Xiong, 2021; Al-Ashmori et al., 2020), AR&VR-based BIM (Elshafey, Chai, Aminudin, Gheisari, & Usmani, 2020; Dallasega et al., 2020; Ratajczak et al., 2019), cloud-based system framework integrated with big data and BIM (Chen, Chang, & Lin, 2016), Blockchain-based sharing platform (Ratajczak et al., 2019), and 3D-printing building (Besklubova et al., 2021; Olsson et al., 2019). Therefore, there is a missing gap in the study: the integration of these 12 emerging technologies in the construction industry. This study aims to integrate these 12 emerging technologies into a viable conceptualized system that can provide the construction industry with better solutions. The proposed system has contributed to the current body of knowledge by providing some guidelines for the construction industry stakeholders, which help to understand and utilize products composed of one or more technologies. Moreover, this system will continuously enhance the project productivity and working efficiency of all construction organizations in the long run since it provides clearer pictures of future technology development for construction companies, especially for small and medium enterprises (SMEs). Also, it lays the foundation for the future development of smart construction and smart cities since it has potentially revolutionized the traditional working methods in the construction industry.

15.2   Twelve Emerging Technologies

The construction industry has benefited from the significant advances in technology and science brought by IR 4.0 (Boyes, Hallaq, Cunningham, & Watson, 2018); hence, the term "Construction 4.0" comes into being. It mainly focuses on the application of digital technologies, also known as industry digitalization (Forcael et al., 2020). The digitalization of the construction industry has introduced a series of advanced technologies to enhance project competitiveness, productivity, and working effectiveness (Beddiar, Grellier, & Woods, 2019). However, the construction industry has still rarely or not fully adopted those advanced technologies, while they have been positively and successfully adopted in other industries (García De Soto et al., 2018). Hence, CIDB Malaysia has designated 12 emerging technologies to help the construction industry adapt more fluidly and quickly forward to digitalization.

15.2.1   Prefabrication and Modular Construction

Driven by the development of the current technological environment, prefabrication building has attracted the interest of the global construction industry due to the growing demand and drastic cost pressure in recent years (Casini, 2022a). Conventional construction projects are usually executed entirely on-site through in-situ casting technologies, which are location-fixed, weather-affected, safety-concerned, and generate extensive construction waste (Lu, Chen, Xue, & Pan, 2018). Prefabrication buildings, on the other hand, allow off-site construction to increase project productivity and working sustainability while reducing safety concerns, construction waste, and project durations (Kamali, Hewage, & Sadiq, 2019). Prefabrication and modular construction are defined as construction projects that either adopt individual prefabricated building components or are entirely assembled by prefabricated building modules (Chen et al., 2016). Prefabrication and modular construction are usually standardized designs, factory prefabricated and modular construction, which is normally manufactured building components and modules at factories or manufacturing workshops, then delivered on-site for assembling.

15.2.2   Building Information Modelling (BIM)

According to Zhao and Taib (2022), building information modelling (BIM) is one of the significant parts of the development of the architecture, engineering, and construction (AEC) industry. BIM has been defined as a digital tool that plays the role of a collaborative platform for the use of processing, generating, evaluating, analyzing, and communicating projects through digital information models (Tam, Toan, Phong, & Durdyev, 2021; Kassem et al., 2020). In other words, BIM is a kind of approach that facilitates project collaboration through visual simulation of computer n-D modellings like the computer-aided design (CAD) model (Vandecasteele, Merci, & Verstockt, 2017). Hence, instead of the traditional sets of separate drawings, it supports many functions for collaborative management and more accurate information sharing throughout the project life cycle (Durdyev, Ashour, Connelly, & Mahdiyar, 2022). Besides, this visual model integrates every single aspect of project information and allows access among all project stakeholders during the entire project life cycle, from the design phase to the marketing phase (Sadeghineko & Kumar, 2021). Therefore, the adoption of BIM has benefited the construction project's productivity and working effectiveness, since it reduces most of the management conflicts and the probability of working mistakes during the project life cycle by providing an integrated information database and allowing all stakeholders to access it (Durdyev, Ashour, Connelly, & Mahdiyar, 2022).

15.2.3   Autonomous Construction

Autonomous construction refers to construction projects adopting the automatic assembly method through fully autonomous, controllable, or programmable robotics (Qi, Razkenari, Costin, Kibert, & Fu, 2021). In the past decades, the application of industrial robots has enhanced project productivity in many industrial fields, which also benefits the introduction of robotics in the construction industry (Vujović, Rodić, & Stevanović,

2017). Autonomous robotics is becoming popular in construction since most tasks related to construction projects are dirty, safety-concerned, dull, and repeatable (Melenbrink, Werfel, & Menges, 2020). Videlicet, a fully autonomous construction project, capable of working without supervision or intervention. The increase in automation implementation in the construction industry has carried many substantial benefits, like the reduction in safety concerns, enhancement in the working efficiency of repetitive tasks, and the ability to construct in environments where humans cannot work, like exoplanet construction or other challenging environments (Gharbia, Chang-Richards, Lu, Zhong, & Li, 2020). Furthermore, autonomous construction provides more options for dealing with the increasing complexity of construction projects such as extensive infrastructure and commercial construction projects (Munoz-Morera, Maza, Fernandez-Aguera, Caballero, & Ollero, 2015).

15.2.4   *Augmented Reality and Virtualization (AR and VR)*

According to Delgado, Oyedele, Demian, and Beach (2020), Augmented Reality and Virtualization (AR and VR) are two critical technologies for the AEC sectors due to the heavy dependency on imagery for communication among people involved in AEC projects. AR refers to the technology that enables overlaying information and computer-generated imagery in the actual world, thereby improving or augmenting the surrounding contextual experience of users (Harikrishnan, Abdallah, Ayer, Asmar, & Tang, 2021). It can be visualised through mobile devices or head-mounted displays (HMD). Conversely, VR refers to the technology that allows users to immerse in the digital imitation of the real world or the imaginative one. This kind of simulation allows interaction between users and the computer-generated environment through devices such as HMDs, VR glasses, etc. (Schiavi et al., 2022b). Basically, the implementation of AR and VR provides an ameliorated connection between the actual world and the virtual world; these two technologies enable people to experience project results in advance of the actual beginning of construction through the assistance of BIM (Li, Yi, Chi, Wang, & Chan, 2018). Moreover, AR and VR have progressively been adopted in most construction projects since they have benefited working productivity and efficiency by enhancing collaboration, detecting errors, and improving the safety of the construction environment (Barazzetti & Banfi, 2017).

15.2.5   *Cloud and Realtime Collaboration*

Construction projects are viewed as highly information-intensive and project-based; thus, project success is heavily dependent on collaborative working among all project members throughout the project lifecycle (Matthews et al., 2015). The development of cloud technology was introduced to address such issues. Implementing cloud technology is crucial since it allows real-time communication and collaborative work among all project stakeholders. It can also provide a platform for storing and sharing all documents and data related to the project (Bello et al., 2021). In short, through Cloud, all stakeholders with authorization within the project are allowed to access the related information and data without the restriction of location and time zone.

15.2.6   *3D Scanning and Photogrammetry*

3D Scanning and Photogrammetry are two types of data collection methods. One is to collect actual data on objects through 3D laser scanners, and another is to generate geometrical properties from the photos taken of objects (El-Omari & Moselhi, 2008). In recent years, an integrated technology of 3D scanning and photogrammetry has emerged with the development of computer science and computer vision technologies (Ulvi, 2021). In short, this technology allows monitoring changes to 3D models generated through data acquisition and photograph mapping. It helps to improve the speed and accuracy of data collection, which helps measure project progress and keep things under control (Yu and Fingrut, 2022).

15.2.7   *Big Data and Predictive Analytic (BDPA)*

Big data is defined as data that is generally associated with volume, velocity, and variety (3Vs) (Bilal et al., 2016). Simply, it is data sets that are more massive and complex than conventional databases, and they have surpassed the capability of the usual traditional database software or tools to capture, store, manage, and analyze them (Chen, Chiang, & Storey, 2012). Predictive analytics uses statistical and data mining approaches to analyze previous performance by censoring huge-volume historical data, determining patterns, inferring relationships, and then forecasting future performance (Wang, Gunasekaran, Ngai, & Papadopoulos, 2016). According to Ngo, Hwang and Zhang (2020), Big Data and Predictive Analytics (BDPA) is known as an application of big data-based predictive analytics, which provides significant potential benefits to the construction industry throughout the project lifecycle. BDPA enables efficient capture, storage, management, and process of vast amounts of project data to generate predictive analysis models to support and guide future decision-making (Grover, Chiang, Liang, & Zhang, 2018). It also improves the project's coordination (Bag, Rahman, Srivastava, Chan, & Bryde, 2022).

15.2.8   *Internet of Things (IoT)*

The Internet of Things (IoT) was initially proposed for collecting, processing, and transmitting data in 1999, which is also considered the beginning of the new generation of the Internet (Lee, Stanley, Spanias, & Tepedelenlioglu, 2016). Besides, IoT is a worldwide infrastructure of the information society that describes surrounding physical objects in virtual form with sensors, software, and interoperable information and communication technologies, then allows for data connection, exchange, generation, and operation with other devices over the Internet or other networks (Berawi, Sunardi, & Ichsan, 2019). The IoT has enormous potential in the construction industry to enhance project productivity and working efficiency. It has usually been used as a technology-based device to optimize the usage of valid resources through proper technological planning with minimal cost and lower risks (Malik et al., 2021b).

15.2.9   *3D Printing and Additive Manufacturing (AM)*

Additive manufacturing (AM) refers to the process of creating objects by joining materials from 3D model data, which is different from subtractive manufacturing (International Organization for Standardization [ISO], 2013). Besides, 3D printing refers to creating objects by depositing materials through printheads, nozzles, or other printer technologies (Farahbakhsh, Rybkowsk, Zakira, Kalantar, & Onifade, 2022). Hence, the integration of 3D printing and AM is to fabricate or re-fabricate physical objects by depositing layers of materials based on the digital model (Ashima, Haleem, Javaid, &

Rab, 2022). Implementing these technologies brings many benefits to the construction industry, like reducing construction duration, waste production, transportation costs, etc. (Carneau, Mesnil, Roussel, & Baverel, 2020).

15.2.10 Advanced Building Materials

Advanced materials refer to materials specifically designed to have new or enhanced technical characteristics or environmental features compared to conventional materials with the same functions (Pacheco-Torgal, Jalali, Labrincha, & John, 2013). In other words, it is used to create and develop new or improved materials and products for the construction industry through technologies and processes (Krupík, 2021). Therefore, the emergence of advanced materials in the construction industry has enabled the project's productivity to be expressively enhanced, provided more secure and comfortable buildings, and enabled a more sustainable construction environment throughout the entire project lifecycle (Casini, 2022b).

15.2.11 Blockchain

A distributed ledger technology database, known as the blockchain, is described as having decentralised activities over a consensus mechanism network, where all data, transactions' records, protocols of the internet, and other items are allowed to be maintained over a network of computers (Elghaish et al., 2021). Besides, all the data will be kept as unchangeable blocks once boned with and authenticated to a chain (Scott, Broyd, & Ma, 2021). As a result, blockchain technology has become an essential prerequisite in the construction industry for increasing value-added and data-acquisition tasks (Hosseini et al., 2020).

15.2.12 Artificial Intelligence (AI)

In the context of Construction 4.0, artificial intelligence (AI) has played as the backbone for the digitalization of the construction industry, which helps to realize the enhancement of automation, productivity, and efficiency (Baduge et al., 2022). AI is a division of computer science; it drives the machine to sense and learn to imitate the cognitive functions of human beings, and then handle complex and ill-defined tasks through perceiving, inferencing, planning, and problem-solving like a human being via a set algorithm (Pan & Zhang, 2021). Therefore, the adoption of AI technology has provided better competitive benefits in automation, risk mitigation, high efficiency, digitalization and computer vision for construction projects (Abioye et al., 2021).

15.3 A Conceptual System to Integrate 12 Emerging Technologies

With the various emerging technologies brought by Construction 4.0, the construction industry's transformation towards digitalization has been able to develop rapidly, as these emerging technologies help improve projects' competitiveness, productivity, and working efficiency (Forcael et al., 2020). Despite academic and industry attention, the rate of emerging technology implementation in the construction industry remains limited (Yap, Lam, Skitmore, & Talebian, 2022). This is because the construction industry is resistant to change, and most stakeholders lack awareness of the emerging technologies' knowledge and experience (Elshafey, Chai, Aminudin, Gheisari, & Usmani, 2020).

Moreover, another issue causing the low adoption of emerging technologies is the lack of study on the integration of emerging technologies, since they are only partially adopted in a few specified fields within construction projects (Kamaruddeen, Rui, Lee, Alzoubi, & Alshurideh, 2022). As previously stated, most previous studies have focused on a single or few technologies, and thus, fragmented introduction and application is one of the major issues for adopting emerging technologies (You & Feng, 2020). Therefore, this chapter aims to provide better solutions for the construction industry by integrating twelve emerging technologies into a conceptual system according to the ANN decision-making models. ANN is a kind of computational methodology that conducts multi-factorial analyses that are inspired by biological neuron networks (Anagnostou, Remzi, & Djavan, 2003). It has been widely touted as a solution to many predictive and decision-modelling problems (Hill, Marquez, O'Connor, & Remus, 1994). Therefore, the proposed conceptual system uses the ANN decision-making model as its foundation, which is shown in Figure 15.1.

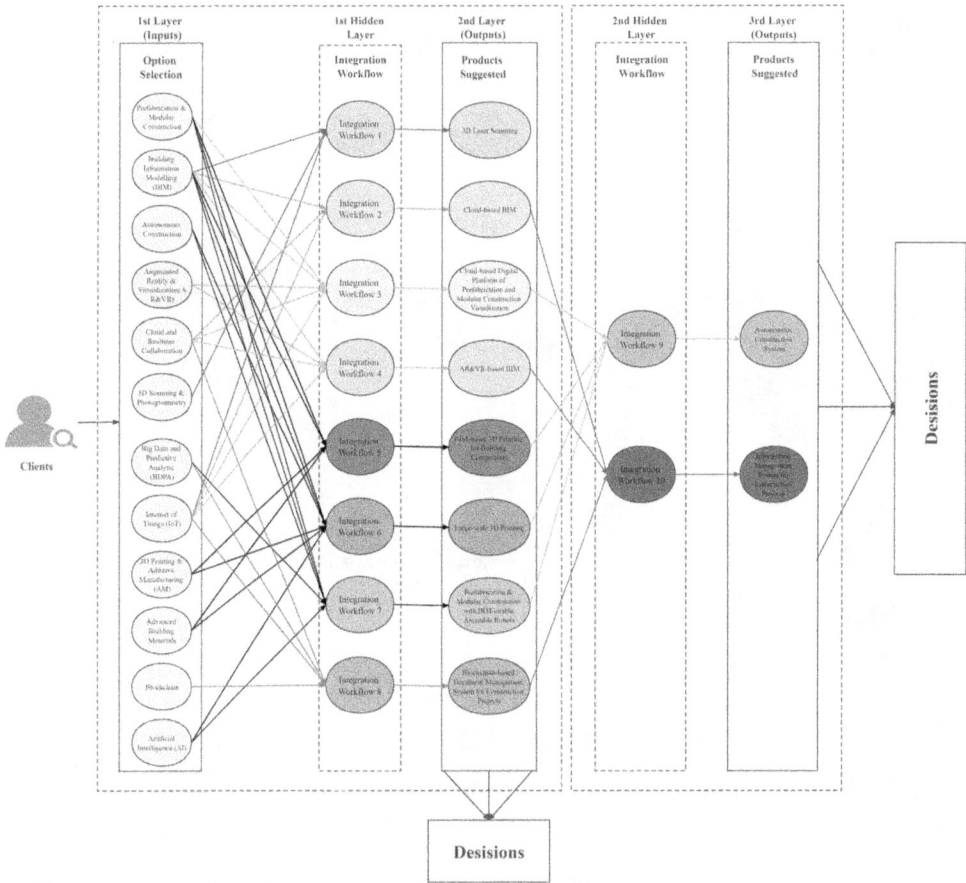

Figure 15.1 ANN-based decision-making tool for digital twinning (ADMT-DT) in the construction industry.

As Figure 15.1 shows, the first layer is an input layer, which is the selection of options composed of 12 emerging technologies. It allows clients to select one or multiple technologies that they or their company/organization own. The first hidden layer is the process of matching the integration workflow of the selected technologies. The second layer is the provided suggestion of construction-related products generated by the system. The suggested product is generated based on the technologies selected by clients, and each product consists of several technologies. The second hidden layer will proceed if selected technologies match the required conditions of the integration workflow, and then the third layer will subsequently show the suggested construction-related products. The second and third layers are output layers, which will show the suggested products and their workflow. Depending on the technology selected in the first layer, the output may be a single product or multiple products. The details of possible products and their integration workflow are as follows.

15.3.1   Integration Workflow 1 – 3D Laser Scanning

3D laser scanning is an accurate reality capture technology that can be used to collect physical object data into a 3D model (Usmani et al., 2020). Besides, it can also be used to scan as-built buildings into a model-based document to streamline the work for as-built verification and model updating. This technique requires 3D scanning and photogrammetry, BIM, and IoT. The workflow for this technique is as follows (Figures 15.2–15.5).

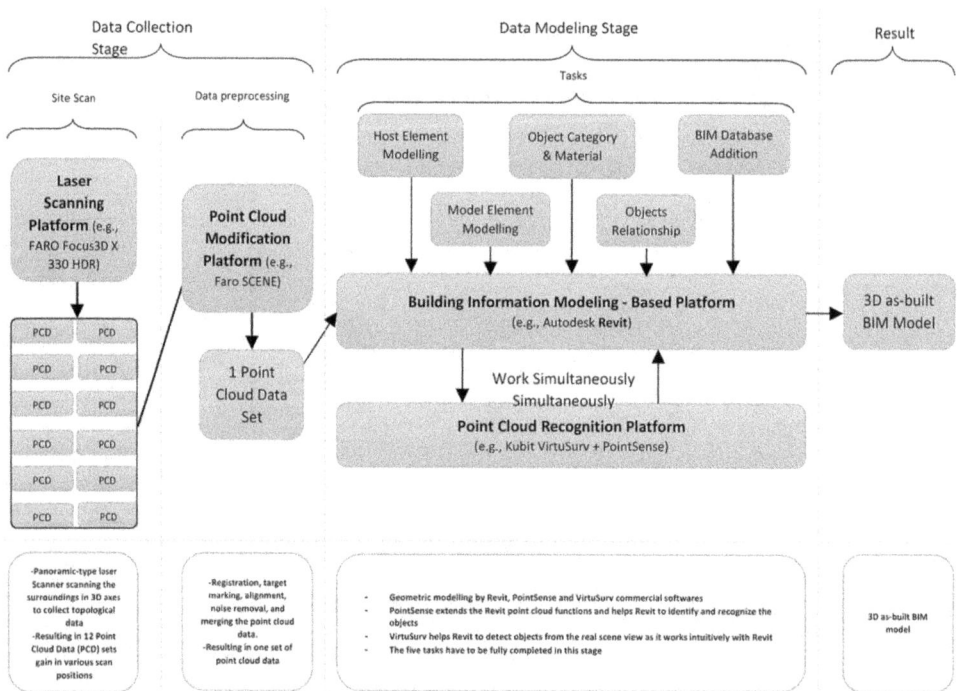

Figure 15.2 Workflow for 3D Laser Scanning of As-built BIM Model. From "As-built building information modeling workflow," by A. R. A. Usmani, A. Elshafey, M. Gheisari, C. S. Chai, E. B. Aminudin and C. S. Tan, 2019, *Journal of Engineering, Design and Technology, 18*(4), pp. 923–940, Copyright [2020] by Emerald Publishing Limited.

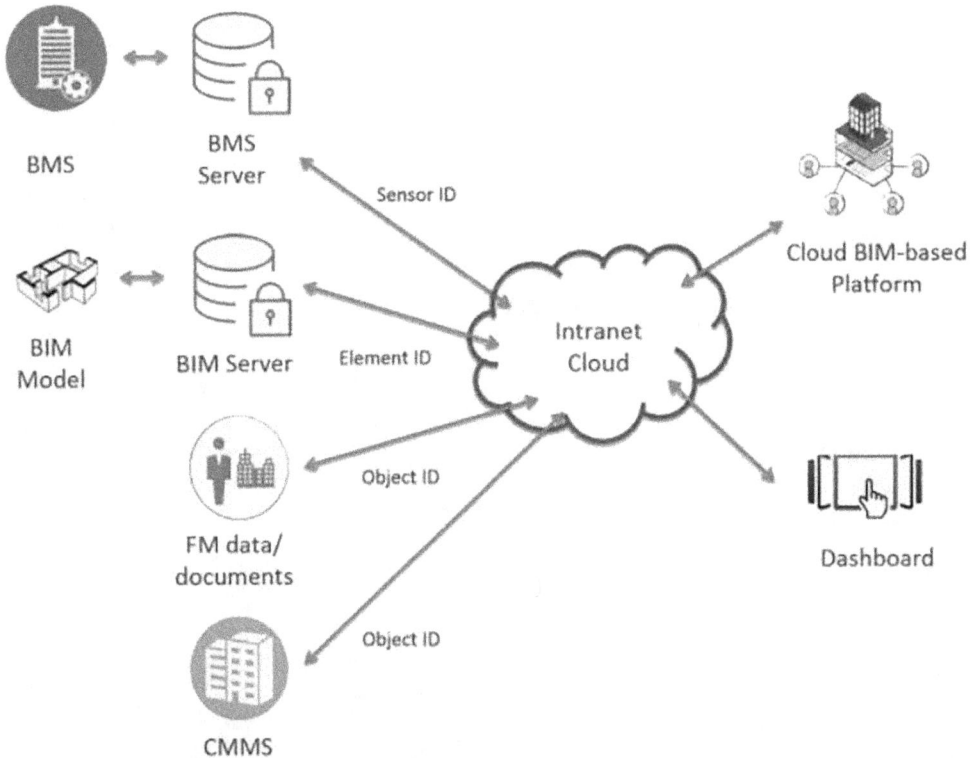

Figure 15.3 Workflow for Cloud-based BIM. From "Implementation of cloud BIM-based platform towards high-performance building services," by S. Chien, T. Chuang, H. Yu, Y. Han, B. Soong and K. J. Tseng, 2017, *Procedia Environmental Sciences, 38*, pp. 436–444, Copyright [2017] by Elsevier.

15.3.2 *Integration Workflow 2 – Cloud-Based BIM*

Shan, Chai, Gui, and Xiong (2021) have mentioned that Cloud-based BIM is the advanced technology of traditional BIM as the integration of BIM, Cloud computing, and IoT. It is mainly used to collaboratively manage construction projects by linking the entire project team cohesively. In short, cloud-based BIM allows all project participants to work on a shared platform without restrictions on location and time zone while retaining the inherent characteristics of a typical BIM platform.

15.3.3 *Integration Workflow 3 – Cloud-Based Digital Platform of Prefabrication and Modular Construction Visualization*

Ezzeddine and García de Soto (2021) have proposed a mixed method to enhance the flexibility of prefabrication and modular construction as a cloud-based digital platform for enhancing collaborative management and decision-making among all project participants. This digital platform allows access and sharing to visualise platform-based 3D models in a virtual environment during the entire project lifecycle. It involved prefabrication and modular construction, BIM, IoT, cloud computing, AR and VR.

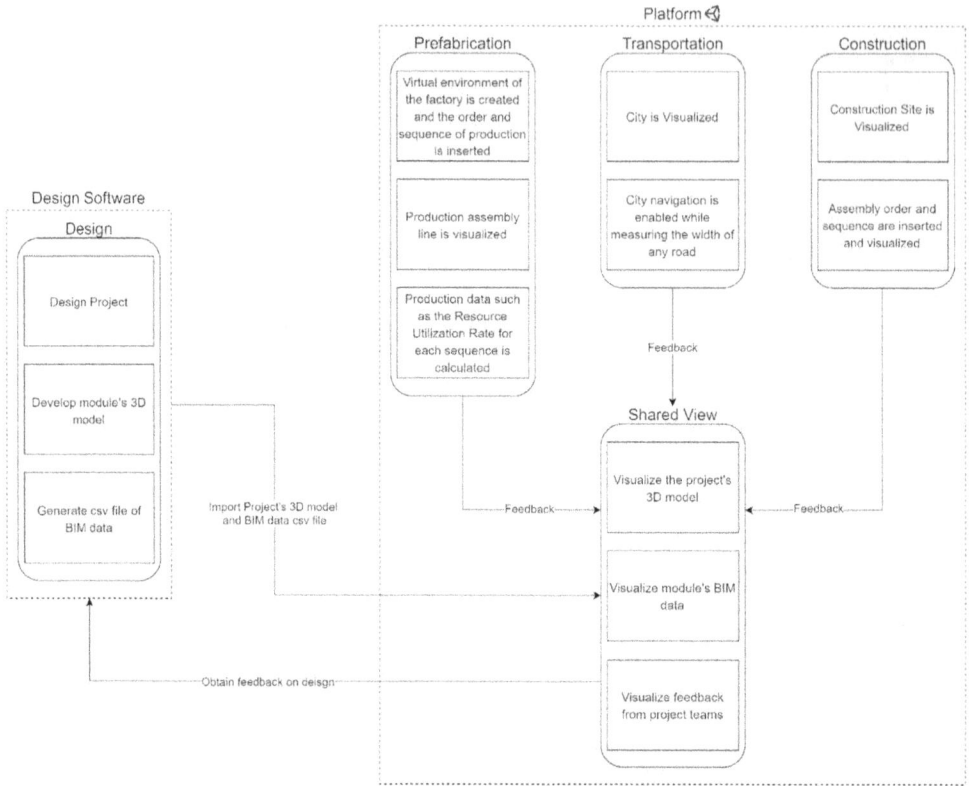

Figure 15.4 Workflow for cloud-based prefabrication and modular construction visualization. From "Connecting teams in modular construction projects using game engine technology," by A. Ezzeddine and B. Garcia de Soto, 2021, *Automation in Construction, 132*, p. 103887, Copyright [2021] by Elsevier.

15.3.4   *Integration Workflow 4 – AR and VR Based BIM*

Schiavi et al. (2022b) have integrated AR and VR technology into the BIM-based construction projects to provide better project collaboration among all project participants by visualising the 3D model. Apart from BIM and AR&VR, cloud computing and IoT also need to be integrated with this technique for the additional 3D content like the capability to process terrain or 3D point cloud data.

15.3.5   *Integration Workflow 5 – BIM-Based 3D Printing for Building Components*

As BIM has advanced traditional 3D modeling, it is now regarded as the necessary foundation for 3D printing strategies in the construction industry (Teizer et al., 2018). The application of BIM-based 3D printing brings many benefits to the construction industry, such as reducing project cost and duration and minimising environmental pollution and safety concerns (Hager, Golonka, & Putanowicz, 2016). Besides, the implementation of 3D printing can be done whether on-site or off-site, and the building materials are also one of the critical parts of 3D printing technology.

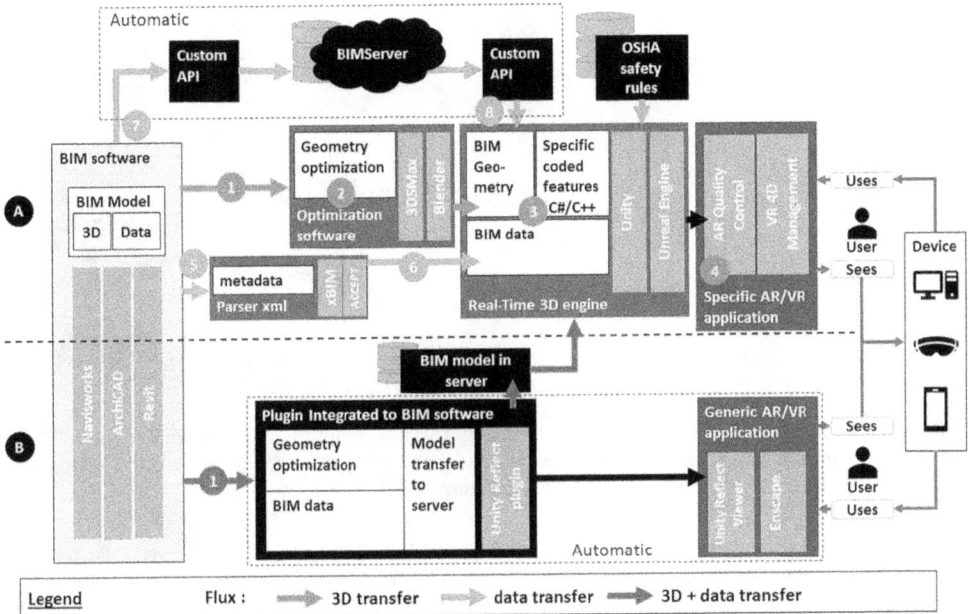

Figure 15.5 Workflow for AR&VR-based BIM and the commercial solutions. From "BIM data flow architecture with AR/VR technologies: Use cases in architecture, engineering and construction," by B. Schiavi, V. Havard, K. Beddiar and D. Baudry, 2022, *Automation in Construction, 134*, p. 104054, Copyright [2022] by Elsevier.

15.3.6 *Integration Workflow 6 – Large-scale 3D Printing*

A Chinese company triggered large-scale 3D printing, an advanced 3D printing technology mainly for constructing structures on large scales, heavy and permanent (Jassmi, Najjar, & Mourad, 2018). Industrial robots have instead of traditional 3D printers due to easier tasks in planning and robotic trajectory control through accessible software and scripting languages. Figure 15.6 shows the workflow of a large-scale 3D printing of a concrete structure, which consists of (0) system command of 3D models, (1) controller of robot, (2) printing controller, (3) robotic arm, (4) printhead, (5) concrete accelerating agent, (6) accelerating agent of peristaltic pump, (7) peristaltic pump for premix, (8) mixer of premix, and (9) 3D printed object (Figures 15.7 and 15.8).

15.3.7 *Integration Workflow 7 – Prefabrication and Modular Construction with BIM-Enable Assemble Robots*

Given the benefits of enhanced efficiency and reduced safety concerns associated with implementing automated robotics in construction projects, the integration of autonomous and prefabricated construction brings new developments to the construction industry (Chea, Bai, Pan, Arashpour, & Xie, 2020). Besides, BIM can be combined to increase the accuracy by collecting data, optimizing construction sequences, monitoring construction activities, and interacting with the robots (Figures 15.9 and 15.10).

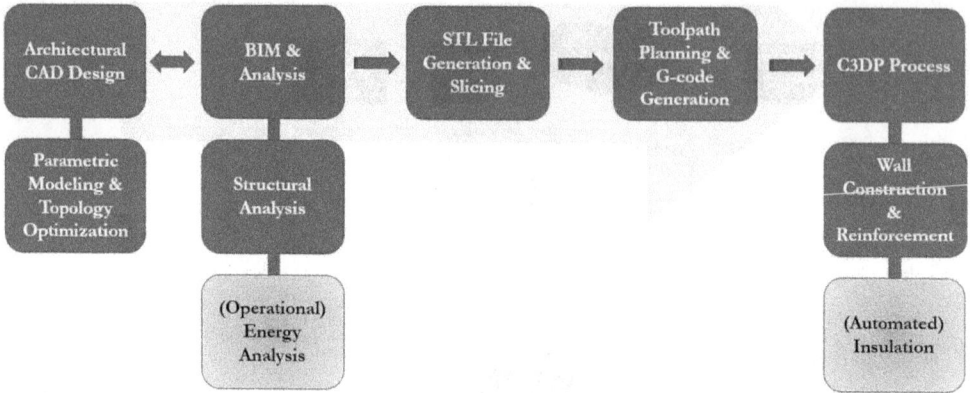

Figure 15.6 Workflow for BIM-based 3D printing for construction components. From "BIM-integrated thermal analysis and building energy modeling in 3D-printed residential buildings," by E. Kamel and A. Kazemian, 2023, *Energy and Buildings, 279*, p. 112670, Copyright [2023] by Elsevier.

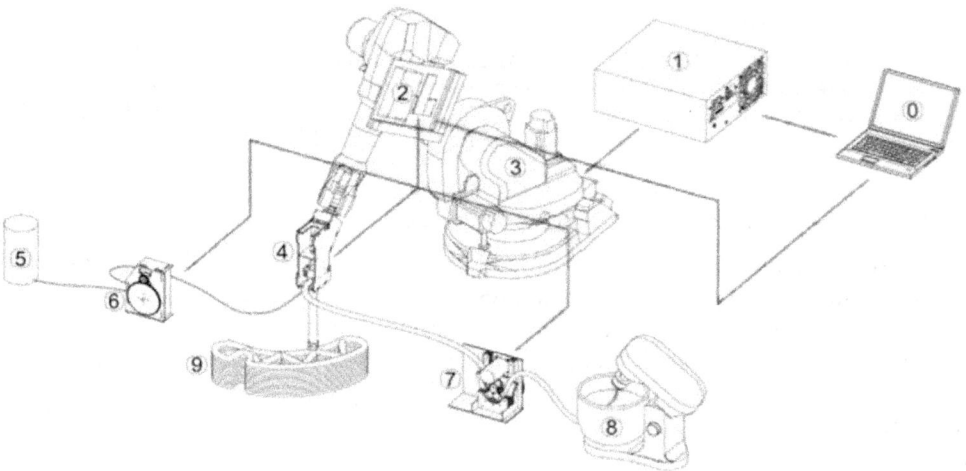

Figure 15.7 Workflow of Large-scale 3D Printing. From "Large-Scale 3D Printing: The Way Forward," by H. A. Jassmi, F. A. Najjar and A. H. I. Mourad, 2018, *IOP Conference Series: Materials Science and Engineering, 324*, p. 012088, Copyright [2018] by IOP Publishing Ltd.

15.3.8   *Integration Workflow 8 – Blockchain-based Document Management System for Construction Projects*

A smart contract is a self-executing digital agreement that can be embedded into the platform of the blockchain system to accomplish a collaborative data network with multiple parties (Ye, 2019). The blockchain-based document management system consists of a peer-to-peer blockchain and a cloud data network that connects project stakeholders

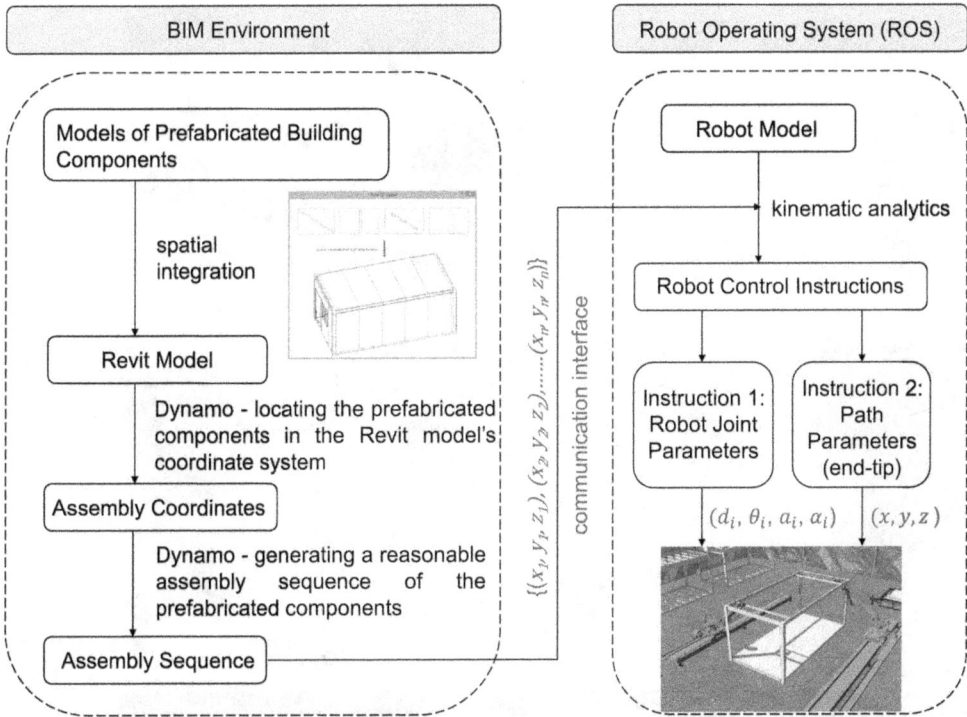

Figure 15.8 Workflow of prefabrication & modular construction with bim-enable assemble robots. From "BIM-based task and motion planning prototype for robotic assembly of COVID-19 hospitalisation light weight structures," by Y. Gao, J. Meng, J. Shu and Y. Liu, 2022, *Automation in Construction, 140*, p. 104370, Copyright [2020] by Elsevier.

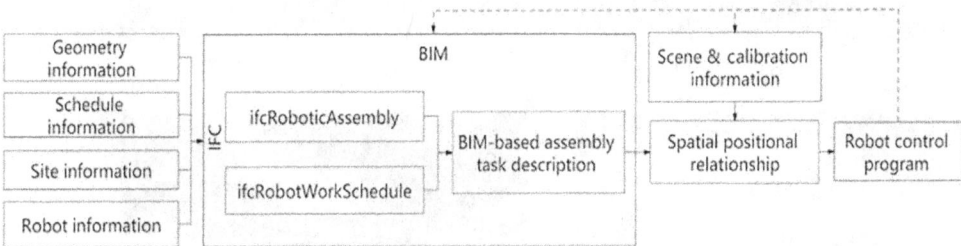

Figure 15.9 Information workflow of BIM-enable robotic assemble. From "Towards fully BIM-enabled building automation and robotics: A perspective of lifecycle information flow," by J. Zhang, H. Luo and J. X, 2022, *Computers in Industry, 135*, p. 103570, Copyright [2022] by Elsevier.

to the functionalities of document management, as shown in Figure 15.11. Besides, technologies like smart contracts and distributed blockchain ledgers are deployed to support document approval procedures, document lifecycle recording, and versioning (Das, Tao, Liu, & Cheng, 2022).

(a)

(b)

(c)

(d)

(e)

(f)

(g)

(h)

Figure 15.10 Examples of BIM-enable robotic assemble in the virtual construction site. From "BIM-based task and motion planning prototype for robotic assembly of COVID-19 hospitalisation light weight structures," by Y. Gao, J. Meng, J. Shu and Y. Liu, 2022, *Automation in Construction, 140*, p. 104370, Copyright [2020] by Elsevier.

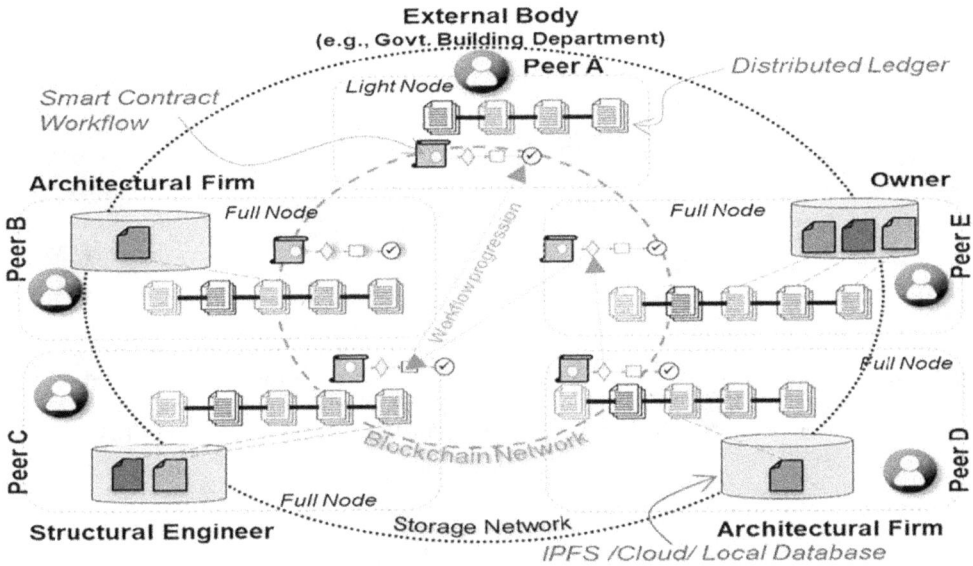

Figure 15.11 Workflow of blockchain-based document management system. From "A blockchain-based integrated document management framework for construction applications," by M. Das, X. Tao, Y. Liu and J. C. Cheng, 2022, *Automation in Construction, 133*, p. 104001, Copyright [2022] by Elsevier.

15.3.9 *Integration Workflow 9 – Autonomous Construction System*

As the implementation of automation and robotics in construction projects reduces the rates of injuries or fatalities and enhances project productivity, those technologies have generated a great interest in the construction field (Xu & Garcia de Soto, 2020). Moreover, the interest in AI is also increasing due to the project size and data size, which have tremendous growth, especially in the analysis, modelling, and building data prediction through Machine Learning (ML) (Seyedzadeh, Rahimian, Oliver, Glesk, & Kumar, 2020). The Autonomous Construction System is an advanced robotic system that combines large-scale 3D printing, prefabrication and modular construction with BIM-enabled assemble robots and AR and VR-based BIM. It realises an automated robotic construction system with high autonomy to conduct specific tasks automatically to replace human efforts and instructions. In other words, this system is to minimise human interactions during the entire construction phase by integrating technologies such as model vision techniques, cloud computing, and train-enable robots (Figures 15.12 and 15.13).

15.3.10 *Integration Workflow 10 – Information Management System for Construction Projects*

With the increasing expansion and complexity of construction projects, the interaction amount and exchange frequency of information among project participants are also increasing (Zhang, Wang, & Yuen, 2021). Nevertheless, fraudulent behaviour is one of the common challenges during digital construction information management, such

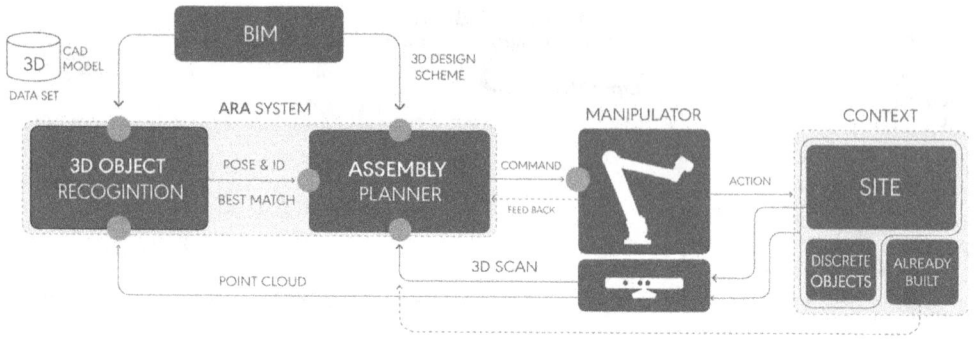

Figure 15.12 Workflow of autonomous construction system. From "BIM-assisted object recognition for the on-site autonomous robotic assembly of discrete structures," by M. Dawod and S. Hanna, 2019, *Construction Robotics, 3*, pp. 69–81, Copyright [2019] by Springer.

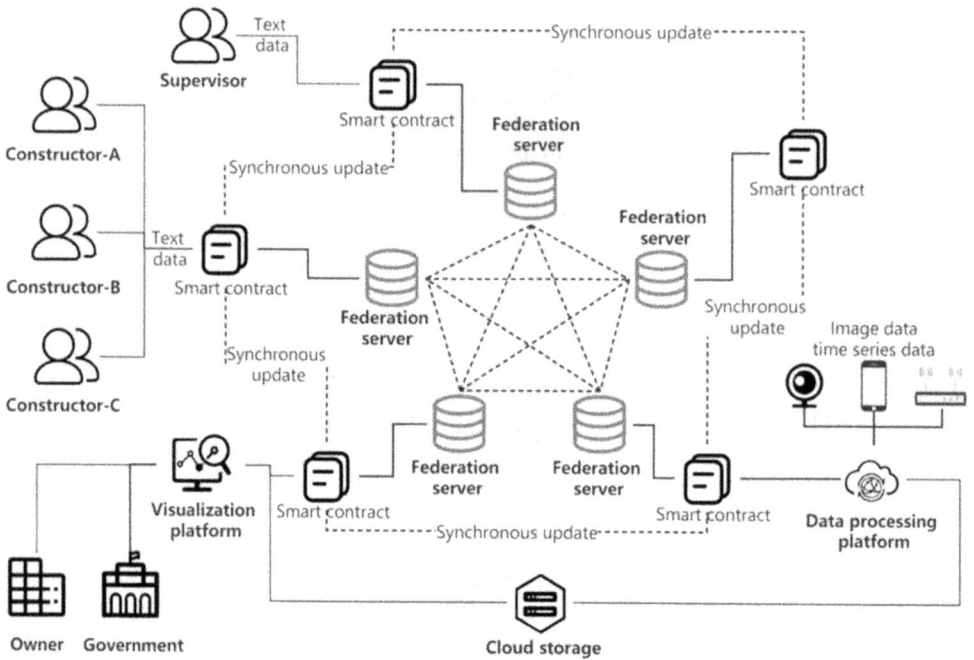

Figure 15.13 Workflow of information management system. From "Construction site information decentralized management using blockchain and smart contracts," by Y. Zhang, T. Wang and K. Yuen, 2020, *Computer-Aided Civil and Infrastructure Engineering, 37*(11), pp. 1450–1467, Copyright [2021] by Wiley.

as forged and modified data (Jessop, 2013). Hence, the information management system has come into being to enhance the security and traceability of construction information by integrating blockchain-based information management systems with cloud-based and AR&VR-based BIM.

15.4 Discussion

In this chapter, a conceptual system with twelve integrated emerging technologies is proposed. These 12 emerging technologies in the construction industry are specified by CIDB Malaysia, which are common and widely known technologies in the Malaysian construction industry. However, the adoption and utilization rate of those emerging technologies is meagre due to a lack of understanding and awareness. Besides, past studies have focused on developing and adopting a single or few technologies, yet, no study has tried to study the integration of all twelve technologies. Hence, this chapter's purpose is to propose a conceptual system to provide better solutions for the construction industry.

The proposed conceptual system, built based on the ANN decision-making model, consists of three display layers and two hidden layers. The first display layer is the input layer, which is composed of twelve emerging technologies. Clients are able to select multi-technologies they own from the first display layer. Then the system will automatically run to match the selected technologies to the integration workflows of possible construction-related products during the first hidden layer, which will be shown in the second display layer afterwards. The second hidden layer will only run if the selected technologies match the required conditions of advanced products. It aims to match the selected technologies to the integration workflow of advanced construction-related products; the matched product will be suggested in the third display layer. Besides, both the second and third display layers are output layers, meaning clients can make decisions based on the suggested products from both the second and third layers. Therefore, the suggestions given by the proposed conceptual system have provided an explicit direction for the future technology development of construction organizations, especially SMEs. For example, most Malaysian construction companies own some emerging technologies such as IoT, Cloud Computing, BIM and Blockchain-based smart contracts, but they are not aware that those technologies can be integrated into an information management system. The primary reason is that most construction organizations with construction-related products are unwilling to share the product's integration process since they treat it as a secret. Another reason is that most studies have focused on developing single technologies while not much study has been done on integrating different technologies. Hence, it is difficult for other organizations to figure out how to integrate their own technologies into a construction-related product, especially SMEs. Therefore, in the long run, the proposed conceptual system will improve project productivity and efficiency for all construction organizations, as it provides a more straightforward future guideline for integrating emerging technologies. In addition, it also lays the foundation for the future development of smart construction and smart cities by promoting digital twinning in the construction industry, and it has the potential to transform the traditional construction working process.

However, there are a few limitations to the proposed conceptual system. Firstly, the twelve emerging technologies of the construction industry are specified by the CIDB Malaysia based on the most common and well-known in the Malaysian construction industry, which does not mean that there are only twelve technologies. Secondly, the suggested products and their workflow are found based on the literature review, which means there are still some products that are not included. It is because not all product owners are willing to share their product and their integration workflow. Lastly, this conceptual system is proposed for the construction industry; hence, all suggested products are construction related. Yet, this system can also be adopted in other industries. The future study of this chapter will focus on investigating the connection weights between the specified technologies and the integrated products, as well as searching for more construction-related products.

References

Abioye, S. O., Oyedele, L. O., Akanbi, L., Ajayi, A., Delgado, J. M. D., Bilal, M., ...Ahmed, A. (2021). Artificial intelligence in the construction industry: A review of present status, opportunities and future challenges. *Journal of Building Engineering*, 44, 103299. 10.1016/j.jobe.2021. 103299

Al-Ashmori, Y. Y., Othman, I., Rahmawati, Y., Amran, Y. H. M., Sabah, S. H. A., Rafindadi, A. D. & Mikic, M. (2020). 'BIM benefits and its influence on the BIM implementation in Malaysia', *Ain Shams Engineering Journal*. 10.1016/j.asej.2020.02.002

Alawag, A. M., Alaloul, W. S., Liew, M. S., Musarat, M. A., Baarimah, A. O., Saad, S., & Ammad, S. (2022). Critical success factors influencing total quality management in industrialised building system: a case of Malaysian construction industry. *Ain Shams Engineering Journal*, 101877. 10.1016/j.asej.2022.101877

Ammar, M., Haleem, A., Javaid, M., Bahl, S., & Verma, A. S. (2021). Implementing Industry 4.0 technologies in self-healing materials and digitally managing the quality of manufacturing. *Materials Today: Proceedings*. 10.1016/j.matpr.2021.09.248

Anagnostou, T., Remzi, M., & Djavan, B. (2003). Artificial neural networks for decision-making in Urologic Oncology. *Reviews in Urology*, 5(1), 15–21. Retrieved from https://www.ncbi.nlm.nih.gov/pmc/articles/PMC1472995/

Ashima, R., Haleem, A., Javaid, M., & Rab, S. (2022). Understanding the role and capabilities of Internet of Things-enabled additive manufacturing through its application areas. *Advanced Industrial and Engineering Polymer Research*, 5(3), 137–142. 10.1016/j.aiepr.2021.12.001

Baduge, S. K., Thilakarathna, S., Perera, J. S., Arashpour, M., Sharafi, P., Teodosio, B., ...Mendis, P. (2022). Artificial intelligence and smart vision for building and construction 4.0: Machine and deep learning methods and applications. *Automation in Construction*, 141, 104440. 10.1016/j.autcon.2022.104440

Bag, S., Rahman, M. S., Srivastava, G., Chan, H. L., & Bryde, D. J. (2022). The role of big data and predictive analytics in developing a resilient supply chain network in the South African mining industry against extreme weather events. *International Journal of Production Economics*, 251, 108541. 10.1016/j.ijpe.2022.108541

Barazzetti, L., & Banfi, F. (2017). Historic BIM for mobile VR/AR applications. *Mixed Reality and Gamification for Cultural Heritage*, 271–290. 10.1007/978-3-319-49607-8_10

Beddiar, K., Grellier, C., & Woods, E. (2019). *Construction 4.0 - Réinventer le bâtiment grâce au numérique: BIM, DfMA, Lean Management*. (Hors Collection) (French Edition). Malakoff, France: Dunod.

Bello, S. A., Oyedele, L. O., Akinade, O. O., Bilal, M., Davila Delgado, J. M., Akanbi, L. A., ... Owolabi, H. A. (2021). Cloud computing in construction industry: Use cases, benefits and challenges. *Automation in Construction*, 122, 103441. 10.1016/j.autcon.2020.103441

Berawi, M. A., Sunardi, A., & Ichsan, M. (2019). Chief-screen 1.0 as the Internet of Things platform in project monitoring & controlling to improve project schedule performance. *Procedia Computer Science*, 161, 1249–1257. 10.1016/j.procs.2019.11.239

Berger, R. (2016). *Digitization in the construction industry: Building Europe's road to "Construction 4.0."* Munich, Germany: ROLAND BERGER GMBH.

Besklubova, S., Skibniewski, M. J., & Zhang, X. (2021). Factors affecting 3D printing technology adaptation in construction. *Journal of Construction Engineering and Management*, 147(5), 04021026. 10.1061/(asce)co.1943-7862.0002034

Bilal, M., Oyedele, L. O., Qadir, J., Munir, K., Ajayi, S. O., Akinade, O. O., ...Pasha, M. (2016). Big data in the construction industry: A review of present status, opportunities, and future trends. *Advanced Engineering Informatics*, 30(3), 500–521. 10.1016/j.aei.2016.07.001

Boyes, H., Hallaq, B., Cunningham, J., & Watson, T. (2018). The Industrial Internet of Things (IIoT): An analysis framework. *Computers in Industry*, 101, 1–12. 10.l016/j.compind. 2018.04.015

Carneau, P., Mesnil, R., Roussel, N., & Baverel, O. (2020). Additive manufacturing of cantilever – From masonry to concrete 3D printing. *Automation in Construction*, 116, 103184. 10.1016/j.autcon.2020.103184

Casini, M. (2022a). Advanced building construction methods. *Construction 4.0*, 405–470. 10.1016/b978-0-12-821797-9.00006-4

Casini, M. (2022b). Advanced construction materials. *Construction 4.0*, 337–404. 10.1016/b978-0-12-821797-9.00005-2

Charehzehi, A., Chai, C. S., Md Yusof, A., Chong, H. Y., & Loo, S. C. (2017). Building information modeling in construction conflict management. *International Journal of Engineering Business Management*, 9, 184797901774625. 10.1177/1847979017746257

Chea, C. P., Bai, Y., Pan, X., Arashpour, M., & Xie, Y. (2020). An integrated review of automation and robotic technologies for structural prefabrication and construction. *Transportation Safety and Environment*, 2(2), 81–96. 10.1093/tse/tdaa007

Chen, C., Ma, Y., Li, Q., Song, G., Pu, Y., & Chen, Z. (2016). Modularization for ground engineering construction in the Anyue Gasfield, Sichuan Basin, with the designed annual gas processing capacity of six billion m^3. *Natural Gas Industry B*, 3(6), 614–620. 10.1016/j.ngib. 2017.05.012

Chen, Chiang, & Storey. (2012). Business intelligence and analytics: From big data to big impact. *MIS Quarterly*, 36(4), 1165. 10.2307/41703503

Chen, H. M., Chang, K. C., & Lin, T. H. (2016). A cloud-based system framework for performing online viewing, storage, and analysis on big data of massive BIMs. *Automation in Construction*, 71, 34–48. 10.1016/j.autcon.2016.03.002

Dallasega, P., Revolti, A., Sauer, P. C., Schulze, F., & Rauch, E. (2020). BIM, augmented and virtual reality empowering lean construction management: A project simulation game. *Procedia Manufacturing*, 45, 49–54. 10.1016/j.promfg.2020.04.059

Das, M., Tao, X., Liu, Y., & Cheng, J. C. (2022). A blockchain-based integrated document management framework for construction applications. *Automation in Construction*, 133, 104001. 10.1016/j.autcon.2021.104001

Delgado, J. M. D., Oyedele, L., Demian, P., & Beach, T. (2020). A research agenda for augmented and virtual reality in architecture, engineering and construction. *Advanced Engineering Informatics*, 45, 101122. 10.1016/j.aei.2020.101122

Durdyev, S., Ashour, M., Connelly, S., & Mahdiyar, A. (2022). Barriers to the implementation of Building Information Modelling (BIM) for facility management. *Journal of Building Engineering*, 46, 103736. 10.1016/j.jobe.2021.103736

El-Omari, S., & Moselhi, O. (2008). Integrating 3D laser scanning and photogrammetry for progress measurement of construction work. *Automation in Construction*, 18(1), 1–9. 10.1016/j.autcon.2008.05.006

Elghaish, F., Hosseini, M. R., Matarneh, S., Talebi, S., Wu, S., Martek, I., Poshdar, M., & Ghodrati, N. (2021). Blockchain and the 'Internet of Things' for the construction industry: Research trends and opportunities. *Automation in Construction*, 132, 103942. 10.1016/j.autcon. 2021.103942

Elshafey, A., Chai, C. S., Aminudin, E. B., Gheisari, M., & Usmani, A. (2020). Technology acceptance model for Augmented Reality and Building Information Modeling integration in the construction industry. *Journal of Information Technology in Construction*, 25, 161–172. 10.36680/j.itcon.2020.010

Ezzeddine, A., & García de Soto, B. (2021). Connecting teams in modular construction projects using game engine technology. *Automation in Construction*, 132, 103887. 10.1016/j.autcon.2021. 103887

Farahbakhsh, M., Rybkowsk, Z. K., Zakira, U., Kalantar, N., & Onifade, I. (2022). Impact of robotic 3D printing process parameters on interlayer bond strength. *Automation in Construction*, 142, 104478. 10.1016/j.autcon.2022.104478

Forcael, E., Ferrari, I., Opazo-Vega, A., & Pulido-Arcas, J. A. (2020). Construction 4.0: A literature review. *Sustainability*, 12(22), 9755. 10.3390/su12229755

Francis, A., & Thomas, A. (2020). Exploring the relationship between lean construction and environmental sustainability: A review of existing literature to decipher broader dimensions. *Journal of Cleaner Production*, 252, 119913. 10.1016/j.jclepro.2019.119913

García De Soto, B., Agustí-Juan, I., Hunhevicz, J., Joss, S., Graser, K., Habert, G., & Adey, B. T. (2018). Productivity of digital fabrication in construction: Cost and time analysis of a robotically built wall. *Automation in Construction*, 92, 297–311. 10.1016/j.autcon.2018.04.004

Gharbia, M., Chang-Richards, A., Lu, Y., Zhong, R. Y., & Li, H. (2020). Robotic technologies for on-site building construction: A systematic review. *Journal of Building Engineering*, 32, 101584. 10.1016/j.jobe.2020.101584

Ghobakhloo, M. (2020). Industry 4.0, digitization, and opportunities for sustainability. *Journal of Cleaner Production*, 252, 119869. 10.1016/j.jclepro.2019.119869

Grover, V., Chiang, R. H., Liang, T. P., & Zhang, D. (2018). Creating strategic business value from big data analytics: a research framework. *Journal of Management Information Systems*, 35(2), 388–423. 10.1080/07421222.2018.1451951

Hager, I., Golonka, A., & Putanowicz, R. (2016). 3D printing of buildings and building components as the future of sustainable construction? *Procedia Engineering*, 151, 292–299. 10.1016/j.proeng.2016.07.357

Harikrishnan, A., Abdallah, A. S., Ayer, S. K., Asmar, M. E., & Tang, P. (2021). Feasibility of augmented reality technology for communication in the construction industry. *Advanced Engineering Informatics*, 50, 101363. 10.1016/j.aei.2021.101363

Hill, T., Marquez, L., O'Connor, M., & Remus, W. (1994). Artificial neural network models for forecasting and decision making. *International Journal of Forecasting*, 10(1), 5–15. 10.1016/0169-2070(94)90045-0

Hosseini, M. R., Jupp, J., Papadonikolaki, E., Mumford, T., Joske, W., & Nikmehr, B. (2020). Position paper: digital engineering and building information modelling in Australia. *Smart and Sustainable Built Environment*, 10(3), 331–344. 10.1108/sasbe-10-2020-0154

IKTISAS ENVIRONMENT SDN BHD. (2019). FOCUS GROUP DISCUSSION (FGD) for the Proposed Development of Solid Waste Transfer Station, on 12.474 Acres of Land on Lots 1336 & 1337, Pekan Nenas, Mukim Jeram Batu, Daerah Pontian, Johor Darul Takzim for Jabatan Pengurusan Sisa Pepejal Negara (JPSPN). https://enviro2.doe.gov.my/ekmc/wp-content/uploads/2019/06/30.-APPENDIX-6.2-FGD-Final-Report.pdf

International Organization for Standardization. (2013). Standard terminology for additive manufacturing — Coordinate systems and test methodologies (ISO Standard No. 52921). Retrieved from https://www.iso.org/standard/62794.html

Jassmi, H. A., Najjar, F. A., & Mourad, A. H. I. (2018). Large-scale 3D printing: The way forward. *IOP Conference Series: Materials Science and Engineering*, 324, 012088. 10.1088/1757-899x/324/1/012088

Jessop, T. (2013). Fundamentals of civil engineering: An introduction to the ASCE body of knowledge. *Proceedings of the Institution of Civil Engineers*, 166(1), 14–15. 10.1680/cien.2013.166.1.14

Kamali, M., Hewage, K., & Sadiq, R. (2019). Conventional versus modular construction methods: A comparative cradle-to-gate LCA for residential buildings. *Energy and Buildings*, 204, 109479. 10.1016/j.enbuild.2019.109479

Kamaruddeen, A. M., Rui, L. S., Lee, K. L., Alzoubi, H. M., & Alshurideh, M. T. (2022). Determinants of emerging technology adoption for safety among construction businesses. *Academy of Strategic Management Journal*, (4), 1–20. Retrieved from https://www.researchgate. net/publication/357699458_Determinants_of_Emerging_Technology_Adoption_for_Safety_ Among_Construction_Businesses

Kassem, M., Li, J., Kumar, B., Malleson, A., Gibbs, D., Kelly, G., & Watson, R. (2020). Building information modelling: Evaluating tools for maturity and benefits measurement. https://www. cdbb.cam.ac.uk/files/bim_evaluating_tools_for_maturity_and_benefits_measurement_report.pdf

Krupík, P. (2021). Advanced building materials – Prefabricated and modular designs as part of Construction 4.0 with a focus on transport constructions. *IOP Conference Series: Materials Science and Engineering*, 1209(1), 012007. 10.1088/1757-899x/1209/1/012007

Lee, J., Stanley, M., Spanias, A., & Tepedelenlioglu, C. (2016). Integrating machine learning in embedded sensor systems for Internet-of-Things applications. 2016 IEEE International Symposium on Signal Processing and Information Technology (ISSPIT). 10.1109/isspit.2016. 7886051

Li, X., Yi, W., Chi, H. L., Wang, X., & Chan, A. P. (2018). A critical review of virtual and augmented reality (VR/AR) applications in construction safety. *Automation in Construction*, 86, 150–162. 10.1016/j.autcon.2017.11.003

Lu, W., Chen, K., Xue, F., & Pan, W. (2018). Searching for an optimal level of prefabrication in construction: An analytical framework. *Journal of Cleaner Production*, 201, 236–245. 10.1016/ j.jclepro.2018.07.319

Malik, P. K., Sharma, R., Singh, R., Gehlot, A., Satapathy, S. C., Alnumay, W. S., ... Nayak, J. (2021b). Industrial Internet of Things and its applications in industry 4.0: state of the art. *Computer Communications*, 166, 125–139. 10.1016/j.comcom.2020.11.016

Matthews, J., Love, P. E., Heinemann, S., Chandler, R., Rumsey, C., & Olatunj, O. (2015). Real time progress management: Re-engineering processes for cloud-based BIM in construction. *Automation in Construction*, 58, 38–47. 10.1016/j.autcon.2015.07.004

Melenbrink, N., Rinderspacher, K., Menges, A., & Werfel, J. (2020). Autonomous anchoring for robotic construction. *Automation in Construction*, 120, 103391. 10.1016/j.autcon.2020.103391

Mohsen, A., Alaloul, W. S., Liew, M., Musarat, M. A., Baarimah, A. O., Alzubi, K. M., & Altaf, M. (2021). Impact of the COVID-19 pandemic on construction industry in Malaysia. 2021 Third International Sustainability and Resilience Conference: Climate Change. 10.1109/ ieeeconf53624.2021.9667984

Munoz-Morera, J., Maza, I., Fernandez-Aguera, C. J., Caballero, F., & Ollero, A. (2015). Assembly planning for the construction of structures with multiple UAS equipped with robotic arms. 2015 International Conference on Unmanned Aircraft Systems (ICUAS). 10.1109/icuas.2 015.7152396

Ngo, J., Hwang, B. G., & Zhang, C. (2020). Factor-based big data and predictive analytics capability assessment tool for the construction industry. *Automation in Construction*, 110, 103042. 10.1016/ j.autcon.2019.103042

Olsson, N. O., Shafqat, A., Arica, E., & Økland, A. (2019). 3D-printing technology in construction: Results from a survey. 10th Nordic Conference on Construction Economics and Organization, 349–356. 10.1108/s2516-285320190000002044

Olsson, N. O., Arica, E., Woods, R., & Madrid, J. A. (2021). Industry 4.0 in a project context: Introducing 3D printing in construction projects. *Project Leadership and Society*, 2, 100033. 10.1016/j.plas.2021.100033

Pacheco-Torgal, F., Jalali, S., Labrincha, J., & John, V. M. (2013). *Eco-Efficient Concrete*. Maarssen, Netherlands: Elsevier Gezondheidszorg.

Pan, Y., & Zhang, L. (2021). Roles of artificial intelligence in construction engineering and management: A critical review and future trends. *Automation in Construction*, 122, 103517. 10.1016/ j.autcon.2020.103517

Petrillo, A., Felice, F. D., Cioffi, R., & Zomparelli, F. (2018). Fourth industrial revolution: Current practices, challenges, and opportunities. *Digital Transformation in Smart Manufacturing.* 10.5772/intechopen.72304

Qi, B., Razkenari, M., Costin, A., Kibert, C., & Fu, M. (2021). A systematic review of emerging technologies in industrialized construction. *Journal of Building Engineering*, 39, 102265. 10.1016/j.jobe.2021.102265

Ratajczak, J., Riedl, M., & Matt, D. (2019). BIM-based and AR application combined with location-based management system for the improvement of the construction performance. *Buildings*, 9(5), 118. 10.3390/buildings9050118

Sadeghineko, F., & Kumar, B. (2021). Application of semantic Web ontologies for the improvement of information exchange in existing buildings. *Construction Innovation*, 22(3), 444–464. 10.1108/ci-03-2021-0058

Schiavi, B., Havard, V., Beddiar, K., & Baudry, D. (2022b). BIM data flow architecture with AR/VR technologies: Use cases in architecture, engineering and construction. *Automation in Construction*, 134, 104054. 10.1016/j.autcon.2021.104054

Scott, D. J., Broyd, T., & Ma, L. (2021). Exploratory literature review of blockchain in the construction industry. *Automation in Construction*, 132, 103914. 10.1016/j.autcon.2021.103914

Seyedzadeh, S., Rahimian, F. P., Oliver, S., Glesk, I., & Kumar, B. (2020). Data driven model improved by multi-objective optimisation for prediction of building energy loads. *Automation in Construction*, 116, 103188. 10.1016/j.autcon.2020.103188

Shan, J., Chai, C. S., Gui, H., & Xiong, Y. (2021). Enhancing project integration using cloud-based building information modelling: A conceptual model. *International Journal of Advanced Research in Technology and Innovation*, 3(3), 12–25. Retrieved from http://myjms.mohe.gov.my/index.php/ijarti

Tam, N. V., Toan, N. Q., Phong, V. V., & Durdyev, S. (2021). Impact of BIM-related factors affecting construction project performance. *International Journal of Building Pathology and Adaptation.* 10.1108/ijbpa-05-2021-0068

Teizer, J., Blickle, A., King, T., Leitzbach, O., Guenther, D., Mattern, H., & König, M. (2018). BIM for 3D Printing in Construction. *Building Information Modeling*, 421–446. 10.1007/978-3-319-92862-3_26

Ulvi, A. (2021). Documentation, three-dimensional (3D) modelling and visualization of cultural heritage by using Unmanned Aerial Vehicle (UAV) photogrammetry and terrestrial laser scanners. *International Journal of Remote Sensing*, 42(6), 1994–2021. 10.1080/01431161.2020.1834164

Usmani, A. R. A., Elshafey, A., Gheisari, M., Chai, C., Aminudin, E. B., & Tan, C. S. (2020). A scan to as-built building information modeling workflow: A case study in Malaysia. *Journal of Engineering, Design and Technology*, 18(4), 923–940. 10.1108/jedt-07-2019-0182

Vandecasteele, F., Merci, B., & Verstockt, S. (2017). Fireground location understanding by semantic linking of visual objects and building information models. *Fire Safety Journal*, 91, 1026–1034. 10.1016/j.firesaf.2017.03.083

Vujović, M., Rodić, A., & Stevanović, I. (2017). Design of modular re-configurable robotic system for construction and digital fabrication. *Advances in Intelligent Systems and Computing*, 550–559. 10.1007/978-3-319-49058-8_60

Wang, G., Gunasekaran, A., Ngai, E. W., & Papadopoulos, T. (2016). Big data analytics in logistics and supply chain management: Certain investigations for research and applications. *International Journal of Production Economics*, 176, 98–110. 10.1016/j.ijpe.2016.03.014

Xu, X., & Garcia de Soto, B. (2020). On-site autonomous construction robots: A review of research areas, technologies, and suggestions for advancement. *The International Association for Automation and Robotics in Construction.* 10.22260/isarc2020/0055

Yap, J. B. H., Lam, C. G. Y., Skitmore, M., & Talebian, N. (2022). Barriers to the adoption of new safety technologies in construction: A developing country context. *Journal of Civil Engineering and Management*, 28(2), 120–133. 10.3846/jcem.2022.16014

Ye, X. (2019). *Combining BIM with smart contract and blockchain to support digital project delivery and acceptance processes (MA Thesis)*. Ruhr-Universität Bochum.

You, Z., & Feng, L. (2020). Integration of Industry 4.0 related technologies in construction industry: A framework of cyber-physical system. *IEEE Access*, 8, 122908–122922. 10.1109/access.2020.3007206

Yu, B., & Fingrut, A. (2022). Sustainable building design (SBD) with reclaimed wood library constructed in collaboration with 3D scanning technology in the UK. *Resources, Conservation and Recycling*, 186, 106566. 10.1016/j.resconrec.2022.106566

Zhang, J., Luo, H., & Xu, J. (2022). Towards fully BIM-enabled building automation and robotics: A perspective of lifecycle information flow. *Computers in Industry*, 135, 103570. 10.1016/j.compind.2021.103570

Zhang, Y., Wang, T., & Yuen, K. (2021). Construction site information decentralized management using blockchain and smart contracts. *Computer-Aided Civil and Infrastructure Engineering*, 37(11), 1450–1467. 10.1111/mice.12804

Zhao, Y., & Taib, N. (2022). Cloud-based building information modelling (Cloud-BIM): Systematic literature review and Bibliometric-qualitative Analysis. *Automation in Construction*, 142, 104468. 10.1016/j.autcon.2022.104468

16 Electromyography-Based Action Recognition of Construction Workers Using and Not Using Wearable Robots

Nihar James Gonsalves[1], *Omobolanle Ogunseiju*[2], *and Abiola Abosede Akanmu*[3]

[1]*Myers-Lawson School of Construction, Virginia Tech, Blacksburg, VA, USA;* [2]*School of Building of Construction, College of Design, Georgia Tech, Atlanta, Georgia, USA;* [3]*Department of Smart Design and Construction, Myers Lawson School of Construction, Virginia Tech, Blacksburg, VA, USA*

16.1 Introduction

Construction projects are labour-intensive, often involving tasks that are physically demanding and repetitive. As a result, construction workers suffer from low productivity and are exposed to safety and health risks (Sherafat et al., 2020). Automated recognition of workers' actions is increasingly recognized as one of the requirements for assessing these risks. Information obtained from the assessment improves awareness of project stakeholders and enhances decision-making. Existing methods for automated recognition of workers' actions and activities can be categorized into vision and wearable sensor-based methods. Vision-based methods infer information about workers' activities from image or video data using computer vision techniques. Such data obtained from cameras, are sensitive to environmental factors and could be computationally expensive to analyze. Wearable sensor-based methods include kinematic (e.g., inertial measurement units, gyroscopes, and accelerometers) and surface electromyography (sEMG) sensors. Previous studies have utilized inertial measurement units (IMU) and smartphones attached to the wrist, arm, waist, and thigh to capture acceleration and angular velocities of workers performing a variety of activities e.g., sawing, hammering, carpentry, bricklaying, and drilling (Akhavian and Behzadan, 2016; Joshua and Varghese, 2014; Joshua and Varghese, 2011; Antwi-Afari et al., 2020). These studies employed supervised and unsupervised machine learning techniques to recognize workers' actions from kinematic data. While good recognition accuracies have been achieved from kinematic data, their potential for inferring safety and health risks is limited (Mudiyanselage et al., 2021; Nouredanesh and Tung, 2019; Bangaru et al., 2021; Diaz, 2020).

sEMG sensors measure the electrical activity of muscles, which can be used to understand movements and investigate triggers of health conditions e.g., muscle disorders and fatigue (Lariviere et al., 2002). Data from sEMG sensors attached to different body parts (e.g., hands, legs, and trunk) have also been used for recognizing activities (Wang et al., 2021; Young et al., 2014). A key motivation for utilizing sEMG data for activity recognition is that specific patterns of electromyographic activity are observed in the muscles when certain body parts are moved (Friedli et al., 1988; Hodges and Richardson, 1996; Zedka and Prochazka, 1997). These patterns have been found to correlate with the direction and extent of movement of the associated body part (Hodges et al., 1999). Sánchez-Zuriaga et al. (2009) provided evidence of activations in

DOI: 10.1201/9781003408949-16

the erector spinae, one of the trunk muscles, resulting from repetitive movement of the hands. Located near sacrum, the erector spinae muscles are a set of long muscles that run vertically up the length of the back. The erector spinae muscles help to rotate the body from side to side and straighten the back. Back spasms and pain could result from an injury or strain to these muscles (Sung et al., 2009). During tasks involving bending and lifting, the erector spinae muscles generate large extending activations to help raise materials and keep the upper body in an upright position (Dolan and Adams, 1993). Despite this evidence, there are few studies exploring the potential of sEMG sensors for classifying workers' actions and activities. Since the hand and arm are key body parts for performing construction tasks and construction tasks involve significant bending and lifting, this study hypothesizes that when construction workers perform work, specific activation patterns are triggered in the erector spinae muscles, which can facilitate detection and classification of construction activities. However, the extent to which these patterns can be distinguished by machine learning algorithms in order to recognize workers' actions is unknown.

According to the United States Bureau of Labor and Statistics (BLS), the rate of non-fatal injuries (i.e. work-related musculoskeletal disorders) among construction workers is about twice the rate in other industries (BLS, 2020). This has triggered increasing awareness and interest in the adoption of wearable robots and postural assist devices (i.e. exoskeletons) for preventing work-related musculoskeletal disorders (WMSDs) amongst construction workers. Kim et al. (2019) reported that the use of an exoskeleton for construction work could improve workforce productivity and safety. In a more recent study, Antwi-Afari et al. (2021) assessed a passive exoskeleton during manual material handling tasks and reported reduced discomfort and muscle activity at the back and improved work performance. Ogunseiju et al. (2021) also investigated a postural assist exoskeleton for manual material handling tasks. The use of the wearable device reduced range of motion and discomfort at the back and also improved task performance. However, studies (Madinei et al., 2021; Gonsalves et al., 2021) have shown that exoskeletons have unintended consequences such as restricting movements. Workers would need to alter their regular work postures to perform work. This makes activity or action recognition models developed for workers not using exoskeletons unsuitable for workers using exoskeletons. In other words, it is possible that the relationship between the activation patterns in the erector spinae muscles and the hand movements will differ for workers wearing and not wearing exoskeletons.

The objective of this research was to investigate automatic recognition of construction workers' activities from activations of the erector spinae muscles measured with a wearable sEMG. To achieve this, an experimental study was performed to collect activations of the erector spinae muscles during a simulated rebar task performed with and without a wearable robot. Rebar task is chosen given that it exposes workers to multiple ergonomic risk factors such as prolonged static and awkward posture (e.g., squatting and back bending), and repetitive movement. Thus, making rebar workers good candidates for exoskeleton use. Supervised machine learning is performed on the muscle activity data to recognize the subtasks involved in rebar work. The main contributions of this research are to (1) propose an automated approach for recognizing construction workers' activities based on muscle activation data; and (2) determinate the extent to which activations of the erector spinae muscles can distinguish actions of workers performing construction work without and with wearable exoskeletons.

16.2   Literature Review

16.2.1   Wearable Robots in the Construction Industry

In the construction industry, workers are often subjected to awkward work postures involving twisting, bending, kneeling, reaching out, lifting, hauling, and squatting (Zhu et al., 2021). These work postures are physically demanding and repetitive which exposes the workers to work-related musculoskeletal disorders (WMSDs) (Sherafat et al., 2020). In the past, various solutions have been explored to address the problem of WMSDs such as workforce training (Albers, 2007), utilization of ergonomic principles for safety manuals (Shoubi et al., 2013), mechanical assistive devices (Akanmu et al., 2020), and real-time monitoring systems (Antwi-Afari et al., 2018), but the effectiveness of these methods is contingent on the workers. Likewise, while real-time warning systems could lead to distractions and possible fatalities on the jobsite, the use of powered assist devices has limited automation, and often requires the physical involvement of construction workers.

Thus, with advancements in wearable technology, there is a growing interest in the potentials of wearable robots such as exoskeleton to protect workers from back-related WMSDs. Wearable robots (exoskeletons) are defined by (Pons, 2008) as 'robotic systems that a person wears to enhance his/her capabilities in some ways'. Wearable systems got recognition from military sectors over a quarter of a century ago (Mudie et al., 2021) and recently have been actively used for rehabilitation purposes to assist recovering human beings (Jarrassé et al., 2014). Considering the promising results of exoskeletons in the military, industrial and health sectors, there is growing interest in exoskeletons for addressing the problem of WMSDs in the construction industry (Kermavnar et al., 2021; Zhu et al., 2021). Generally, exoskeletons are classified based on power systems such as active and passive. The active systems use electrical means to enhance the capabilities of the workers while passive systems utilize non-electrical means (Lowe et al., 2019). Even though passive wearable systems lack the enhancing capabilities of active exoskeletons, they are lighter in weight and less expensive. Bearing in mind the size of the construction industry, the adoption of passive wearable robots appears to be rationale. Although passive exoskeletons can be used for back, shoulder, leg, and full-body support (Zhu et al., 2021), the high occurrences of back disorders amongst construction workers make it imperative to investigate the potentials of back support exoskeletons.

16.2.2   Action Recognition in Construction

Traditionally, action recognition in construction involved manual observations which are time-consuming, costly, and error-prone. With advancements in technology, automated solutions are increasingly receiving traction. The technology-based action recognition techniques can broadly be classified into vision-based and kinematic methods. In the construction industry, extant studies have explored vision-based methods for action recognition. For example, Luo et al. (2018b) utilized site surveillance videos to identify construction workers' continuous and diverse activities using convolutional networks with an accuracy of 80.5%. A wireless real-time video monitoring system was employed by Bai et al. (2020) to determine the efficiency of the workforce for bridge construction. Using artificial neural network (ANN) human pose analysis was conducted to classify workers' actions into effective work, contributory work, and ineffective work which attained an accuracy of up to 81%. Luo et al. (2019) achieved 84% accuracy for workers' activity recognition using far-field surveillance videos by employing deep learning-based hierarchical method. A real-world data set with 11 common actions across five different construction crews was created by Yang et al. (2016), and action recognition was carried out by employing the dense trajectory method. Yang et al. (2016) further employed support vector machine for data classification and reported an average accuracy of 59%. Likewise, Luo et al. (2018a) used computer vision for action recognition of rebar workers by employing a convolutional neural network. Even though vision-based methods can provide successful results and reliable documentation, limitations poses barrier to widespread use. For example, action recognition with vision-based methods requires the installation of multiple cameras which comes with initial cost, requires constant line-of-sight, and is prone to environmental factors.

These barriers made several authors explore the potentials of kinematic methods involving wearable sensors for action recognition of workers and equipment. Wearable kinetic sensors like the Inertial Measurement Unit (IMU) consist of sensors such as accelerometer and gyroscope that have been found effective in identifying human postures. For example, Joshua and Varghese (2011) used accelerometer to detect workers' actions for masonry work. Using IMU sensors, Conforti et al. (2020) carried out automated posture detection and ergonomic evaluation for lifting tasks and by employing support vector machine classifier, attained an accuracy of 99.4% and 76.9% respectively. Likewise, recent studies have identified smartphones as wearable technology for understanding construction workers' activity. For example, Akhavian and Behzadan (2016) used smartphones to measure diverse activities performed by construction workers such as sawing, hammering, wrench turning, pushing, and dumping loaded wheelbarrows and trained five machine learning algorithms. The results indicate that neural network outperformed other classifiers by yielding an accuracy of up to 97%. Hence, with these promising results, and the low cost, high accuracy, and user-friendly nature of kinetic sensors, it is viable to explore kinematic methods for construction action recognition.

16.3 Methodology

The method adopted in this study is shown in Figure 16.1. Muscle activity data was collected from participants performing simulated rebar tasks with and without an exoskeleton using a wearable sensor. Collected muscle activity data were segmented and several features were calculated. Each segment was labeled based on the corresponding

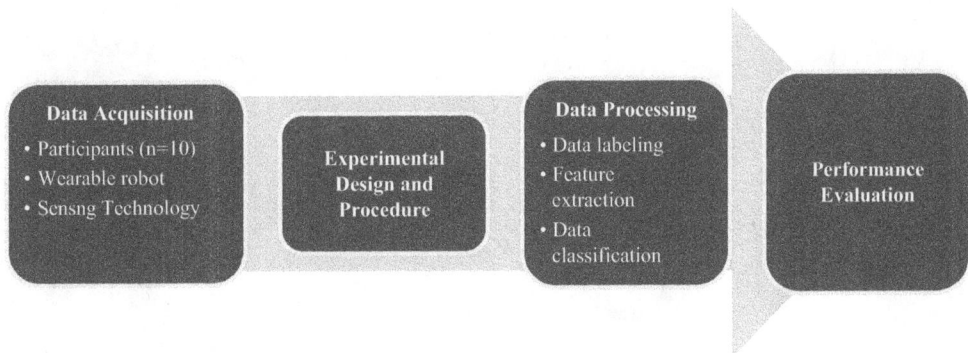

Figure 16.1 Overview of methodology.

rebar subtasks. Nine supervised machine learning classifiers were trained to detect and classify subtasks involved in rebar work performed in the laboratory. All data processing was performed using MS Office Excel 2021 and MATLAB Release 2021a (Matlab, The MathWorks Inc., MA, USA).

16.3.1   Data Acquisition

16.3.1.1   Participants

The study adopted a convenience sample size of ten participants from XXX. All the participants were male with no prior experience of the task and signed the informed consent form approved by Virginia Tech Institutional Review Board (IRB). None of the participants reported any health issues or prior musculoskeletal disorders which could impact their task performance. The demographics of the students in mean and standard deviation is age = 23yrs ± 1.99, weight = 155.70 lbs. ± 22.51 and height = 173.40 cm ± 4.97.

16.3.1.2   Wearable Robot

A passive back support exoskeleton, BackX™ S from SuitX industries (https://www.suitx.com/backX), was employed in this study. BackX is designed to reduce the stress from the workers' lower back during forward bending, squatting, lifting, and stooping tasks. The exoskeleton weighs 3.4 kg and comprises a harness and a frame that provides up to 13.6 kg of support to the lower back. The harness consists of straps (shoulder, chest, and leg), a chest pad, and a hip belt whereas the frame has a chest plate, thigh pad, and a torque generator which acts as the actuation point for the exoskeleton. The frame connects with the harness via the hip support as shown in Figure 16.2.

16.3.1.3   Sensing Technology

A wearable device called Cricket (Figure 16.3a), developed by Somaxis Inc. (https://www.somaxis.com/), was used to collect EMG data from the erector spinae muscle group

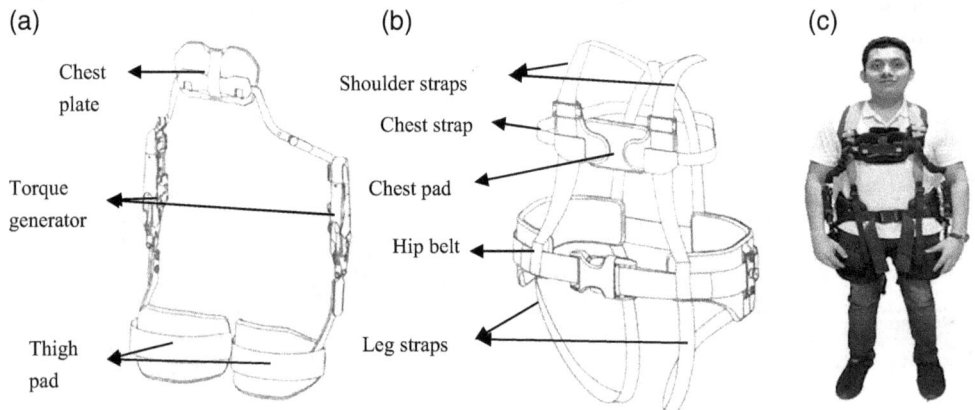

Figure 16.2 BackX exoskeleton (a) frame (left), (b) harness (middle), and (c) complete exoskeleton (right).

(a)

(b)

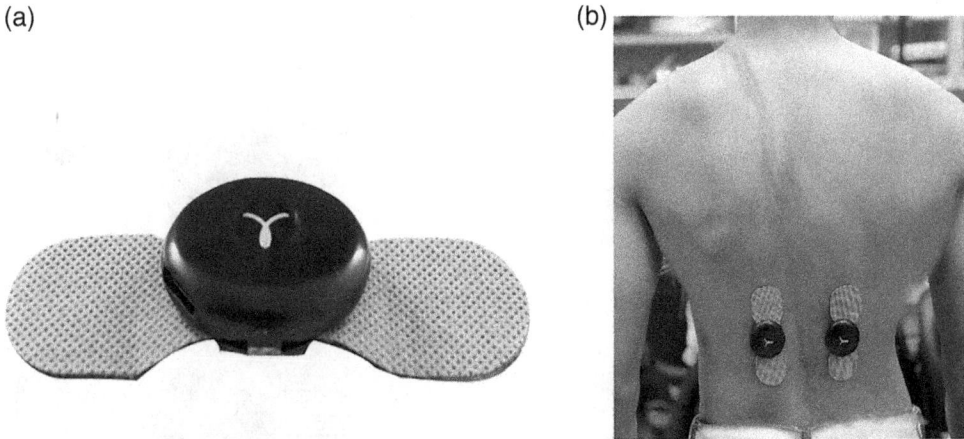

Figure 16.3 (a) Cricket wearable device (left); (b) Cricket devices on the left and right erector spinae muscles (right).

(i.e., left and right erector spinae muscles) (Figure 16.3b). Cricket is a wearable device that consists of an EMG, electrocardiogram (EKG), electroencephalogram (EEG), and one 6-axis IMU sensor (3-axis for acceleration and 3-axis for gyroscope). The device weighs approximately 0.18 kg and contains a rechargeable lithium-ion battery, which according to the manufacturer, can last for 11 hours at continuous live-streaming at 1000 samples/sec (for EMG, EKG, and EEG data) and 30 samples/sec (for IMU data). The device is attached to the body using a disposable patch as shown in Figure 16.3a. The patch also serves as an electrode that provides access to one channel of data from the sensors. The data from the EMG is transmitted in real-time to local or cloud storage via Bluetooth Low Energy (BLE) wireless connection and recorded by an application developed by Somaxis Inc. The Cricket device has been reported as having an acceptable accuracy when employed for other applications such as recreation (Boddy et al., 2018).

16.3.2 *Experimental Design and Procedure*

Post signing the informed consent, the rebar task was explained to the participants since they had no prior experience, and they were allowed to practice multiple cycles of the task. Once comfortable with the task, the Cricket devices were placed directly on each participant's skin bilaterally on the erector spinae muscle group by exercising the placement technique identified in Florimond (2009). Prior to placing the devices, each participant's skin was cleaned with alcohol to remove impurities. Thereafter the participants performed the rebar task without the exoskeleton (Figure 16.4a) and were allowed to rest for 15 mins to avoid fatigue. Subsequently, the functioning of the exoskeleton was introduced to the participants. Once comfortable with the exoskeleton, the participants performed the rebar task using BackX (Figure 16.4b). A prefabricated assembly was utilized to simulate the rebar task. The assembly comprises #6 rebars placed at 127mm on center (both ways) with 600mm by 400mm cross-sectional dimension. Four cycles involving placing and tying subtasks were performed with and without the exoskeleton (i.e., Exo and NoExo conditions). For each cycle, the participants placed one

(a)   (b)

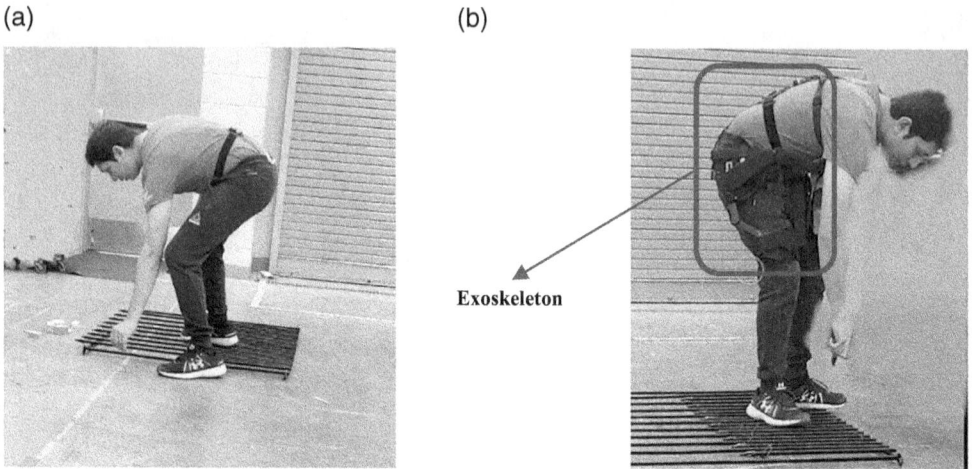

Figure 16.4 Performing simulated rebar task: (a) NoExo condition; (b) Exo condition.

prefabricated assembly on the ground during the placing subtask and subsequently they tied the joints using a plier and pre-cut ties. The participants were video recorded while performing the experimental tasks to identify the durations of the subtasks.

16.3.3   Data Analysis

16.3.3.1   Data Preparation and Labeling

During the rebar task, the time-stamped raw EMG data was collected at a frequency of 500 Hz and extracted using the accompanying device. The timings of the placing and tying subtasks were identified by comparing the time-stamped EMG data with the recorded videos. Thereafter, the data was sorted into subtasks using a Python script and the non-productive data was discarded. The data for the placing and tying subtasks were separately combined for all the participants for both the left and right erector spinae muscles. The total number of data samples is 2,729,400 for the NoExo condition and 2,141,500 for the Exo condition. The raw EMG data was progressively filtered between 20–450 Hz with fourth-order Butterworth band-pass filter (De Luca et al., 2010). The filtered data was subsequently transposed into rows, with each row having 100 columns. Once transposed, the data for the placing and tying subtasks were structured using concatenation and each row was labeled to correspond to the correct subtask. All the data extraction, structuring, and processing were performed using MS Office Excel 2021 and MATLAB Release 2021a.

16.3.3.2   Feature Extraction

Feature extraction was performed to transfigure the raw input data into features that contain useful information about the data. Time-domain features were employed for the feature extraction. Time-domain features are widely used for classifying EMG data because the resulting classifications contain a reduced amount of noise and the processing time is lower than that of the frequency-domain features (Toledo-Pérez et al., 2019).

From literature (Spiewak et al., 2018; Phinyomark et al., 2012; Chowdhury et al., 2013), seven common features such as mean absolute value (MAV), integrated EMG (iEMG), variance (VAR), standard deviation (STD), simple square integral (SSI), kurtosis (KT) and average energy (AE) were extracted from the filtered EMG data. This study only employed seven features because the use of additional features (i.e., root mean square and coefficient of variance) significantly increased computational time (212 times) and led to reduction in accuracy of the models (approximately 1.5–2%). MAV is expressed as the moving average of the EMG rectified signal (Zhang et al., 2019). iEMG enables signal recognition but without a pattern and is related to the trigger point of the EMG signal sequence. It is defined as the sum of the absolute values of each EMG sample (Alkan and Günay, 2012). Both the MAV and iEMG are related to the strength of the muscle contraction. VAR is derived from the EMG power while the STD is defined as the square root of the VAR. SSI is the sum of squared values of the EMG signal amplitude (Spiewak et al., 2018). KT helps to evaluate the effect of changes in VAR on the shape of the EMG distribution. KT is determined by comparing the peak of the curve inclination data distribution and the normal curve. AE is the average power of the EMG signal for a given period of time.

16.3.3.3 Data Classification

To classify the rebar subtasks, supervised machine learning classifiers were used to learn patterns from the extracted features. After all the features were extracted, all the algorithms in the MATLAB classifier application were trained to find the classification model best suited using all the seven features combined. Researchers have used a variety of supervised machine learning classifiers for recognizing activities from EMG data, including Decision tree (DT), Support Vector Machine (SVM), Ensemble, K-nearest neighbor (KNN), Naïve Bayes (NB), Neural Network (NN), and Logistic Regression (Chan et al., 2022). There is evidence that there is no specific best-performing classifier for all the applications (Murthy, 1998). As such, it would be beneficial to compare the performance of the different types of supervised machine learning classifiers to achieve the best-performing model. This study selected the following classifiers: SVM, KNN, Ensemble, NN, NB, Logistic Regression, DT, Linear Discriminant, and Kernel from the MATLAB Toolbox. Furthermore, default hyperparameters (mentioned in Section 16.4.1) from MATLAB were employed for training all the models. Hyperparameters are a set of 'high level' factors which impact the performance of an algorithm (Duan et al., 2003). Prior to training the classifiers, the extracted feature dataset was split into a training dataset (80%) and a testing dataset (20%). The training dataset was trained with the nine classifiers and thereafter validated using a holdout validation. Generally, with a large dataset holdout validation can save on the computational cost (Wang and Fey, 2018).

16.3.4 Performance Evaluation

After training the classifiers, the performance of the resulting models needs to be evaluated to understand their efficacy for future applications. Four common performance metrics were used to measure and compare the performance of the classifiers, namely, accuracy, precision, recall, and F1 score. The accuracy of a classification model can be computed by dividing the total number of correctly predicted classes by the total number of class samples as shown in Equation 16.1. However, the accuracy could be biased towards a class. For example, if the training dataset for both classes are different and the correctly predicted values of the class with a higher training dataset is more, then the accuracy of the model will be biased towards the class with the larger training dataset. Thus, to overcome this bias, it is necessary to employ other metrics such as precision, recall, and F1 score. These metrics were computed for both classes i.e., the placing and tying subtasks. Precision and recall account for the implications of the misclassification. Precision indicates the number of positive class predictions (i.e., True positive) that belong to the positive class (i.e., True positive + False positive) (Equation 16.2) while the recall signifies the ability of the model to correctly identify the true positive or negative class. That is, the recall is the number of positive class predictions (i.e., True positive) made from the total class samples (i.e., True positive + False negative) (Equation 16.3). The F1 score is the harmonic mean of the precision and recall (Equation 16.4). The higher the F1 score, the better the model of a given class performs.

$$\text{Accuracy} = \frac{\text{Total correctly predicted classes}}{\text{Total class samples}} = \frac{TP + TN}{TP + FP + TN + FN} \tag{16.1}$$

$$\text{Precision} = \frac{\text{Correct class predictions}}{\text{Total positive class predictions}} = \frac{TP}{TP + FP} \tag{16.2}$$

$$\text{Recall} = \frac{\text{Correct class predictions}}{\text{Total class sample}} = \frac{TP}{TP + FN} \tag{16.3}$$

$$\text{F1 score} = \frac{2 \times \text{Recall} \times \text{Precision}}{\text{Recall} + \text{Precision}} \tag{16.4}$$

Note. TP = True positive, TN = True negative, FP = False positive, and FN = False negative.

16.4   Results and Discussion

This study is focused on understanding the extent to which muscle activity obtained from the erector spinae muscle groups can be used to detect and classify activities performed by construction workers. While performing the rebar task, EMG data was collected from all participants. The data were combined and used to detect and classify rebar subtasks i.e., placing and tying subtasks. In this study, the activities of the erector spinae muscles of all the participants were collected while performing the rebar subtasks with and without a back-support exoskeleton. Considering both conditions in the study is significant given the growing interest in exoskeletons to address WMSDs in the construction industry. The performance of the classifiers was evaluated on 20% of the datasets (i.e., 20% holdout) using the performance metrics described in Section 16.3.4. The performance evaluation allows for investigating the extent to which construction activities can be detected and classified from new data collected from future instances. The extent to which the activities of the erector spinae muscles can be used to recognize rebar subtasks is discussed by comparing the with and without exoskeleton experimental conditions (i.e., Exo and NoExo conditions) for the best classifier and the top-performing classifiers.

16.4.1 Performance of the Best Classifier

Table 16.1 shows the accuracy of the classifiers employed in training the datasets for both Exo and NoExo conditions. The confusion matrixes of the classifier with the highest accuracy are presented in Figure 16.5. The confusion matrixes illustrate the correctly and wrongly classified classes of the classification model in the form of a matrix. The dark blue diagonal boxes in the confusion matrix represent the correctly predicted classes whereas the pink boxes represent the wrongly predicted class. For both experimental conditions, the SVM algorithm had the highest accuracy in recognizing the actions of construction workers during rebar tasks. This is supported by similar studies (Antwi-Afari et al., 2018) that highlighted the effectiveness of SVM for classifying construction workers' postures. The default hyperparameters employed in this study for SVM model are (1) gaussian kernel function; (2) kernel scale = 10; (3) box constraint level = 1; (4) multiclass method = one-vs-one; and (5) standardized data = true.

For the Noexo condition, the holdout validation utilized a total of 545,800 datasets, transposed into 5,458 rows out of which 2,728 and 2,730 rows represented placing and tying subtasks respectively. The classifier with the highest accuracy is the SVM with an accuracy of 74.10% whereas the classifier with the lowest accuracy is KNN with an accuracy of 69%. The confusion matrix for SVM (Figure 16.5a) indicates that the model correctly classified 1,702 and 2,343 rows of samples as placing and tying subtasks respectively. The model misclassified 1,026 and 387 rows of data as placing and tying subtasks respectively. The precision signifies that from the total number of placing predictions, only 81.5% are correct, whereas, from the tying predictions, only 69.5% are correct. The recall indicates that the model can accurately identify 62.4% and 85.8% of placing and tying subtasks respectively. Compared with the F1 score of the placing subtask that was 0.707, the F1 score of the tying subtask is 0.768.

Table 16.1 Accuracy of all the trained algorithms for both experimental conditions

	Decision Tree	Linear Discriminant	Logistic Regression	NB	SVM	KNN	Ensemble	NN	Kernel
Noexo	73.6%	73.4%	73.5%	73.7%	**74.1%**	69.0%	73.5%	73.7%	72.9%
Exo	81.7%	78.8%	82.4%	82.1%	**83.8%**	78.0%	82.9%	83.7%	82.4%

(a)   (b)

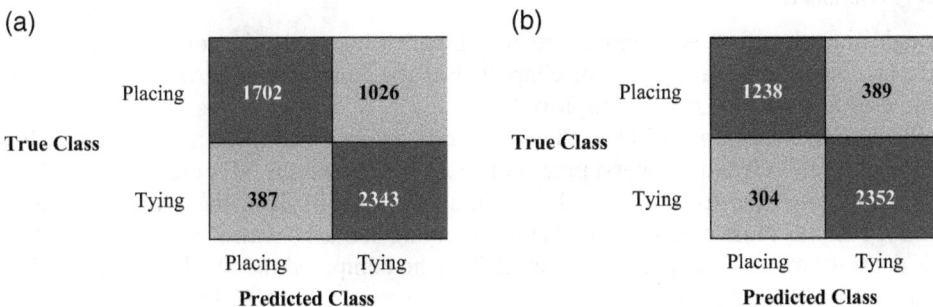

Figure 16.5 (a) NoExo confusion matrix (SVM); (b) Exo confusion matrix (SVM).

Table 16.2 Performance of the top three classifiers for both experimental conditions

Experimental Condition	Classifier		Accuracy	Precision	Recall	F1 score
NoExo	Support Vector Machine	Placing	74.10%	0.815	0.624	0.707
		Tying		0.695	0.858	0.768
	Neural Network	Placing	73.70%	0.786	0.651	0.712
		Tying		0.702	0.823	0.758
	Naïve Bayes	Placing	73.70%	0.767	0.679	0.720
		Tying		0.712	0.794	0.751
Exo	Support Vector Machine	Placing	83.80%	0.803	0.761	0.781
		Tying		0.858	0.886	0.872
	Neural Network	Placing	83.70%	0.801	0.758	0.779
		Tying		0.857	0.884	0.870
	Ensemble	Placing	82.90%	0.762	0.801	0.781
		Tying		0.874	0.847	0.860

The total number of training samples for the Exo condition was 428,300. This was transposed into 4,283 rows comprising 1,627 and 2,656 rows of placing and tying subtasks respectively. SVM also outperformed the other classifiers with an accuracy of 83.80% while KNN had the least accuracy of 78%. Figure 16.5b shows the confusion matrix of SVM for the Exo condition. The model accurately predicted 1238 rows of the testing dataset as placing subtask while 2,352 rows were accurately predicted as tying subtask. The recall and precision for the tying subtasks are higher than the placing subtask. This means that the model can predict and identify the tying subtasks than the placing sub-tasks. This is also reflected in the F1 score, which is higher for the tying subtask (0.872) than the placing subtask (0.781).

Compared with the NoExo condition, the classifiers for the Exo condition performed better and yielded a higher accuracy as shown in Table 16.2. For both conditions, the classifier with the highest accuracy was SVM while the classification algorithm with the least accuracy was the KNN (Table 16.1). From the total number of placing and tying predictions, the Exo condition had the highest precision (Figure 16.6). The ability of the model to correctly identify either subtask (recall) was higher for the Exo condition (Figure 16.6). The overall performance of the model, considering the precision and recall (F1 score), was highest for the Exo condition as shown in Figure 16.6.

16.4.2   Performance of Top Classifiers

Although nine classifiers were employed for classification, only the top three are compared as a convenience sample in this chapter. For the comparison, performance metrics described in Section 16.3.4 were employed.

In the NoExo condition, SVM was the most accurate in predicting placing subtask with a precision of 0.815. NB had the worst precision i.e., 0.767. However, NB outperformed SVM and NN in being able to correctly identify the placing subtask with a recall of 0.679. Similarly, NB was most precise in predicting the tying subtasks with a precision of 0.712. This is contrary to the findings of Narayan (2021) who compared the SVM and NN models for classifying EMG data and found SVM to be better than NB. Furthermore, SVM correctly predicted the tying subtasks with a recall of 0.858 than NN and NB. SVM had the highest F1 score for predicting tying subtask compared with placing subtask (Figure 16.7a).

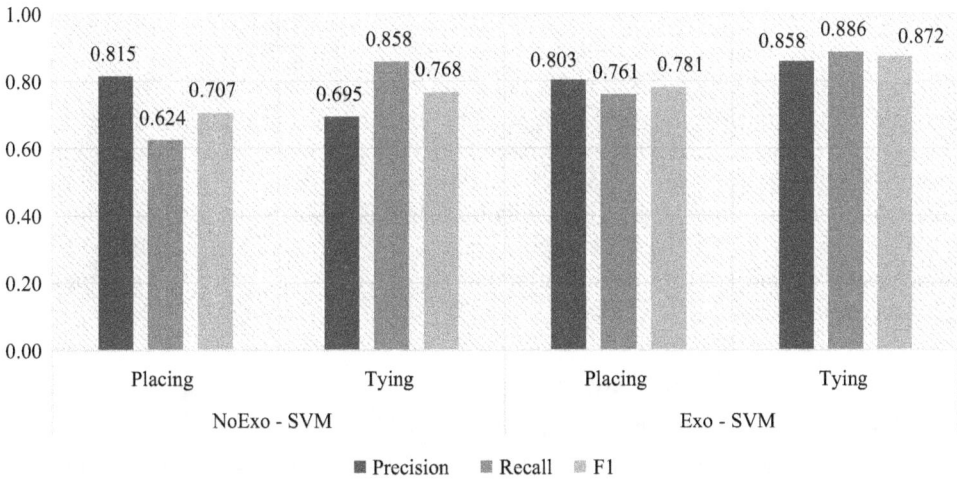

Figure 16.6 Comparison between Exo and NoExo conditions for SVM.

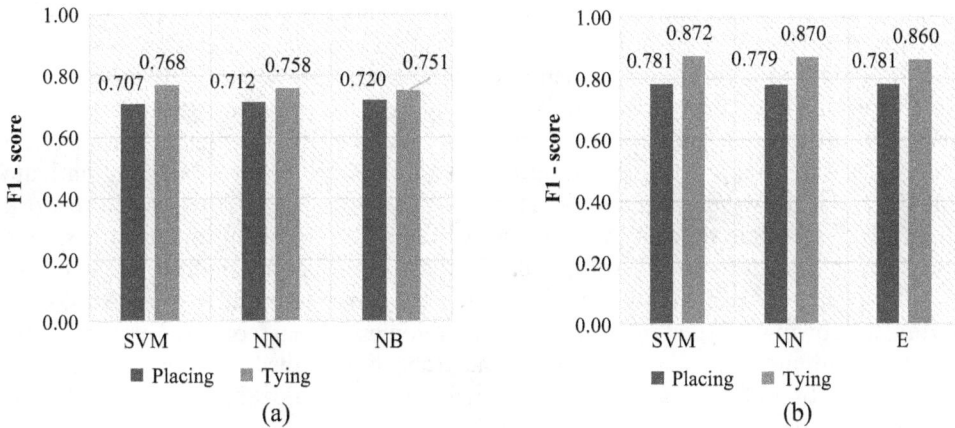

Figure 16.7 (a) F1 score graph – Noexo; (b) F1 score graph – Exo.

SVM also performed best in accurately predicting the placing subtasks (0.803) for the Exo condition. The performance of SVM for correctly identifying the placing subtasks was not far behind the highest recall value. The highest recall was registered by Ensemble (0.801) followed by SVM (0.761). During the Exo condition, Ensemble was the most precise in predicting tying subtask (0.874). The potential of the models in correctly predicting the tying subtask (recall) was greatest for SVM (0.886). Similar to the NoExo condition, SVM also outperformed NN and Ensemble in predicting tying subtasks compared with the placing subtask during the Exo condition (Figure 16.7b). The performance of the SVM model in this study is in line with the findings of Li et al. (2020) who observed that the SVM classifier attained an accuracy of 96.27% for prosthetic action recognition using data from surface EMG. Furthermore, considering the precision, recall, and F1 score for the NoExo condition during placing and tying subtasks, the model yielded better results for the tying subtask than the placing subtask. This may imply that there are more movements in the hand during the tying subtask that could trigger more activations in the erector spinae muscles than during the placing subtask. Bangaru et al. (2021) also revealed the efficacy of EMG data for recognizing construction activities involving more muscle activations. Furthermore, this could be attributed to similar postures being assumed by the participants since they were not experienced rebar workers. It is evident that the performance of the Exo classification model is better than that of the NoExo condition. This reveals the potential of the classification model for recognizing rebar tasks during the utilization of wearable robots, which can culminate in improved workers' safety.

16.5   Conclusions and Future Work

This study evaluated the potential of muscle activity acquired from the erector spinae muscles to automatically detect actions of construction workers wearing and not wearing an exoskeleton. A simulated rebar task was performed in a laboratory to examine the feasibility of the approach. The performance of supervised machine learning classifiers was compared in order to select the most suitable. Overall, SVM obtained the highest accuracy of correctly classifying placing and tying subtasks in both experimental conditions (i.e., Exo and NoExo). The performance of the SVM algorithm showcases the efficacy of erector spinae muscle activity for the recognition of workers' tasks. It is expected that the implementation of the classification models will enhance workers' productivity analysis and human resource allocation. Likewise, understanding workers' trunk muscle activity can support ergonomic risk assessments, which can improve decision-making for the adoption of a wearable robot system to mitigate WMSDs in the construction industry. This study also sets precedence for future work that explores real-time models for detecting workers' actions using the erector spinae muscle activity, such as the investigation of smart exoskeletons that are triggered in response to the required muscle activations during construction activities. Also, considering a growing interest in wearable technologies as part of workers' personal protective equipment, this study suggests the efficacy of embedded EMG sensors for real-time detection of workers' muscle activities, which is critical for reducing WMSDs and crew management.

Although the findings of this study reveal the effectiveness of detecting rebar tasks from activations of the erector spinae muscles which can be utilized for improving construction workers' safety, some limitations exist and should be addressed in future studies. Firstly, muscle activations from only ten participants who are students and not construction professionals were employed. Hence, future work will entail a field study with more participants who are actual construction workers to achieve a robust training dataset that is representative of the construction workers' actions. Furthermore, data augmentation methods will be employed to create synthetic datasets for improving the accuracy of the classifiers. Also, the classification algorithm was trained using seven time-domain features, but the impact of each feature on the classification model was not explored in this study. Future work will investigate the models' performance for different sets of data features to understand which combination yields the highest performance. This will inform the choice of the dataset for classification models when limited data is available. Also, the models were trained using default hyperparameters without considering their impact on the model performance. As part of future studies, hyperparameter tuning will be conducted to improve model accuracy. Two separate classification models were trained for both the experimental conditions in this study. Thus, future studies will explore the suitability of recognizing workers' actions during both experimental conditions using the same classifier.

References

Akanmu, A. A., Olayiwola, J., Ogunseiju, O. & Mcfeeters, D. (2020). Cyber-physical postural training system for construction workers. *Automation in construction*, 117, 103272. 10.1016/j.autcon.2020.103272

Akhavian, R. & Behzadan, A. H. (2016). Smartphone-based construction workers' activity recognition and classification. *Automation in Construction*, 71, 198–209. 10.1016/j.autcon.2016.08.015

Albers, J. T. (2007). *Simple solutions: Ergonomics for construction workers*, US Department of Health and Human Services, Public Health Service, Centers.

Alkan, A. & Günay, M. (2012). Identification of EMG signals using discriminant analysis and SVM classifier. *Expert Systems with Applications*, 39, 44–47. 10.1016/j.eswa.2011.06.043

Antwi-Afari, M. F., Li, H., Yu, Y. & Kong, L. (2018). Wearable insole pressure system for automated detection and classification of awkward working postures in construction workers. *Automation in Construction*, 96, 433–441. 10.1016/j.autcon.2018.10.004

Antwi-Afari, M. F., Li, H., Umer, W., Yu, Y. & Xing, X. (2020). Construction activity recognition and ergonomic risk assessment using a wearable insole pressure system. *Journal of Construction Engineering and Management*, 146, 04020077. 10.1061/(ASCE)CO.1943-7862.0001849

Antwi-Afari, M. F., Li, H., Anwer, S., Li, D., Yu, Y., Mi, H.-Y. & Wuni, I. Y. (2021). Assessment of a passive exoskeleton system on spinal biomechanics and subjective responses during manual repetitive handling tasks among construction workers. *Safety Science*, 142, 105382. 10.1016/j.ssci.2021.105382

Bai, Y., Huan, J. & Kim, S. (2020) Measuring bridge construction efficiency using the wireless real-time video monitoring system. *American Society of Civil Engineers*. 10.1061/(ASCE)ME.1943-5479.0000061

Bangaru, S. S., Wang, C., Busam, S. A. & Aghazadeh, F. (2021). ANN-based automated scaffold builder activity recognition through wearable EMG and IMU sensors. *Automation in Construction*, 126, 103653. 10.1016/j.autcon.2021.103653

BLS. (2020). *Nonfatal Occupational Injuries and Illnesses Requiring Days Away from Work* [Online]. Available: https://data.bls.gov/PDQWeb/cs [Accessed December 12, 2020].

Boddy, K., Rogers, K. & Scheffey, J. (2018). A surface electromyographic examination of the serratus anterior during exercise, weight, and order-dependent variations of the bench press. 10.31236/osf.io/84gsf

Chan, V. C., Ross, G. B., Clouthier, A. L., Fischer, S. L. & Graham, R. B. (2022). The role of machine learning in the primary prevention of work-related musculoskeletal disorders: A scoping review. *Applied Ergonomics*, 98, 103574. 10.1016/j.apergo.2021.103574

Chowdhury, R. H., Reaz, M. B., Ali, M. A. B. M., Bakar, A. A., Chellappan, K. & Chang, T. G. (2013). Surface electromyography signal processing and classification techniques. *Sensors*, 13, 12431–12466. 10.3390/s130912431

Conforti, I., Mileti, I., Del Prete, Z. & Palermo, E. (2020). Measuring biomechanical risk in lifting load tasks through wearable system and machine-learning approach. *Sensors*, 20, 1557. 10.3390/s20061557

De Luca, C. J., Gilmore, L. D., Kuznetsov, M. & Roy, S. H. (2010). Filtering the surface EMG signal: Movement artifact and baseline noise contamination. *Journal of Biomechanics*, 43, 1573–1579. 10.1016/j.jbiomech.2010.01.027

Diaz, C. (2020). Combination of IMU and EMG for object mass estimation using machine learning and musculoskeletal modeling. Published in Digitala Vetenskapliga Arkivet, Available online at: https://www.diva-portal.org/smash/record.jsf?pid=diva2%3A1526030&dswid=2407

Dolan, P. & Adams, M. (1993). The relationship between EMG activity and extensor moment generation in the erector spinae muscles during bending and lifting activities. *Journal of Biomechanics*, 26, 513–522. 10.1016/0021-9290(93)90013-5

Duan, K., Keerthi, S. S. & Poo, A. N. 2003. Evaluation of simple performance measures for tuning SVM hyperparameters. *Neurocomputing*, 51, 41–59.

Florimond, V. 2009. *Basics of surface electromyography applied to physical rehabilitation and biomechanics*. Montreal, Canada: Thought Technology Ltd.

Friedli, W., Cohen, L., Hallett, M., Stanhope, S. & Simon, S. 1988. Postural adjustments associated with rapid voluntary arm movements. II. Biomechanical analysis. *Journal of Neurology, Neurosurgery & Psychiatry,* 51, 232–243. 10.1136/jnnp.51.2.232

Gonsalves, N. J., Ogunseiju, O. R., Akanmu, A. A. & Nnaji, C. A. 2021. Assessment of a passive wearable robot for reducing low back disorders during rebar work. *Journal of Information Technology in Construction (ITCON)*, 26, 936–952. 10.36680/j.itcon.2021.050

Hodges, P., Cresswell, A. & Thorstensson, A. 1999. Preparatory trunk motion accompanies rapid upper limb movement. *Experimental Brain Research*, 124, 69–79. 10.1007/s002210050601

Hodges, P. W. & Richardson, C. A. 1996. Inefficient muscular stabilization of the lumbar spine associated with low back pain: a motor control evaluation of transversus abdominis. *Spine*, 21, 2640–2650. ISSN: 0362-2436

Jarrassé, N., Proietti, T., Crocher, V., Robertson, J., Sahbani, A., Morel, G. & Roby-Brami, A. 2014. Robotic exoskeletons: a perspective for the rehabilitation of arm coordination in stroke patients. *Frontiers in Human Neuroscience*, 8, 947. 10.3389/fnhum.2014.00947

Joshua, L. & Varghese, K. 2011. Accelerometer-based activity recognition in construction. *Journal of Computing in Civil Engineering*, 25, 370–379. 10.1061/(ASCE)CP.1943-5487.0000097

Joshua, L. & Varghese, K. 2014. Automated recognition of construction labour activity using accelerometers in field situations. *International Journal of Productivity and Performance Management*. ISSN: 1741-0401

Kermavnar, T., De Vries, A. W., De Looze, M. P. & O'sullivan, L. W. 2021. Effects of industrial back-support exoskeletons on body loading and user experience: An updated systematic review. *Ergonomics*, 64, 685–711. 10.1080/00140139.2020.1870162

Kim, S., Moore, A., Srinivasan, D., Akanmu, A., Barr, A., Harris-Adamson, C., Rempel, D. M. & Nussbaum, M. A. 2019. Potential of exoskeleton technologies to enhance safety, health, and performance in construction: Industry perspectives and future research directions. *IISE Transactions on Occupational Ergonomics and Human Factors*, 7, 185–191. 10.1080/24725838.2018.1561557

Lariviere, C., Arsenault, A., Gravel, D., Gagnon, D. & Loisel, P. 2002. Evaluation of measurement strategies to increase the reliability of EMG indices to assess back muscle fatigue and recovery. *Journal of Electromyography and Kinesiology*, 12, 91–102. 10.1016/S1050-6411(02)00011-1

Li, C., Li, G., Jiang, G., Chen, D. & Liu, H. 2020. Surface EMG data aggregation processing for intelligent prosthetic action recognition. *Neural Computing and Applications*, 32, 16795–16806. 10.1007/s00521-018-3909-z

Lowe, B. D., Billotte, W. G. & Peterson, D. R. 2019. ASTM F48 formation and standards for industrial exoskeletons and exosuits. *IISE Transactions on Occupational Ergonomics and Human Factors*, 7, 230–236. 10.1080/24725838.2019.1579769

Luo, H., Xiong, C., Fang, W., Love, P. E., Zhang, B. & Ouyang, X. 2018a. Convolutional neural networks: Computer vision-based workforce activity assessment in construction. *Automation in Construction*, 94, 282–289. 10.1016/j.autcon.2018.06.007

Luo, X., Li, H., Cao, D., Yu, Y., Yang, X. & Huang, T. 2018b. Towards efficient and objective work sampling: Recognizing workers' activities in site surveillance videos with two-stream convolutional networks. *Automation in Construction*, 94, 360–370. 10.1016/j.autcon.2018.07.011

Luo, X., Li, H., Yang, X., Yu, Y. & Cao, D. 2019. Capturing and understanding workers' activities in far-field surveillance videos with deep action recognition and Bayesian nonparametric learning. *Computer-Aided Civil and Infrastructure Engineering*, 34, 333–351. 10.1111/mice.12419

Madinei, S., Kim, S., Srinivasan, D. & Nussbaum, M. A. 2021. Effects of back-support exoskeleton use on trunk neuromuscular control during repetitive lifting: A dynamical systems analysis. *Journal of Biomechanics*, 123, 110501. 10.1016/j.jbiomech.2021.110501

Mudie, K., Billing, D., Garofolini, A., Karakolis, T. & Lafiandra, M. 2021. The need for a paradigm shift in the development of military exoskeletons. *European Journal of Sport Science*, 1–8. 10.1080/17461391.2021.1923813

Mudiyanselage, S. E., Nguyen, P. H. D., Rajabi, M. S. & Akhavian, R. 2021. Automated workers' ergonomic risk assessment in manual material handling using sEMG wearable sensors and machine learning. *Electronics*, 10, 2558. 10.3390/electronics10202558

Murthy, S. K. 1998. Automatic construction of decision trees from data: A multi-disciplinary survey. *Data Mining and Knowledge Discovery*, 2, 345–389. 10.1023/A:1009744630224

Narayan, Y. 2021. Comparative analysis of SVM and Naive Bayes classifier for the SEMG signal classification. *Materials Today: Proceedings*, 37, 3241–3245. 10.1016/j.matpr.2020.09.093

Nouredanesh, M. & Tung, J. 2019. IMU, sEMG, or their cross-correlation and temporal similarities: Which signal features detect lateral compensatory balance reactions more accurately? *Computer Methods and Programs in Biomedicine*, 182, 105003. 10.1016/j.cmpb.2019.105003

Ogunseiju, O., Olayiwola, J., Akanmu, A. & Olatunji, O. A. 2021. Evaluation of postural-assist exoskeleton for manual material handling. *Engineering, Construction and Architectural Management*. ISSN: 0969-9988

Phinyomark, A., Phukpattaranont, P. & Limsakul, C. 2012. Feature reduction and selection for EMG signal classification. *Expert Systems with Applications*, 39, 7420–7431. 10.1016/j.eswa.2012.01.102

Pons, J. L. 2008. *Wearable robots: biomechatronic exoskeletons*, John Wiley & Sons. ISBN: 978-0-470-98765-0

Sherafat, B., Ahn, C. R., Akhavian, R., Behzadan, A. H., Golparvar-Fard, M., Kim, H., Lee, Y.-C., Rashidi, A. & Azar, E. R. 2020. Automated methods for activity recognition of construction workers and equipment: state-of-the-art review. *Journal of Construction Engineering and Management*, 146, 03120002. 10.1061/(ASCE)CO.1943-7862.0001843

Shoubi, M. V., Barough, A. S. & Rasoulijavaheri, A. 2013. Ergonomics principles and utilizing it as a remedy for probable work related injuries in construction projects. *International Journal of Advances in Engineering & Technology*, 6, 232. ISSN: 2231-1963

Spiewak, C., Islam, M., Zaman, A. & Rahman, M. H. 2018. A comprehensive study on EMG feature extraction and classifiers. *Open Access Journal of Biomedical Engineering and Biosciences*, 1, 1–10.

Sung, P. S., Lammers, A. R. & Danial, P. 2009. Different parts of erector spinae muscle fatigability in subjects with and without low back pain. *The Spine Journal*, 9, 115–120. 10.1016/j.spinee.2007.11.011

Sánchez-Zuriaga, D., Vera-Garcia, F. J., Moreside, J. M. & Mcgill, S. M. 2009. Trunk muscle activation patterns and spine kinematics when using an oscillating blade: Influence of different postures and blade orientations. *Archives of Physical Medicine and Rehabilitation*, 90, 1055–1060. 10.1016/j.apmr.2008.12.015

Toledo-Pérez, D. C., Rodríguez-Reséndiz, J., Gómez-Loenzo, R. A. & Jauregui-Correa, J. 2019. Support vector machine-based EMG signal classification techniques: A review. *Applied Sciences*, 9, 4402. 10.3390/app9204402

Wang, J., Dai, Y. & Si, X. 2021. Analysis and recognition of human lower limb motions based on electromyography (EMG) signals. *Electronics*, 10, 2473. 10.3390/electronics10202473

Wang, Z. & Fey, A. M. 2018. Deep learning with convolutional neural network for objective skill evaluation in robot-assisted surgery. *International Journal of Computer Assisted Radiology and Surgery*, 13, 1959–1970. 10.1007/s11548-018-1860-1

Yang, J., Shi, Z. & Wu, Z. 2016. Vision-based action recognition of construction workers using dense trajectories. *Advanced Engineering Informatics*, 30, 327–336. 10.1016/j.aei.2016.04.009

Young, A., Kuiken, T. & Hargrove, L. 2014. Analysis of using EMG and mechanical sensors to enhance intent recognition in powered lower limb prostheses. *Journal of Neural Engineering*, 11, 056021. 10.1088/1741-2560/11/5/056021

Zedka, M. & Prochazka, A. 1997. Phasic activity in the human erector spinae during repetitive hand movements. *The Journal of Physiology*, 504, 727–734. 10.1111/j.1469-7793.1997.727bd.x

Zhang, J., Ling, C. & Li, S. 2019. EMG signals based human action recognition via deep belief networks. *IFAC-PapersOnLine*, 52, 271–276. 10.1016/j.ifacol.2019.12.108

Zhu, Z., Dutta, A. & Dai, F. 2021. Exoskeletons for manual material handling – A review and implication for construction applications. *Automation in Construction*, 122, 103493.

17 Design and Development of Virtual Reality Environment for Human-Robot Interaction on Construction Site

Adetayo Onososen[1], Innocent Musonda[1], Christopher Dzuwa[1], and Ramabodu Molusiwa[2]
[1]*Center of Applied Research and Innovation in the Built Environment (CARINBE) Department of Construction Management and Quantity Surveying Faculty of Engineering and the Built Environment University of Johannesburg, Johannesburg, South Africa;* [2]*Department of Construction Management and Quantity Surveying Faculty of Engineering and the Built Environment University of Johannesburg, Johannesburg, South Africa*

17.1 Introduction

It has been crucial for the construction sector to increase performance in terms of productivity and safety, given the high incidence of hazards and low productivity in infrastructure delivery (Onososen & Musonda, 2022). Hence, recent studies have prioritised developing robotic systems as a solution to the challenges of safety and productivity (Golizadeh et al., 2019; Eiris et al., 2020). Such robotic development has seen the applications of robots in road construction, drone deployment, autonomous transport, automated infrastructure inspection and maintenance, 3D concrete printing, road pavement and painting, amongst others (Bock, 2015; Gheisari & Esmaeili, 2016). Unmanned aerial vehicles (UAVs), commonly referred to as drones, are among the most adopted robotic systems currently being deployed on construction sites(Wen & Gheisari, 2020; Elghaish et al., 2021). Drones are frequently utilised in the construction sector for tasks including aerial site mapping, quality control, logistics on the project site, assessing structural integrity and damage at construction sites, maintenance evaluations, and more (Bogue, 2018; Khan et al., 2020; Aiyetan & Das, 2022). Drones are also being explored further for tasks such as delivery and transportation of materials and equipment on-site (Mendes et al., 2022). This, however, has severe implications for training given the low available expertise in drone flying and human-drone interaction arising from increased interaction between human workers and drones.

The shared workplace between humans and robots is already a reality given the increasing adoption of drones on construction sites, thus making it even more crucial to train workers on using and interacting with drones. Techniques based on virtual reality (VR) are the best for creating such training because VR-based simulations reduce the cost of training with real drones, are safer, and allow repeatability to hone skills and inform on the disposition of workers to how the presence of robots (i.e., drones) will be adopted in such environments (Kang & Miranda, 2006; Prabhakaran et al., 2022; Matsas & Vosniakos, 2015). Such training is impractical to carry out in real-world settings and exposes workers to more hazardous circumstances involving robots, which worsens their safety performance. VR simulations allow for the effective use of cutting-edge building and construction technologies before actual deployment in an interactive space with human workers.

DOI: 10.1201/9781003408949-17

There have only been a few test uses of drones for material handling in the built industry, which the future is predicted to see more of. Therefore, this study advances a virtual-reality application of delivery drone training to reduce the cost of drone training and safety hazards in drone applications and improve human-robot teaming for industrialised construction. To achieve this, the chapter presents the tools, design philosophy, implementation details, and tests of a DJI Phantom 4 drone clone. It must be noted that some features of the real drone were not implemented. However, the implementation added a gripping mechanism, which is not part of the physical DJI Phantom 4 drone.

17.2   Drone Design and Development Method

17.2.1   Design Philosophy

The design parameters of the DJI Phantom 4 clone considered the use case of the drone, i.e., human-robotic interaction in construction environments. As such, design parameters were chosen to meet his requirement. The design philosophy integrates the design criteria and choice adopted in developing the virtual environments and drone models, the controls, the need for data parameters and how they improve human-robot or human-drone collaboration.

17.2.2   Design Details

To arrive at good design specifications for the drone, the following parameters were scored using a Pugh scoring matrix: The rating scale is between 1 and 3: 3 for good, 2 for average and 1 for bad. Fixed-wing hybrid vertical take-off and landing (VTOL) drones were not included since only a handful of fixed-wing hybrid VTOLs are currently on the market, and the technology used in these drone types is still in the nascent stage (Rennie, 2016).

Table 17.1 Drone Pugh scoring matrix

Design Criteria	Weight (%)	Multirotor		Fixed Wing		Single Rotor	
		Rating	Weight Score	Rating	Weight Score	Rating	Weight Score
Speed	5	1	0.05	3	0.15	2	0.1
Range	7	1	0.07	3	0.21	2	0.14
Light Weight	10	3	0.3	1	0.1	2	0.2
Hovering Capability	6	3	0.18	1	0.06	2	0.12
VTOL Capability	10	3	0.3	1	0.1	2	0.2
Endurance	8	1	0.08	3	0.24	2	0.16
Ease of operation	15	3	0.45	2	0.3	1	0.15
Stability	12	3	0.36	2	0.24	1	0.12
Carrying capacity	12	1	0.12	2	0.24	3	0.36
Safety	15	3	0.45	2	0.3	1	0.15
	Sum		2.36		1.94		1.7
	Rank		1		2		3
	Continue		Yes		No		No

17.2.3   Choice Solution

Based on the Pugh matrix ranks, the multi-rotor drone passed the design matrix score check and was selected for implementation. Since pugh matrices are accurate with ±0.2, this was checked to see if any other solution could be considered (Wettergreen, 2014). The closest solution was the fixed-wing drone but is still outside the allowable consideration limit, i.e., $1.94 + 0.2 = 2.16 < 2.36$.

17.2.4   Drone Model

The drone was modelled in Blender 3.1 on a Windows 10 machine. No special textures were applied to the drone's body except specific colours to give the drone a nice look and feel. Figure 17.1 shows the drone's side view, plan, and 3D model.

17.2.5   Implementation of the simulator

The drone simulator was implemented in Unity 2022.1.16 using the C# programming language. Following the importation of the model into unity, design parameters, i.e.,

Side View-Conceptual view   Drone Plan-2D Wireframe

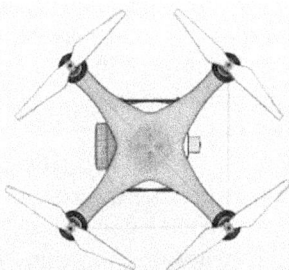

3D model-Conceptual View

Figure 17.1 Drone model.

weight, gripping mechanism joint friction, rotation, retraction and expansion speed, were assigned to the drone. To enable collisions with other physical objects, the exposed parts of the drone were assigned box colliders. Although mesh colliders could fit the irregular shape of the drone, box colliders were chosen due to their inexpensive operation compared to mesh colliders. Scripting involved simulating the rotor animations, object gripping, grip retraction and extension, switching between cameras, drone motion control, drone powering and starting, collision tracking, and payload delivery logging. Control of the drone was made possible using a controller with a display screen.

17.2.6 Flight Environment

The flight environment was a construction site with four uncompleted buildings. Other objects commonly found on construction sites were also added to the scene.

17.3 Experiment Procedure

17.3.1 Procedural Order for Flying the Drone

The following approach was adopted in flying the drone in the virtual construction environment. This was made human-friendly and easy to use through many iterations to reduce negative user feedback and enable smooth interaction between humans and the drone in the VR. Since most of the users were new to VR and usage of the VR touch controls, they were first taken through a short training to have knowledge of the drones and how to minimise motion sickness in the virtual environment. The following order was adopted in the experiment:

a The virtual environment is launched in Oculus HMD, as shown in Figures 17.3–17.5.
b The remote drone controller shown in Figure 17.2 is switched on to enable activating the drone for flight operations.
c The user switches on the drone after the remote controller is active and can also be switched off by the user at any moment.
d The user then begins to fly the drone, as shown in Figure 17.3, to learn the drone controls, stabilise the drone, familiarise with the virtual environment and pick the payloads (Bricks as shown in Figure 17.4) with an identified drop-off point on the building adjacent the bricks.
e The remote controller enables beyond visual line of sight (BVLOS) operations helping the user view the environment whether the drone is within or off sight. This is essential to train drone pilots in BVLOS operations and BVLOS delivery tasks.
f The user subsequently picks the payload, ensuring it is firm to avoid falling objects and controlling the drone to deliver the item to the designated delivery points (Figure 17.5).
g Drone pilots can switch between cameras to ensure the flight is well-directed and avoid collision during flight.
h The experimental process takes 10 minutes, with warning indicators showing battery levels. The drone performs an emergency landing once the battery level is depleted and users are to exit the environment.

17.3.2 Data Collection

The following data were collected during each flight session: payload, collision object, Time, Drone speed, Drone Position, Drone Orientation, Safe Distance, Pilot Position, Command, Drone Range, and Battery life. These markers are described next.

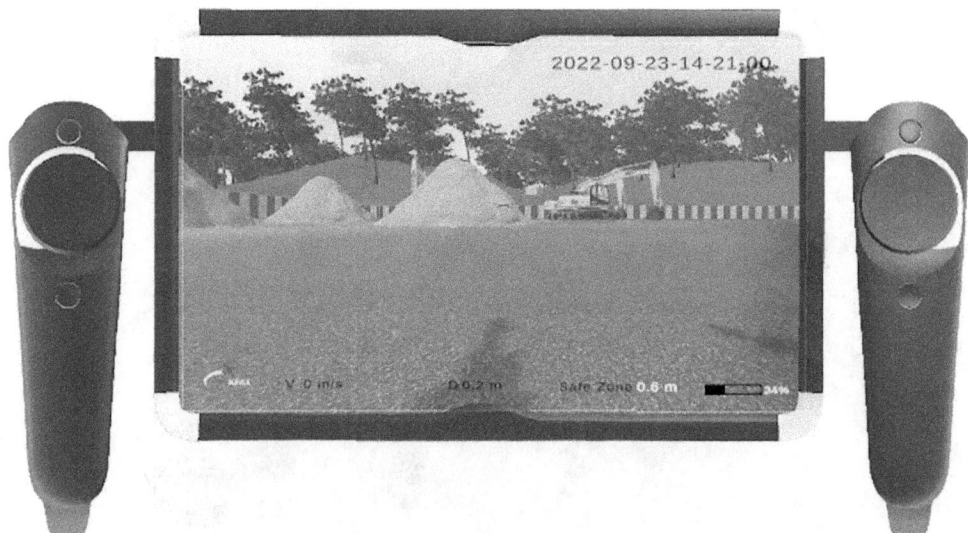

Figure 17.2 Drone controller with camera feedback of the site.

Figure 17.3 Drone in flight in virtual environment construction site.

Figure 17.4 Bricks in VR environments ready to be picked.

Figure 17.5 Drone mid-flight with the object to deliver.

Table 17.2 Drone data parameters

Parameter	Data Type	Description	Use	Frequency
Payload	Text	A brick was chosen as the payload for the delivery tasks	This was used to test the ability of drones to be used in material handling and transport	No of payloads picked & successfully delivered
Collision object	Text	Measurement of collision of drones with objects and humans for	Safety indicators to measure the propensity of drones to crash into objects and humans	No of Collisions made
Time	Time	Time-taken to perform the experiment and to deliver tasks	To determine training periods required by different users in successful human-drone interaction to deliver tasks	The average time-taken in the environment and to deliver tasks
Drone speed	Numeric	To measure the speed of drone during flight	To measure drone flight patterns	Experiment Duration
Drone Position	Numeric	To measure drone position during flight operations	Vital to track the activities of the user in the environment, such as idleness	Experiment Duration
Drone Orientation	Numeric	To measure drone orientation pre-fight, during and post-flight	To measure drone flight patterns	Experiment Duration
Safe Distance	Numeric	To measure human hazards issues in human-drone interaction	To measure healthy and safe human-drone interaction	Experiment Duration
Pilot Position	Numeric	To study users' flight patterns and behaviour	To measure users' movement patterns	Experiment Duration
Command	Text	To study users' drone communication patterns and behaviour	To improve training development	Experiment Duration
Drone Range	Numeric & Text	To examine the propensity of users to use drones outside specified uses	To train users to avoid spying/ privacy issues	Experiment Duration
Battery life	Numeric	Battery levels from 100% reducing depending on the use of drone in the virtual environment	To train users on battery level consciousness	Experiment Duration

17.3.3   User Task

We identified drone delivery tasks in a typical construction environment and evaluated the risks involved based on the markers in 17.3.1. The development first focused on ensuring a typical construction site with plants, materials, site fencing, ongoing construction, etc. The design of the VR procedure for flying the drone is discussed in 17.3.1 above.

17.3.4   Testing

To make the experience as real-life-like as possible, the simulation was run in VR. The XR interaction toolkit provided VR capabilities for the simulation. This includes motion, turning around, and interacting with other objects in the scene. The XR interaction toolkit was chosen due to its support for multiple VR platforms such as SteamVR, Oculus Quest 2, Oculus Rift, HTC Vive, etc. Motion and turning speeds were left at 1.5 m/s and 1.0 degree/s, respectively, as these played out well in reducing motion sickness in VR. Tests with users were done using the Oculus Quest 2. Each user was given 10 minutes to deliver items to a designated spot. It was noted that most of the users found it difficult to deliver items on their first attempt. However, this improved with further attempts. Each flight test recorded several parameters which act as primary data to test the interaction happening in the environment. These parameters are saved to a CSV file for the total flight duration (Figure 17.6).

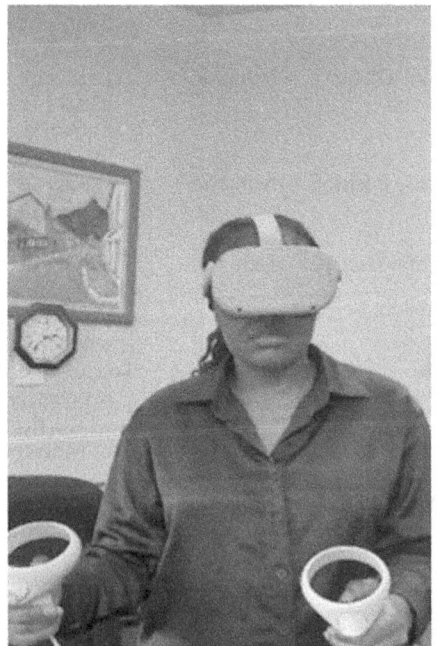

Figure 17.6 Test images of users in the VR environment.

17.4   Discussion

Drones greatly aid infrastructure delivery and construction management, but like technological innovations, they also bring emerging issues in safety hazards, privacy and training needs. These risks might result from drones physically coming into touch with employees, dropping materials from heights, spying on workers/surrounding environment, losing drones, etc. However, previous studies have revealed that with enhanced training techniques, these risks can be mitigated by virtual training to enhance drone flying expertise, ensure safe applications, and reduce training costs to enhance adoption.

The use of VR-based simulations for construction training has been analysed in several studies. For instance, Sakib et al. (2021) created a VR training system for drone operators. The study employed VR to gather physiological indicators from the drone operators during virtual flights, including mental workload (MWL), stress level, trait, and state anxiety. VR offers several benefits over conventional safety training methods by replicating high-risk interaction settings without subjecting learners to dangers (Wen & Gheisari, 2020). Therefore, this approach was considered effective in drone training for users. To further improve the application of VR-based human-drone interaction studies, qualitative feedback from the users was further integrated to improve the design and development of the virtual drone flight environment. The following observations and recommendations were made by users who participated in the pilot study.

The participants stated that the virtual environment should have the presence of multiple users for collaborative task participation and to enable supervised learning and training. Stabilising the drone was considered the task's most difficult aspect, and users suggested pre-training familiarisation with the controls. The pre-experiment training should consider familiarising users new to VR with the virtual environment to enable comfort within the environment. Also, it was mentioned that the pre-experiment training should allow users to familiarise themselves with XR touch controls.

Flying the drone in Attitude mode (ATTI) was considered essential by drone pilots participating in the study. This is a flight mode where GPS and other visual positioning systems are disabled. The pilots stated that over-reliance on automatic positioning systems in place of manual systems could put users in trouble when the drone suddenly loses the ability to hover in real-life situations. This is important as the drone might lose GPS reception when flying near a strong source of signal interference, resulting in the loss of GPS stabilisation. The advent of indoor drones also, as well as the ATTI mode skill requirement before pilot certification in some countries, makes this feature important.

Also important is to note that the users considered the flight simulation realistic with a difficulty level rightly mimicking real-life drone flight. For drone flight training purposes, the pilots recommended the drone camera-direction/nosing to train on drone control. The users also stated the usage of the flight simulation is essential for night operations, and the "safe zone" indicator in the remote drone controller was also considered important for safety.

17.5   Conclusion and Future Improvements

Despite the progress made so far, there are a lot of improvements which can be made to the current simulation. These include making the simulation multiplayer, giving users options to choose a drone based on the task, and enhancing the drone's autonomy using

Artificial Intelligence (AI) and Geographical Information Systems (GIS) tracking. The study's contributions are vital to efficiently utilising VR for human-robot interaction training. The design and development process presented is vital for future development studies in similar applications in the built environment. Also, as research in the usage of drones for material handling and human-robot interaction is underexamined, the study provides practical insights into processes to measure these dynamics associated with the adoption of robots in the built environment.

Acknowledgements

The drone was modelled by Mr Samuel Moyo Ndovie, a student at the University College of Engineering, Osmania University. This research is funded by the National Research Foundation, NRF (Grant Number-129953). Opinions and conclusions are those of the authors and are not necessarily attributable to the NRF. The work is supported and part of collaborative research at the Centre of Applied Research and Innovation in the Built Environment (CARINBE).

References

Aiyetan, A. O., & Das, D. K. (2022). Use of Drones for construction in developing countries: Barriers and strategic interventions. *International Journal of Construction Management*, 0(0), 1–10. 10.1080/15623599.2022.2108026

Bock, T. (2015). The future of construction automation: Technological disruption and the upcoming ubiquity of robotics. *Automation in Construction*, 59, 113–121. 10.1016/j.autcon.2015.07.022

Bogue, R. (2018). What are the prospects for robots in the construction industry? *Industrial Robot: An International Journal*, 45(1), 1–6. 10.1108/IR-11-2017-0194

Eiris, R., Benda, B., & Faris, R. (2020). Indrone: Visualizing Drone flight patterns for indoor building inspection tasks. *Enabling The Development And Implementation of Digital Twins – Proceedings of the 20th International Conference on Construction Applications of Virtual Reality – (30th Sep–2nd Oct 2020)*, 3, 281–290.

Elghaish, F., Matarneh, S., Talebi, S., Kagioglou, M., Hosseini, M. R., & Abrishami, S. (2021). Toward digitalisation in the construction industry with immersive and drones technologies: A critical literature review. *Smart and Sustainable Built Environment*, 10(3), 345–363. 10.1108/SASBE-06-2020-0077

Gheisari, M., & Esmaeili, B. (2016). Unmanned aerial systems (UAS) for construction safety applications Masoud. *Construction Research Congress*, 2016, 2642–2650.

Golizadeh, H., Hosseini, M. R., Martek, I., Edwards, D., Gheisari, M., Banihashemi, S., & Zhang, J. (2019). Scientometric analysis of research on "remotely piloted aircraft": A research agenda for the construction industry. *Engineering, Construction and Architectural Management*, 27(3), 634–657. 10.1108/ECAM-02-2019-0103

Kang, S., & Miranda, E. (2006). Planning and visualisation for automated robotic crane erection processes in construction. *Automation in Construction*, 15, 398–414. 10.1016/j.autcon.2005.06.008

Khan, N. A., Jhanjhi, N. Z., & Brohi, S. N. (2020). *Emerging use of UAV's: Secure communication protocol issues and challenges. In Drones in Smart-Cities.* Elsevier Inc. 10.1016/B978-0-12-819972-5/00003-3

Matsas, E., & Vosniakos, G. C. (2017). Design of a virtual reality training system for human–robot collaboration in manufacturing tasks. *International Journal on Interactive Design and Manufacturing (IJIDeM)*, 11(2), 139–153. 10.1007/s12008-015-0259-2

Mendes, E., Albeaino, G., Brophy, P., Gheisari, M., & Jeelani, I. (2022). Working safely with Drones: A virtual training strategy for workers on heights. *Construction Research Congress*, 3–C, 964–973. 10.1061/9780784483978.098

Onososen, A. O., & Musonda, I. (2022). Research focus for construction robotics and human-robot teams towards resilience in construction: Scientometric review. *Journal of Engineering, Design and Technology*. 10.1108/jedt-10-2021-0590

Prabhakaran, A., Mahamadu, A., & Mahdjoubi, L. (2022). Understanding the challenges of immersive technology use in the architecture and construction industry: A systematic review. *Automation in Construction, 137*(March), 104228. 10.1016/j.autcon.2022.104228

Rennie, J. (2016). Drone Types: Multi-Rotor vs Fixed-Wing vs Single Rotor vs Hybrid VTOL. *AUAV.Com.* https://www.auav.com.au/articles/drone-types/

Sakib, N., Asce, S. M., Chaspari, T., & Behzadan, A. H. (2021). Physiological data models to understand the effectiveness of Drone operation training in immersive virtual reality. *Journal of Computing in Civil Engineering*, 35(1), 1–13. 10.1061/(ASCE)CP.1943-5487.0000941

Wen, J., & Gheisari, M. (2020). Using virtual reality to facilitate communication in the AEC domain: a systematic review. In *Construction Innovation* (Vol. 20, Issue 3, pp. 509–542). Emerald Group Publishing Ltd. 10.1108/CI-11-2019-0122

Wettergreen, M. (2014). Engineering Decision Making 7: Running Scoring Matrices. https://www.youtube.com/watch?v=ePMV-NrmUXo

Appendix

Table 17.3 Asset sources

Model	Source
Building model 1	Construction site – modular exterior and props
Building model 2	Modular Construction Site
Trees	Realistic trees
Construction Toolkit	Construction Package
Terrain	Terrain Sample Asset
Rocks	Rock Pack

Table 17.4 Platform specifications

Platform/Tool	Specifications
Unity	Free version, v. 2022.1.16
Visual studio (IDE)	Community edition, v.2022
Computer/OS	Dell Inspiron G15 15 5515-3472 AMD Ryzen 5 5600H up to 4.20GHz Processor, 16MB Cache, 6x Cores, 12x Threads/16GB DDR4 RAM/ 256GB Ultra-Fast NVME SSD/15.6" FHD (1920 x 1080) Anti-glare 120Hz Refresh Rate With IPS-Level Display/NVIDIA GeForce RTX 3050 4GB GDDR6 Dedicated Graphics/Windows 10 Home (64bit)
VR Toolkit	XR Interaction toolkit, v. 2.1.1
HMDs	Oculus Quest 2, 64 GB HDD, 1832×1920 per eye, 6GB RAM, 2–3 hours battery life

18 Application of Blockchain Technology in the Engineering and Construction Sector

A State-of-the-Art Review

Arturo De Jesús Miranda, Emmanuel Daniel, and Hamlet Reynoso Vanderhorst

University of Wolverhampton, Wolverhampton, United Kingdom

18.1 Introduction

The architecture, engineering and construction (AEC) industry accounts for approximately 13 per cent of the world GDP, but their productivity growth does not reflect its relevance to the global economy in the last 20 years (Barbosa et al., 2017). The productivity of the sector is often seen as laggard or stagnated (Fulford and Standing, 2014). AEC has aimed to hyper-automate its process for supporting and reacting to technology developments during the last decade, however, there are limitations in terms of comprehending the role of emerging technologies, their problems and applications in general. To address this issue, the sector is evolving by implementing productivity-enhancing technology. There have been a few published studies that compare the importance of blockchain in the (AEC) business to other developing technologies like drones or virtual reality (VR).

Some researchers from the AEC field (Mudan Wang, Cynthia Changxin Wang, Samad Sepasgozar, and Sisi Zlatanova, 2020) mentioned that some of the current adoptions of technologies and approaches towards digitalisation in construction are related to "building information modelling" (BIM), radio frequency identification devices (RFID), global positioning systems (GPS), the Internet of Things (IoT), geographic information systems (GIS), sensors, augmented reality (AR), virtual reality (VR), photogrammetry, laser scanning, artificial intelligence (AI), 3D printing, robotics, big data and blockchain. Although digital technologies are rapidly being utilised in construction, the repercussions of their deployment in the engineering and construction industry are yet unknown, particularly in the case of blockchain adoption, which is one of the most recent technologies discussed within the research community.

Blockchain technology may have a pivotal role in the transformation of multiple sectors including architecture AEC (Perera, et al., 2020; Xu, L. D., Xu, & Li, 2018). Even though its role, a recent research has concluded that the blockchain application in AEC is still relatively fresh and dispersed across numerous subjects (Xu, Chong and Chi, 2022). Transparency, speed in transactions and security are some of the actual subjects of this technology for the sector, in summary, but researchers have identified areas of application in order to define the trends despite the resistance to change from the organisations.

The consistent six areas of applications of blockchain are related to AEC are "BIM and Computer Aided Design (CAD), contract management and smart contracts, construction project management, smart buildings and smart cities, construction

DOI: 10.1201/9781003408949-18

supply chain management, and property ownership, land titles, asset management and maintenance in real estate" (Plevris, Lagaros and Zeytinci, 2022; Zhaojing Wang et al., 2020a; Xu et al., 2022). Some of these areas of applications are still in proof of concepts that could be classified as future works, in others there are significant number of tests that can provide credibility for investment (Perera et al., 2020). However, based on this area of applications, extension into other realms of applications has presented frameworks that properly arrange the key potentials and convergence of the blockchain technology. Perera et al. (2020) believe that using these technologies has enormous potential. As a technological convergence, blockchain has the potential to combine various technologies and eliminate some of their limitations, resulting in a massively increased potential for current applications in the industry for construction by identifying terms of drivers, features and barriers. Similarly, Plevris, Lagaros and Zeytinci (2022) conducted research focused on blockchain in the built environment, intending to adopt a unified framework to encourage its use, particularly in the construction industry. In this study, smart energy, smart cities and the sharing economy, smart governance, smart housing, intelligent transportation, BIM, construction management, business models and organisational structures are the seven themes highlighted. Then, the investigations in blockchain are willing to resolve issues regarding fraud, transparency of the transactions, rapid payment methods, digital records and others.

However, there are social challenges that inflict directly in the application of blockchain according to Li, Greenwood and Kassem (2019). In their research findings, the lack of communication, cooperation, trust between parties, poor productivity, late payments, a lack of policy compliance and issues surrounding ownership and intellectual property rights were identified as significant challenges causing slow technological adoption in the construction industry.

Overall, there seems to be some evidence to indicate that the application of blockchain technology in the construction and built environment is evolving rapidly but social and technical barriers should be addressed for its massive adoption. To deal with this, some systematic studies have been carrying out an extensive elaboration of the problem to understand the present state of this technology but detailed answers have not yet been found (Cheng, Liu, Xu, and Chi, 2021). Although prior studies in this research area have contributed significantly, several limitations justify the need for more research. In the past five years, some blockchain studies have been conducted to provide an understanding and identify the gaps for investigations on the topic. These investigations have been adequately helpful for the state of the art of the technology but (i) do not provide an overall direction of where the databases and journals are appointing towards the blockchain technology, (ii) quantitatively do not provide evidence of lack of studies and finally (iii) do not show the suitable keywords for exploring blockchain technology in the AEC industry. Breaking down large amounts of information into manageable units from two databases (Scopus and Web of Science) will allow us to visualise the trends and clusters emerging from the scientific community. Furthermore, there is a need for more academic research on the subject to understand further the direction, challenges and opportunities of applying blockchain technology in the (AEC). Hence, this research aims to discover and analyse the research trends, challenges and advancements in blockchain use in engineering and construction.

18.2   Methodology and Data Collection

To assist analyze and identify the present knowledge on blockchain in the construction and engineering industries, a bibliometric research of the literature was conducted utilising best practices in scientific knowledge and a systematic literature review approach. Wallin (2005) described a systematic review as a document identification procedure that may discover, assess and analyse foundational research to answer specific issues. The data selection process is straightforward and repeatable, which can serve researchers in increasing the validity and dependability of their findings. According to Kitchenham (2004), there are many reasons to conduct a systematic review, summarise empirical evidence of technology with the benefits and limitations, and identify the gaps in the current research to correctly place new research initiatives with a solid background and framework. Thus, we can imply that to address this research, the best methodology is a systematic literature review along with a bibliometric analysis and content analysis because it helps to identify the field's trends, gaps and challenges. This systematic review was carried out in three phases using secondary data: (1) bibliometric analysis, (2) paper retrieval and screening method, and (3) content analysis.

The bibliometric analysis incorporates quantitative data concepts. This method is useful for business research in engineering and construction, as well as other sectors that require sufficient breadth in a topic. The most important information is key nations, places, terminology, years and an outline of the issue. It is often used to retrieve databases such as Scopus, Web of Science, PubMed and Scival. Each database has a mechanism for interoperability that is compatible with specialised software for interpretation (VOSviewer, Gephi, Nvivo, Citenet Explorer and others). This type of analysis supports the importance and impact of a topic; reveals the strength and weakness of the current research trends, locates knowledge within the scientific sphere and presents in a timeline the trends emerging in the topic. Therefore, this research will provide a broader understanding of the current knowledge of blockchain technology applied to construction and all the other integrations across the engineering and built environment. For these reasons, breadth and depth are pursued in the bibliometric analysis obtained through Scopus database and Web of Science utilising the tools of VOS Viewer. However, depth explication for the research is intended to convey a selection criterion for screening in two parts to comprehend and examine the interrelation of the blockchain with multiple technologies and other areas of study. Nvivo software was used to identify the key application, challenges and contributions throughout the content analysis.

18.2.1   Bibliometric Methodology

The Scopus database was chosen as the primary retrieval source due to its wide coverage of scholarly articles. Additionally, the Web of Science database was used to compare and confirm the location of the researchers' expertise and trends relating to blockchain technology. The steps taken were as follows: First, a fast assessment of the most recent scientific literature indicates how important blockchain application in building and engineering has grown in recent years for the scientific community. The first phase of the search was conducted by searching:

Table 18.1 Codes utilised for retrieving articles

For Scopus	Web of Science
TITLE-ABS-KEY (blockchain AND construction) AND (LIMIT TO (PUBSTAGE, "final")) AND (LIMIT-TO (OA, "all")) AND (LIMIT-TO (PUBYEAR, 2023) OR LIMIT-TO (PUBYEAR, 2022) OR LIMIT-TO (PUBYEAR, 2021) OR LIMIT-TO (PUBYEAR, 2020) OR LIMIT-TO (PUBYEAR, 2019) OR LIMIT-TO (PUBYEAR, 2018) OR LIMIT-TO (PUBYEAR, 2017)) AND (LIMIT-TO (DOCTYPE, "ar")) AND (LIMIT-TO (SUBJAREA, "ENGI")) AND (LIMIT-TO (LANGUAGE, "English"))	WEB OF SCIENCE CORE COLLECTION FOR: BLOCKCHAIN CONSTRUCTION (ALL FIELD); REFINED BY: PUBLICATION YEARS: 2023 OR 2022 OR 2021 OR 2020 OR 2019 OR 2018 OR 2017; DOCUMENT TYPES: ARTICLES; OPEN ACCESS: ALL OPEN ACCESS; RESEARCH AREAS: ENGINEERING; LANGUAGES: ENGLISH

"Organisations construction" OR aec OR aeco OR "Build Environment" OR "Engineering OR Civil Engineering" were combined in order to produce results with each of the combinations. Then, the query returned 2,225 unique manuscripts in which 139 are unique from Scopus and 2,091 for Web of Science and 80 journals as intersection between the two databases as presented in the tables below. Further analyses (performance, co-word, cluster, visualisation and network) were used to assess the present and/or future relationships between the topics within the field by focusing on the title, abstract, key-words, full text, graph, year of publication, authors, countries or index. It supports the researcher's interpretations of the field of investigation. Furthermore, the VOS Viewer supports retrieving files from Web of Science, Scopus, PubMed, Lens and Dimensions to produce the visualisation map.

The process in merging two different databases was managed meticulously. The Web of Science database manages another set of data that limits the merging of different resources easily. Therefore, the details of authors, Article Title, Source Title, languages, Document Type, abstract were combined with the Scopus database in order to produce a VOS viewer map. Furthermore, criterium of citation, relevancy to the topic and other fundamental aspects to include literature to the study were considered previous this action.

The years provide an idea of when it happened, authors/countries give the idea who/m are contributing, and the graph brings a perspective of the links involved in the topic. In the analysis, the words are clustered according to the commonalities in publications or themes. These thematic clusters enrich the breadth of the field and are later useful for in-depth investigation with other software such as Nvivo.

Table 18.2 Keywords search and number of publications for Scopus and Web of Science database

No.	Keywords Combination	Keywords 2nd	Criteria	SCOPUS	Web of Science
1	Blockchain	Construction	Open Access, 2017–2023,	105	110
2		Civil Engineering	AR, ENGI, ENGLISH,	1	44
3		Engineering	FINAL	88	1637
4		AEC		5	5

(*Continued*)

Table 18.2 (Continued)

No.	Keywords Combination	Keywords 2nd	Criteria	SCOPUS	Web of Science
5		AECO		1	–
6		Build Environment		37	73
7	Digital Ledger	Construction		7	8
8		Civil Engineering		–	1
9		Engineering		6	63
10		AEC		–	–
11		AECO		–	–
12		Build Environment		3	3
13	Smart Contract	Construction		37	50
14		Civil Engineering		–	55
15		Engineering		40	935
16		AEC		2	2
17		AECO		2	1
18		Build Environment		11	62
19	dlt	Construction		8	7
20		Civil Engineering		–	7
21		Engineering		10	75
22		AEC		1	1
23		AECO		–	–
24		Build Environment		–	4
25	decentralised AND	Construction		–	–
26	autonomous AND	Civil Engineering		–	–
27	organisations	Engineering		3	13
28		AEC		–	–
29		AECO		–	–
30		Build Environment		–	–
TOTAL				367	3156
UNIQUE MANUSCRIPTS				219	2171

18.2.2 *Paper Retrieval Strategy and Screening Method for Systematic Review*

But nonetheless, reducing the amount of articles can be a time-consuming and laborious effort. As a result, these datasets were subjected to additional filtering. The screening approach was then applied, with keywords from the bibliometric analysis identified in the title and/or abstract in connection to the goal being selected. This stage's scope was narrowed by picking 37 journal articles based on citation number and relevance to the topic. These articles were found in both databases in order to assess the sector's highlights. Ultimately, for the systematic review, Nvivo software was utilised to gather details about the application and difficulties through content analysis.

18.3 Results Describing the Bibliometric Analysis and Content Analysis

18.3.1 *Bibliometric Analysis*

18.3.1.1 *Performance Analysis*

18.3.1.1.1 PUBLICATION YEARS

After the initial search, the results of paper publication per year clearly show the interest from the academia in blockchain technology for construction and engineering Figure 18.1. This graph displays the search results by year from 2017 to 2023. In the figure, we can see the trend and an uptrend, starting from the year 2021. However, the application of this in

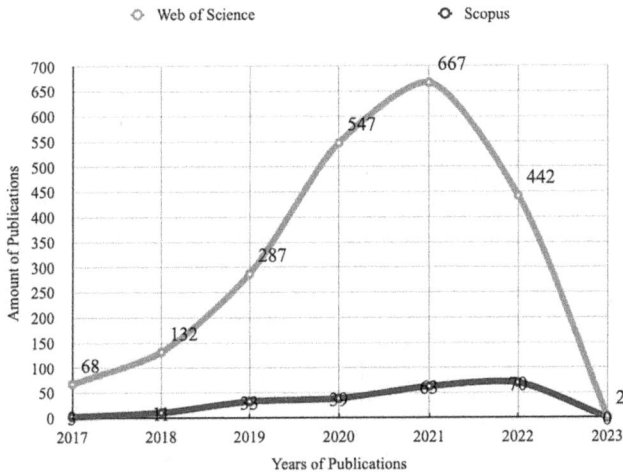

Figure. 18.1 Amount of manuscripts for a year according to the databases. According to SCOPUS and Web of Science database, this figure presents the recent rise of blockchain publications. Web of Science showed an increase in 2021 before Scopus.

AECO sector began at an earlier stage in 2009 with the creation of Bitcoin (Böhme, Christin, Edelman, and Moore, 2015). But, significant interest started in 2017 as researchers may have improved their outcomes in the last recent years.

18.3.1.1.2   TOP COUNTRIES

The production rate by country seen in Figure 18.2 used the same data from Scopus for the keyword analysis. After analysing the data, the five countries with the highest production can be seen. The country with the most literary production on the subject is China (720 and 73), followed by the UK (27 in Scopus), but South Korea presented 323 publications on Web of Science. Furthermore, India has the fourth place in publications homogeneously, but in Scopus, India shares this place with the United States. The implications these countries have published on this topic mean that financial implications and regulations are considered to advance the scientific knowledge and best practice of the technology.

18.3.1.1.3   AUTHORS AND JOURNALS

The data relating to authors was complex to compare as the database has different citation criteria. However, the authors provide an idea of who/whom has impacted the topic related to the amount of publication. Still, the amount may not represent citation, or in other words, a shared perspective of the scientific community that a paper provides real insights into a topic. Therefore, it was focused on citations and sources, as presented in Table 18.3.

The blockchain subject is centred on engineering and computer science journals rather than building magazines. Although some of these contributions are not directly relevant to the construction industry, their algorithm and technical approach are applicable to the sector's practice. The most cited journals include technical topics such as (1) integration with other technologies, (2) guarantee of privacy and security, (3) supply chain, (4) transactions

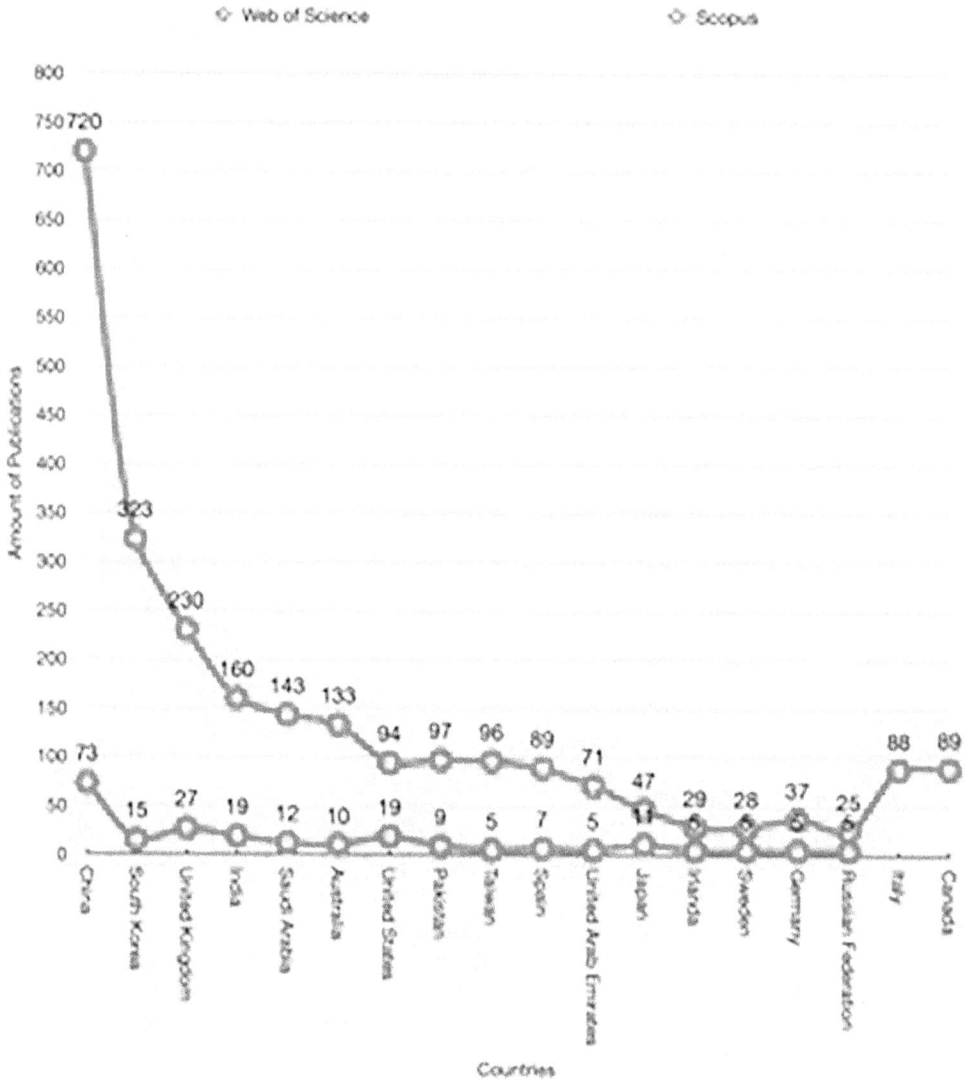

Figure 18.2 Countries with publication records between 2017 and 2023. The figure shows that China has the highest publication number in both databases. But South Korea and the United Kingdom, as well as India and the United States, are the ones who differ in the following place in a publication from both databases. Furthermore, there are countries only in specific databases such as Italy, Canada and others.

and (5) algorithms and education. Also, both the database IEEE Access journal and the other publications have indicated a higher prevalence on the topic.

18.3.1.2   Visualisation and Interpretation of the Clustering Analysis and Co-word Analysis

There were three VOS maps developed. Two were mined utilising Scopus and Web of Science databases, and a third was mined using a combination of both. Observing the

Table 18.3 Top authors and journals Scopus

No.	Authors	Title	Year	Source Title	Cited By
1	Turkanović, M. Hölbl, Heričko, & A. Kamišalić	EduCTX: A blockchain-based higher education credit platform	2018	IEEE Access	264
2	Guido Perboli; Stefano Musso; Mariangela Rosano	Blockchain in Logistics and Supply Chain: A Lean Approach for Designing Real-World Use Cases	2018	IEEE Access	219
3	Greenwood; Mohamad Kassem	Blockchain in the built environment and construction industry: A systematic review, conceptual models, and practical use cases	2019	Automation in Construction	164

Table 18.4 Top authors and journals Web of Science

No.	Authors	Title	Year	Source Title	Cited By
1	Zhetao Li; Jiawen Kang; Rong Yu; Dongdong Ye; Qingyong Deng; Yan Zhang	Consortium Blockchain for Secure Energy Trading in Industrial Internet of Things	2018	IEEE TRANSACTIONS ON INDUSTRIAL INFORMATICS	525
2	Daniel K. Molzahn; Florian Dörfler; Henrik Sandberg; Steven H. Low; Sambuddha Chakrabarti; Ross Baldick; Javad Lavaei	A Survey of Distributed Optimization and Control Algorithms for Electric Power Systems	2018	IEEE TRANSACTIONS ON SMART GRID	470
3	Qi Xia; Emmanuel Boateng Sifah; Kwame Omono Asamoah; Jianbin Gao; Xiaojiang Du; Mohsen Guizani	MeDShare: Trust-Less Medical Data Sharing Among Cloud Service Providers via Blockchain	2019	IEEE ACCESS	414

number of clusters, Scopus (by its small number of articles) produced three clusters covering the topics of (1) cloud services and AI referring to enterprise infrastructure, (2) security within the blockchain system and the different layers of them (decentralised, algorithms, devices, etc.), (3) the applicability for the construction industry based on frameworks, policies and cryptos. The predominant discussion occurring on this database is around the social implications of scientific knowledge produced in blockchain.

On the other hand, the Web of Science map reported other topics related more to technical aspects, which expanded the breadth of the blockchain application. The clusters refer to the topic of (1) intelligent transport system, (2) consensus algorithms, (3) authentication protocols, (4) vulnerability within the IT infrastructure, (5) Application of supply chain system in construction, (6) Honeypot instances, (7) smart contracts, (8) issues in energy consumption by mining and smart buildings, (9) control of the datasets. This database provided enough topics to discuss and evaluate blockchain for

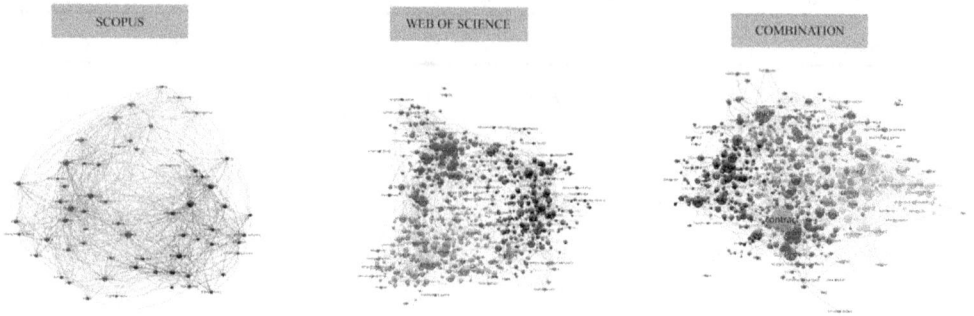

Figure 18.3 VOS viewer maps generated from Scopus, Web of Science and combination of both.

engineering and construction. Finally, the conjunction between the two databases summarises the topic in four major pillars: (1) Security of the system (authentication, traceability, servers, security analysis, etc.), (2) transformation of the supply chain and buildings operations (business models, standards, stakeholder management, big data, interoperability, etc.), (3) Energy consumptions and resources to make it sustainable, (4) Paradigm change in the transportation system (Figure 18.3).

The combination or conjunction of the database reveals eight clusters that significantly map the research in terms of challenges in blockchain. Consensus protocols for decision-making process in blockchain systems, cybersecurity issues, the risks in the application of supply chain management with BIM, tokenisation of sustainable practices in smart cities, challenge in best practices for appropriate electricity distribution systems, adequacy of base station and path trajectory platform for Unmanned Aerial and ground vehicles tasks are some of the remarkable practical problems that the clusters underlined. However, a further analysis was made to understand in-depth the challenges.

18.3.2   *Content Analysis*

After screening the data, the 37 papers were subjected to content analysis, allowing the categorisation of papers into four sections of priority to apply and further research blockchain technology: (1) Document management and quality of the information, (2) Smart contracts and payments, (3) Supply chain management in construction and (4) Other technologies for construction like BIM, IoT, big data and other technology applications in the industry 4.0. However, there is necessary to have an overview of blockchain technology to understand the context of the technology and how the blockchain can transform the construction and engineering industry.

18.4   Literature Review of the Blockchain Technology

Zheng et al. (2020) described a blockchain as a system that enables information and transaction exchanges to be executed without the involvement of a third party. This database is shared between the network nodes and stores the information digitally. The data is structured in groups or blocks with a specific size or storage capacity, and when the block is completed, another block is added to the chain. Also, this data structure creates an irreversible data chronology when implemented in a decentralised manner. The adoption of blockchain significantly increases traceability, transparency, and information sharing throughout project development, improving cooperation and ensuring the trustworthiness of information exchanges (Sheng et al., 2020; Wang et al., 2020; Zheng et al., 2020). The accountability and trust come from the nature of the blockchains because the transaction of information is confirmed by all the participants in the node network to be executed by the blockchain technology or Distributed Ledger Technology. Furthermore, Zheng et al. (2019) explain that the word "distributed" is not only to point out that the data is distributed between the participants, but at the same time a record is kept among the participants. Blockchain technology at its fundamentals can depend on the extent of its application can be public or private. A public Blockchain provides read and write permissions and access to all users. Although, some public Blockchains restrict access to either reading or writing. On the other hand, a private Blockchain restricts access to a small number of trusted participants to keep user information secret (Miraz and Ali, 2018).

18.4.1 Application of Blockchain Technologies in Engineering and Construction

18.4.1.1 Document Management and Quality of the Information

Information security, integrity and data transparency were among the most studied topics in this category. To give context to a document management system, Das et al. (2022) categorised the primary functions of a document management system in the AECO sector, which are: "(1) approval workflow management i.e., facilitating customisable document approval workflows in which project participants are connected using pre-defined business rules to support processes such as design review and RFI (Request for Information) management, (2) 'document lifecycle recording', i.e. establishing 'audit logs' to help project comply with standard guidelines and monitor potential problems, (3) 'document version management, i.e. storing, categorising'". Furthermore, although all these documents are handled digitally, there is a lack of coordination and integration between all the processes that produce these deliverables since they result from many isolated processes. The engineering and construction sector faces many challenges, collaboration and workflows, integrity of the data, the intellectual property of the documents, the trust, traceability, and security of the data are among the things. The property of the information was another topic mentioned by (Perera et al., 2020; Xu et al., 2022; Z. Zheng et al., 2020), they pointed out that traceability of the blockchain creates a distributed record and has all the signatures, this is on the basis of a property ownership exchange protocol, securing the data in all the process and keeping a trustworthy record for all the stakeholders. Wang et al. (2022) proposed the use of intelligent keyless signature, based on parallel and edge computing to ensure data security. This uses one of the properties of blockchain, the consensus mechanism to validate the data between the network peers. Liu et al. (2019) proposed a framework combining blockchain and BIM for sustainable building design coordination and collaboration in multiple building life cycle stages, by capturing the data exchange at three user-driven levels: user, system and transaction. Zheng et al. (2019) suggested the application of a cloud-based platform BCBIM to solve the problems of the cloud base information transactions in the BIM environment, based on private and public blockchain. Sheng et al. (2020) stated that the lack of a uniform and open process for handling quality information in the construction sector jeopardises information certification and may lead to conflicts among stakeholders when problems arise. He created a hyper-ledger-fabric blockchain architectural solution to decentralise information management, resulting in more reliable and secure information management systems. Perera et al. (2020) in their research, they talked about the amount of data produced in the sector and the problems of storing the data. They proposed Blockchain file storage sharing, as a solution for construction practitioners to store data and manage documentation in a blockchain that solves the problem of the unused data storages.

18.4.1.2   *Smart Contracts and Payments*

Most of the papers mention the smart contracts in some form applying them in combination with other technologies like BIM, IoT, with supply chain, contract management, document management (Sonmez, Ahmadisheykhsarmast and Güngör, 2022; Gourisetti et al., 2021; Hamledari and Fischer, 2021a; Ciotta et al., 2021). Hu et al. (2020) described a smart contract as an algorithm that automates contractual transactions predefined by the parties. A smart contract function is programming their contract clauses and giving if-else-if flow statements, allowing the blockchain to execute the conditions automatically and take the corresponding actions in each case. (Z. Zheng et al., 2020). Several studies which presented frameworks or systems for automated payments and highlighted the importance to solve the cashflow and insolvency problems that attacks contractors, subcontractors and suppliers caused by late payments or unsolved disputes have been noted by a few researchers (Ahmadisheykhsarmast and Sonmez, 2020; Sigalov et al., 2021; Sonmez et al., 2022). Sigalov et al. (2021) presented a concept in combination of BIM with blockchain-based smart contracts is adopted to ensure traceability, trust and transparency in the enforcement of the contracts and payments. The framework used is a container-based data exchange, and the digital contract management workflow. On the other hand, he stated that due to the lack of legal precedents and rules, binding construction contracts pose a hurdle. In the meanwhile, a feasible solution is a semi-automated method of providing a legally compliant and practical answer. (Hamledari and Fischer, 2021b; Saygili, Mert, and Tokdemir, 2022; Sigalov et al., 2021). Moreover, Hamledari and Fischer (2021b) add that using smart contracts in the industry depends on the adoption of the technology by all the participants with no exception, the clients and contractors, to make the technology work appropriately. Sonmez, Ahmadisheykhsarmast and Güngör (2022) conducted research and identify that connecting real-world data to the blockchain might be difficult for building projects. The proposal to solve this problem was to integrate an as-built model with blockchain; by applying the concept and conducting a survey of professionals in the area, most of them recognised the potential of blockchain with lump-sum contracts. Even though some of the participants raised their concerns about the data malleability of the blockchain clauses in case of changes and modifications, this was mentioned by a few articles due to the complexity of the problem in relation of the immutability of the blockchain (Das, Luo and Cheng, 2020; Nawari and Ravindran, 2019; Sonmez, Ahmadisheykhsarmast and Güngör, 2022).

The data indicates the advantages of applying this technology to construction, removing the trust issues by augmenting transparency, reducing costs by solving problems in an automatic form and eliminating third-party actors and large fees with the dispute resolutions (Saygili, Mert and Tokdemir, 2022; Ahmadisheykh sarmast and Sonmez, 2020).

Dakhli, Lafhaj and Mossman (2019) conducted a study to determine the potential cost savings in real estate using blockchain and eliminating unnecessary parts of the construction chain value. The study looked at 56 residential buildings, and the results suggested that blockchain using smart contracts might save 8.3 per cent of the entire cost of building a house, with a standard deviation of 1.26 per cent. The savings are primarily from transactional expenses entrenched in a traditional organisation (Dakhli et al., 2019).

Hu et al. (2020) studied the main challenge of the smart contracts, which is the "correctness and unambiguity are its essential formal properties, but also conformance to any legislation governing the matter of the transaction", the study addresses these challenges by developing a suitable engineering process to satisfy the rigorous requirements while allowing for mass production and distribution. Hu et al. (2020) described the framework as "Smart contract engineering (SCE) is a systematised, modularised, and judgmental process for the smart contract that is based on development, maintenance, and execution, and that integrates with software engineering, intelligent methods as well as legal code technology". This aims to decrease mistakes and increase efficiency throughout the contract development process while also encouraging the standardisation of contract design approaches. The system verifies that the contract code and wording are consistent and compliant with all the rules and properties of the smart contract during the value transfer process, mitigating the losses (Hu et al., 2020).

18.4.1.3 Supply Chain Management in Engineering and Construction

Christopher (2011) described supply chain management as "the upstream and downstream relationships with suppliers and customers to deliver superior customer value at less cost to the supply chain as a whole". There is where the blockchain enters, offering multiple solutions for the chain created between the suppliers and the customers. From the design specifications and documents to the acquisition of equipment, materials, human resources and engineering equipment are all part of supply chain management. For instance, Wang et al. (2020b) developed a framework combining BIM and blockchain to address real-world difficulties in precast supply chain automation, information traceability and transparency. They conducted a pilot study with a precast plant in China to demonstrate the effectiveness of the system, Shanghai Jiangong's management team also provided encouraging comments, agreeing that the blockchain-based solution can improve precast supply chain management in practice by improving the sharing of the information between all the participants, facilitating real-time schedule control and execution, which allows them to react and control in a more precise manner the supply chain and real traceability of the information during all the process. Moreover, Bakhtiarizadeh et al. (2021) identify the importance of the information for the prefabrication supply chain industry. Hence, he concludes with the adoption of blockchain as an information integration tool in the prefabrication industry in New Zealand.

18.4.1.4 Other Technologies for Engineering and Construction

Liu et al. (2019) identify the challenge BIM has in sustainable building design and how blockchain in conjunction with other technologies will help overcome the BIM adoption challenges. They describe that the problems presented are financial, technical and legal risks, the resistance presented by the stakeholders in the adoption of the strategies, the intellectual property and cyber security and the responsibility and trust of all the parties involved in the BIM environment. Moreover, according to Liu et al. (2019) the main application of blockchain in sustainable design and construction is the smart contracts to manage the information of the project, enforce responsibility, give transparency and trust and enhance collaboration and dispute resolution, which Teisserenc and Sepasgozar also mentioned (2021) that developed a framework for a sustainable blockchain digital-twins for all the life cycle and dimensions of BIM from 3D,4D,5D,6D and 7D. The research proposes a new category, a contractual dimension (8DcD), that uses smart contracts to help close the BIM gap and improve the data value chain's trust, data integrity and security over time. In the same vein, Zheng et al. (2019) explained that "BIM can be summarised into four aspects, namely, integrating with various databases, facilitating document management, visualising analytical processes and results, and providing sustainability analyses and simulation". As a result, they argue that the future development of BIM must be the application of advanced communication technology to improve the efficiency of the construction industry. He presents the combination of IoT sensors to collect information in real-time wirelessly, big data to store and process the data and blockchain to improve the security and integrity of BIM data and the security concerns associated with updating BIM models and parameters in intelligent structures (R. Zheng et al., 2019).

Similarly, M. Wang, Wang, Sepasgozar, and Zlatanova (2020) conducted a systematic review to identify the technology adoption in Industry 4.0 for off-site construction. The trend identified by the study suggests the most used digital technologies in the construction industry adoption of BIM in conjunction with technologies like IoT to collect information in real-time and photogrammetry, the use of Radio Frequency Identification Devices (RFID) to help track the labour, materials and equipment, as well as the application of artificial intelligence (AI) to increase the efficiency and accuracy of other technologies. Tiwari and Batra (2021) proposed the creation of a platform for facility management uses with the combination of IoT sensors and blockchain smart contracts with proof-of-concept, to automate the maintenance works. Hence ensuring automation, encouraging transparency and giving security to the whole process. Nawari and Ravindran (2019) evaluated the potential integration of blockchain with the BIM process in the design review process with smart contracts and Hyperledger fabric technologies. According to Nawari and Ravindra (2019), a distributed ledger blockchain system generation platform with a modular architecture that provides a flexible and adaptable digital foundation with significant levels of secrecy and scalability. It has a ledger, employs smart contracts and is a mechanism through which players control their transactions, like blockchain technology. This also aids the BIM process by allowing participants to create subgroups in the network and create a separate ledger of transactions while being part of the Hyperledger Fabric network and sharing that transaction with the rest of the nodes.

Lee et al. (2021) designed and tested a blockchain and digital twin infrastructure for traceable data transmission to close the gap between BIM technology and the digital twin, using IoT technology to capture the data on-site and blockchain to authenticate and give transparency and traceability. The study shows that the blockchain helped achieve near real-time data, sharing the information to the stakeholders within 3.42 seconds in a reliable, transparent and secure way the information, without any intermediaries.

18.4.2   *Roadblocks and Challenges in the Adoption of Blockchain in the Engineering and Construction Industry*

Although the implementation of all of these technologies in the engineering and construction industry appears promising, there were typical constraints mentioned in the studies, such as a lack of regulations and legal implications, privacy and security issues with data, high implementation costs, the inflexibility of blockchain technology due to the immutability of the blocks (malleability or adaptability) and a lack of adaptability with other technologies.

Table 18.5 Challenges in the adoption of blockchain technology in construction

Article	Challenges							
	Legal Implications and Regulations	*Privacy and Security of the Data*	*Adoption Resistance*	*Implementation Cost*	*Modification and Adaptability to the Blockchain Code (Immutability)*	*Lack of Development and Adaptability with Other Technologies (Interoperability)*	*Validity of Data Uploaded to the Blockchain*	*The Actual State of the Technology (Maturity)*
A blockchain-based integrated document management framework for construction applications	1	1					1	
Construction quality information management with blockchains	1	1	1	1	1		1	1
Applications of Blockchain Technology beyond Cryptocurrency						1		1
A Systematic Review of Digital Technology Adoption in Off-Site Construction: Current Status and future Direction towards Industry 4.0	1	1		1				1
Digital building twins and blockchain for performance-based (smart) contracts	1			1				1
Building Information Management (BIM) and Blockchain (BC) for Sustainable Building Design Information Management Framework			1	1			1	
Evaluation of Production of Digital Twins Based on Blockchain Technology	1	1		1				

(Continued)

Table 18.5 (Continued)

Article	Challenges							
	Legal Implications and Regulations	Privacy and Security of the Data	Adoption Resistance	Implementation Cost	Modification and Adaptability to the Blockchain Code (Immutability)	Lack of Development and Adaptability with Other Technologies (Interoperability)	Validity of Data Uploaded to the Blockchain	The Actual State of the Technology (Maturity)
Public and private blockchain in construction business process and information integration		1		1	1		1	1
The Potential of Blockchain in Building Construction	1	1		1	1		1	
Securing interim payments in construction projects through a blockchain based framework		1			1			
Measuring the impact of blockchain and smart contracts on construction supply chain visibility	1	1	1			1		1
Construction payment automation using blockchain-enabled smart contracts and robotic reality capture technologies						1	1	1
Blockchain smart contract reference framework and program logic architecture for transactive energy system		1				1		1
Blockchain Enabled Reparations in Smart Buildings-Cyber Physical System		1				1		1
BIM integrated smart contract for construction project progress payment administration					1	1		1

Automated Payment and Contract Management in the Construction Industry by Integrating Building Information Modeling and Blockchain-Based Smart Contracts		1			1		1	
A smart contract system for security of payment of construction contracts		1				1	1	
Blockchain in the built environment and construction industry: A systematic review, conceptual models and practical use cases		1	1					1
Exploring the Barriers against Using Cryptocurrencies in Managing Construction Supply Chain Processes		1	1				1	1
Total	10	13	7	7	7	6	9	12

18.5   Discussion of the Challenges

As the challenges influence the adoption of blockchain in the AECO industry different solutions should be established as a guidance for initial point to defeat them. In the following list:

18.6   Trends and Future Research Areas

Future research must address technological problems linked with the security, scalability and interoperability of blockchain systems. Hence, the study deduces the following research fields based on the researcher's obstacles and opportunities: Further empirical and case studies in the study domain that illustrate the usage of blockchain in a real-world setting are required. Furthermore, the technology's scalability must be addressed in order to manage the vast quantity of data and transactions that a real-world application will have. Similarly, data security was a major worry throughout all of the publications. Certain forms of security vulnerabilities continue to plague blockchain. While it is difficult to hack a blockchain system, it is critical to keep in mind that the technology is not without flaws. As a result, additional research should be conducted to address these security restrictions. Another topic of investigation is information privacy. Smart contracts cannot be encrypted in the same manner that plain text can due to their intrinsic nature as executable programs. Although there are private and public blockchains, further study is required to establish a viable solution that takes use of the benefits of the two forms of blockchain encryption. Future research should look on techniques to disguise information for certain stakeholders in network transaction data. Lastly, the immutability of the blockchain increases trust and security; yet, it poses issues in the engineering and construction environments since projects change and, as a result, contracts do as well. Therefore, this issue needs to be investigated further to find a solution.

18.7   Conclusion

The research goal is to identify and analyse new research subjects and advancements in blockchain use in engineering and construction and provide a general overview of the current state of blockchain technology application in the engineering and construction industry, as well as roadblocks and future research opportunities. This was achieved using a mixed-method research approach, bibliometric analysis and a systematic approach. The investigation found that many studies have explored that the use of blockchain in the AECO sector to improve its productivity by improving the collaboration, traceability and transparency in the business process of the AECO sector. However, it has been observed that because of the nature of blockchain technology, there is a need to provide up-to-date information systematically that summarises the current trends and implementations of the technology and provides directions for future work.

The originality of this study in providing where and what knowledge is focused provides academics with a method to easily search for construction topics in the blockchain. Additionally, the study contributes to the field in the following ways: Firstly, the bibliometric study provided a fast overview of the literature published in the AECO industry in connection to blockchain adoption, including the most relevant literature, research production per year, top authors, nations and author keywords. Additionally, the keyword analysis helps us to understand more clearly the technologies that are currently being researched in the field for blockchain applications.

Table 18.6 Critical discussion of the challenges of blockchain in the AECO industry

No.	Challenges	Discussion	Example of the Solution
1	Current lack and unharmonised standardisation between blockchain systems	Companies may find it challenging to use blockchain technology due to a lack of standards and compatibility among different blockchain platforms. This is due to a lack of knowledge about which platform to employ, its suitability of the purpose or how to connect it with their existing workflows and systems. To overcome this issue, key stakeholders in the AEC industry might collaborate to create common standards and protocols for blockchain-based systems. This might assist to guarantee that multiple platforms are interoperable, making it easier for businesses to use blockchain technology.	Evaluation of the standardisation of payments with cryptocurrencies in the AEC Industry.
2	Data privacy and security	The blockchain is frequently promoted as a safe platform for data sharing, there are still concerns regarding how and which sensitive data will be handled on these networks. An approach of managing this perspective might be solved by implementing strong security measures such as encryption, multi-factor authentication, zero knowledge proof systems, and access limitations. Additionally, the system might also investigate the usage of public, private, or permissioned blockchains, which restrict access to only approved users.	Accessing to personal information for enforcing the compliance of anti-money laundering acts.
3	Integration with the existing systems	In the industry there are some invested in legacy systems that may not be compatible with new technologies like blockchain. To address this challenge, a solution could be explored in ways to integrate their existing systems with new technologies through APIs or other integration tools. The consideration of using hybrid solutions that combine elements of both legacy systems and new technologies like blockchain are welcomed.	The gap closed by the blockchain in carrying out bank transaction and identification of beneficiary in a sole unified system.

(Continued)

Table 18.6 (Continued)

No.	Challenges	Discussion	Example of the Solution
4	Investment cost of implementation	Implementing new technologies like blockchain can be expensive, particularly for small size businesses. To address this challenge, industry stakeholders could work together to develop cost-effective solutions that make it easier for smaller businesses to adopt new technologies. This could include developing open-source software or offering financing options for companies looking to invest in new technologies.	Assessment of liquidity pools for business payments to supply chain beneficiaries inside of a centralised or hybrid system
5	Lack of expertise	Many engineers and construction professionals may not have experience working with blockchain-based systems, which can make it difficult to implement these technologies effectively. To address this challenge, open-source online resources should be developed by the academia that compromise the trends and education of employees in developing the skills and knowledge about emerging technologies.	They could consider partnering with universities or other educational institutions to develop specialised training programs for professionals in the engineering and construction industries.
6	Scalability of the articulated blockchain-based systems	As more businesses adopt blockchain technology, there is a risk that the system could become overloaded and slow down. This is because each transaction on the blockchain requires verification by multiple nodes, which can be time-consuming. To address this challenge, business could explore ways to improve the scalability of blockchain-based systems,	Utilising sharding, other scaling solutions, apply solidity and other platforms as a standard for the industry.
7	Regulatory uncertainty	Uncertainty in the regulatory environment is another obstacle to the implementation of blockchain technology in engineering and construction. Due to the fact that blockchain is a relatively new technology, it may be challenging for businesses to abide by current rules and regulations due to regulatory gaps or discrepancies. Industry participants might collaborate with authorities to create clear rules and regulations for the use of blockchain in engineering and construction in order to overcome this problem.	discussing how taxing may impact in blockchain solutions for cashflow problems in public and private partnership projects.

8	Lack of trust	The trustworthiness and quality of the data kept on these platforms may still be an issue, despite the fact that blockchain is sometimes hailed as a trustless system. This is especially true in fields where accuracy and dependability are crucial, like engineering and construction. Companies should look into ways to increase the traceability and transparency of data on blockchain-based systems to address this issue.	Identification and Evaluation of successful cases in applying blockchain for AEC.
9	Resistance to change	Due to their comfort with their current systems or procedures, many businesses may be reluctant to accept new technology like blockchain. Businesses could attempt to raise knowledge of the potential advantages of blockchain technology and offer rewards to staff members who adopt new technologies in order to overcome this issue.	Rewards in fomenting new effective workflows with blockchain technology.

We may use such data to determine the present state and trends in blockchain use in the AECO industry. Secondly, after conducting the screening, we submitted the papers to a content analysis based on the information obtained from the bibliometric analysis with the keyword concurrence, allowing the categorisation of chapter into four sections of priority to apply and further research blockchain technology: (1) Document management and quality of the information, (2) Smart contracts and payments, (3) Supply chain management in construction and (4) Other technologies for construction like BIM, IoT, big data and other technology applications in the industry 4.0. Thirdly, the analysis identified the current applications that have been developed so far concerning the use of the blockchain in the AECO sector and the current challenges in using the technology in the sector. The research showed that the most studied application of blockchain in the sector solves the lack of coordination and integration between all the processes with document management and other applications, such as IoT, big data and BIM. Likewise, the use of smart contracts is the most mentioned application of blockchain, and this is a solution for the trust problem in some way, combining them with other technologies such as BIM, IoT, supply chain, contract management and document management. Also, coordination and information quality are addressed with blockchain document management tools. The most mentioned challenge is the lack of privacy when the public type of blockchain is used, the maturity of the technology, and the legal implications and regulations. This reflects the obstacles associated with these technologies and points the academics to continue developing and investigating them to answer the problems reflecting the need for more research and development. Finally, we discussed the challenges and trends and future opportunities in adopting blockchain technologies for the AECO sector. Hence, this will give researchers a better idea of what applications need future work.

References

Ahmadisheykhsarmast, S., & Sonmez, R. (2020). A smart contract system for security of payment of construction contracts. *Automation in Construction,* 120. 10.1016/j.autcon.2020.103401

Bakhtiarizadeh, E., Shahzad, W. M., Poshdar, M., Khalfan, M., & Bamidele Rotimi, J.O. (2021). Blockchain and information integration: Applications in New Zealand's prefabrication supply chain. *Buildings*, 11(12), 608. 10.3390/buildings

Barbosa, F., Woetzel, J., Mischke, J., João Ribeirinho, M., Sridhar, M., Parsons, M., … Brown, S. (2017, February 27). Reinventing construction through a productivity revolution | McKinsey. Retrieved 18 April 2022, from https://www.mckinsey.com/business-functions/operations/our-insights/reinventing-construction-through-a-productivity-revolution

Böhme, R., Christin, N., Edelman, B., & Moore, T. (2015). Bitcoin: Economics, technology, and governance. *Journal of Economic Perspectives,* 29(2), 213–238. 10.1257/jep.29.2.213

Cheng, M., Liu, G., Xu, Y., & Chi, M. (2021, August 1). When blockchain meets the aec industry: Present status, benefits, challenges, and future research opportunities. *Buildings*, 11. MDPI AG. 10.3390/buildings11080340

Christopher, M. (2011). Logistics & Supply Chain Management. Retrieved from www.pearson-books.comwww.pearson-books.com

Ciotta, V., Mariniello, G., Asprone, D., Botta, A., & Manfredi, G. (2021). Integration of blockchains and smart contracts into construction information flows: Proof-of-concept. *Automation in Construction*, 132. 10.1016/j.autcon.2021.103925

Dakhli, Z., Lafhaj, Z., & Mossman, A. (2019). The potential of blockchain in building construction. *Buildings*, 9(4). 10.3390/buildings9040077

Das, M., Luo, H., & Cheng, J. C. P. (2020). Securing interim payments in construction projects through a blockchain-based framework. *Automation in Construction*, 118. 10.1016/j.autcon.2020.103284

Das, M., Tao, X., Liu, Y., & Cheng, J. C. P. (2022). A blockchain-based integrated document management framework for construction applications. *Automation in Construction*, 133. 10.1016/j.autcon.2021.104001

Fulford, R., & Standing, C. (2014). Construction industry productivity and the potential for collaborative practice. *International Journal of Project Management*, 32(2), 315–326. 10.1016/J.IJPROMAN.2013.05.007

Gourisetti, S. N. G., Sebastian-Cardenas, D. J., Bhattarai, B., Wang, P., Widergren, S., Borkum, M., & Randall, A. (2021). Blockchain smart contract reference framework and program logic architecture for transactive energy systems. *Applied Energy*, 304. 10.1016/j.apenergy.2021.117860

Hamledari, H., & Fischer, M. (2021a). Measuring the impact of blockchain and smart contracts on construction supply chain visibility. *Advanced Engineering Informatics*, 50. 10.1016/j.aei.2021.101444

Hamledari, H., & Fischer, M. (2021b). The application of blockchain-based crypto assets for integrating the physical and financial supply chains in the construction & engineering industry. *Automation in Construction*, 127. 10.1016/j.autcon.2021.103711

Hu, K., Zhu, J., Ding, Y., Bai, X., & Huang, J. (2020). Smart contract engineering. *Electronics (Switzerland)*, 9(12), 1–26. 10.3390/electronics9122042

Kitchenham, B. (2004). Procedures for Performing Systematic Reviews. *Keele University*, 33, 1–26. Retrieved from https://www.researchgate.net/publication/228756057

Lee, D., Lee, S. H., Masoud, N., Krishnan, M. S., & Li, V. C. (2021). Integrated digital twin and blockchain framework to support accountable information sharing in construction projects. *Automation in Construction*, 127. 10.1016/j.autcon.2021.103688

Li, J., Greenwood, D., & Kassem, M. (2019). Blockchain in the built environment and construction industry: A systematic review, conceptual models and practical use cases. *Automation in Construction*, 102, 288–307. 10.1016/j.autcon.2019.02.005

Liu, Z., Jiang, L., Osmani, M., & Demian, P. (2019). Building information management (BIM) and blockchain (BC) for sustainable building design information management framework. *Electronics (Switzerland)*, 8(7). 10.3390/electronics8070724

Miraz, M. H., & Ali, M. (2018). Applications of blockchain technology beyond cryptocurrency. *Annals of Emerging Technologies in Computing*, 2(1), 1–6. 10.33166/AETiC.2018.01.001

Nawari, N. O., & Ravindran, S. (2019, September 1). Blockchain and the built environment: Potentials and limitations. *Journal of Building Engineering*, 25. Elsevier Ltd. 10.1016/j.jobe.2019.100832

Perera, S., Nanayakkara, S., Rodrigo, M. N. N., Senaratne, S., & Weinand, R. (2020, March 1). Blockchain technology: Is it hype or real in the construction industry? *Journal of Industrial Information Integration*, 17. Elsevier B.V. 10.1016/j.jii.2020.100125

Plevris, V., Lagaros, N. D., & Zeytinci, A. (2022). Blockchain in civil engineering, architecture and construction industry: State of the art, Evolution, Challenges and opportunities. *Frontiers in Built Environment*, 8. 10.3389/fbuil.2022.840303

Saygili, M., Mert, I. E., & Tokdemir, O. B. (2022). A decentralised structure to reduce and resolve construction disputes in a hybrid blockchain network. *Automation in Construction*, 134. 10.1016/j.autcon.2021.104056

Sheng, D., Ding, L., Zhong, B., Love, P. E. D., Luo, H., & Chen, J. (2020). Construction quality information management with blockchains. *Automation in Construction*, 120. 10.1016/j.autcon.2020.103373

Sigalov, K., Ye, X., König, M., Hagedorn, P., Blum, F., Severin, B., ... Groß, D. (2021). Automated payment and contract management in the construction industry by integrating building information modeling and blockchain-based smart contracts. *Applied Sciences (Switzerland),* 11(16). 10.3390/app11167653

Sonmez, R., Ahmadisheykhsarmast, S., & Güngör, A. A. (2022). BIM integrated smart contract for construction project progress payment administration. *Automation in Construction,* 139, 104294. 10.1016/j.autcon.2022.104294

Tiwari, A., & Batra, U. (2021). Blockchain enabled reparations in smart buildings-cyber physical system. *Defence Science Journal,* 71(4), 491–498. 10.14429/DSJ.71.16454

Wallin, J. A. (2005). Bibliometric methods: Pitfalls and possibilities. *Basic & Clinical Pharmacology & Toxicology,* 97(5), 261–275. 10.1111/J.1742-7843.2005.PTO_139.X

Wang, M., Wang, C. C., Sepasgozar, S., & Zlatanova, S. (2020, November 1). A systematic review of digital technology adoption in off-site construction: Current status and future direction towards industry 4.0. *Buildings,* 10, 1–29. MDPI AG. 10.3390/buildings10110204

Wang, Z., Wang, T., Hu, H., Gong, J., Ren, X., & Xiao, Q. (2020). Blockchain-based framework for improving supply chain traceability and information sharing in precast construction. *Automation in Construction,* 111. 10.1016/J.AUTCON.2019.103063

Wang, Z., Wang, K., Wang, Y., & Wen, Z. (2022). A data management model for intelligent water project construction based on blockchain. *Wireless Communications and Mobile Computing,* 2022, 1–16. 10.1155/2022/8482415

Xu, Y., Chong, H. Y., & Chi, M. (2022, February 1). Blockchain in the AECO industry: Current status, key topics, and future research agenda. *Automation in Construction,* 134. Elsevier B.V. 10.1016/j.autcon.2021.104101

Zheng, R., Jiang, J., Hao, X., Ren, W., Xiong, F., & Ren, Y. (2019). Bcbim: A blockchain-based big data model for BIM modification audit and provenance in mobile cloud. *Mathematical Problems in Engineering,* 2019. 10.1155/2019/5349538

Zheng, Z., Xie, S., Dai, H.-N., Chen, W., Chen, X., Weng, J., & Imran, M. (2020). An overview on smart contracts: Challenges, advances and platforms. *Future Generation Computer Systems,* 105, 475– 491. 10.1016/j.future.2019.12.019

Index

Pages in *italics* refer to figures and pages in **bold** refer to tables.

For Product Safety Concerns and Information please contact our EU
representative GPSR@taylorandfrancis.com
Taylor & Francis Verlag GmbH, Kaufingerstraße 24, 80331 München, Germany